SPACEFLIGHT
A HISTORICAL ENCYCLOPEDIA

Spaceflight

A HISTORICAL ENCYCLOPEDIA

Patrick J. Walsh

GREENWOOD
An Imprint of ABC-CLIO, LLC

Santa Barbara, California • Denver, Colorado • Oxford, England

Library of Congress Cataloging-in-Publication Data

Walsh, Patrick J., 1963–
Spaceflight : a historical encyclopedia / Patrick J. Walsh.
p. cm.
Includes bibliographical references and index.
ISBN 978-0-313-37869-0 (set : print : alk. paper) — ISBN 978-0-313-37870-6 (set : ebook) — ISBN 978-0-313-37871-3 (volume 1 : print : alk. paper) — ISBN 978-0-313-37872-0 (volume 1 : ebook) — ISBN 978-0-313-37873-7 (volume 2 : print : alk. paper) — ISBN 978-0-313-37874-4 (volume 2 : ebook) — ISBN 978-0-313-37875-1 (volume 3 : print : alk. paper) — ISBN 978-0-313-37876-8 (volume 3 : ebook)
1. Space flight—Encyclopedias. 2. Outer space—Exploration—Encyclopedias. I. Title.
TL788.W35 2010
629.403—dc22 2009037030

14 13 12 11 10 1 2 3 4 5

This book is also available on the World Wide Web as an eBook.
Visit www.abc-clio.com for details.

ABC-CLIO, LLC
130 Cremona Drive, P.O. Box 1911
Santa Barbara, California 93116-1911

This book is printed on acid-free paper ∞
Manufactured in the United States of America

This work is dedicated with love and gratitude to
my parents, Helen E. and John E. Walsh.

CONTENTS

VOLUME ONE

VOLUME TWO

VOLUME THREE

LIST OF ENTRIES

SAKIGAKE

(January 7, 1985–November 15, 1995)

Japanese Scientific Satellite

Living up to its name, the Sakigake (Pioneer) spacecraft was the first Japanese craft to achieve a planet swingby successfully, and it also encountered the comet Halley.

Sakigake was a project of the Institute of Space and Aeronautical Science at the University of Tokyo. It was identical to the Suisei spacecraft, which launched on August 18, 1985. Sakigake launched on January 7, 1985, and achieved the flyby of Halley on March 11, 1986, at a distance of 7 million kilometers.

The first Earth swingby occurred on January 8, 1992, followed by two more, on June 14, 1993 and October 28, 1994. During the course of its flight, Sakigake also allowed mission controllers to study the Earth's geotail, and it successfully conducted measurements of plasma wave spectra, solar wind ions, and interplanetary magnetic fields.

Although its mission profile envisioned continued operations through 1998, including additional comet flybys, the Sakigake mission came to an end with the exhaustion of the spacecraft's hydrazine fuel supply. Mission controllers lost contact with the craft on November 15, 1995.

SALYUT SPACE STATION PROGRAM

(1971–1991)

Soviet Manned Space Station Program

In the early years of the space race between the Soviet Union and the United States, both nations explored the concept of placing a manned observation post in orbit to conduct space-based reconnaissance. As each nation developed unmanned satellite spying capabilities, the idea of manned spy missions seemed a logical extension to military planners in both the Soviet Union and the United States.

Soviet space station development began with the Almaz military program, which was first proposed in 1964. Derived from the same technology that would serve as the basis for the Soyuz spacecraft, the Almaz concept called for separate, linked compartments that would facilitate high resolution photographic surveillance of Earth-based objects as small as one foot wide. In addition to photo reconnaissance, the Almaz station was reportedly equipped with a satellite-destroying device that was used in various tests.

The station's propulsion system and its related electronics would reside in one module; a second compartment would house the camera equipment the crew would use to conduct the desired surveillance, along with the work area and living quarters necessary to support the crew during a mission; and the third Almaz component would consist of a craft essentially identical to the Soyuz Descent Module, which could be used to return film and other data and equipment to Earth at the end of a mission.

At about the same time that the original Soviet Almaz plans were taking shape in the mid-1960s, the United States Air Force was pursuing a similar concept with its Manned Orbiting Laboratory (MOL) program. Envisioned essentially as a military version of the civilian Gemini spacecraft, the MOL concept would have launched military personnel on 30-day flights designed solely for surveillance purposes. Although the program progressed to the point where the Air Force selected astronauts to train for the flights, a combination of technical and financial difficulties hindered its development, and by the end of the 1960s the MOL project was abandoned entirely.

In retrospect, the idea of American spies in space seems an anomaly for the early years of the U.S. space program, given the heavy emphasis on civilian development and control of the nation's space efforts. The tightly controlled totalitarian Soviet political regime had no similar concerns to contend with as they pursued the Almaz station concept in the mid-1960s—and yet, as the decade passed, the Soviets would find themselves translating their military space station plans into a civilian design, at just about the same historical moment that the United States gave up the idea of a manned military station entirely.

A key element in the decision-making process on both sides was the American triumph in the race to be first to land human beings on the Moon. Manned space stations in Earth orbit seemed a logical next step for both the Russian and American space programs in the post-Apollo period, and the Soviets particularly hoped that a string of successful space station missions would soften the impact of the American lunar landings, and help re-establish some of the glory the Soviet program had enjoyed in the early years of its space competition with the United States.

Aside from the substantial technical difficulties and crew safety issues involved in the development of an orbiting station, the Soviets were also faced with the necessity of adapting their Almaz hardware and systems to the new, nonmilitary design. The civilian version of the Almaz was intended for widespread public display, while, by definition, the military station program was a classified, secret project. The dilemma of protecting the Almaz development while openly touting the details of the new program was solved when the civilian station was given a name: Salyut.

Because the first launch was scheduled to take place in April 1971, in the midst of the nation's celebrations marking the tenth anniversary of Yuri Gagarin's epic first spaceflight, the civilian station was named Salyut ("salute" in Russian) in Gagarin's honor. Both the launch and the anniversary celebrations bore heavy significance for ordinary Russians who had exulted at Gagarin's 1961 triumph, then mourned his loss when he was killed in the 1968 crash of a MiG 15 airplane, just eight days before the seventh anniversary of his historic spaceflight.

In addition to honoring the nation's first space hero, the Salyut designation also provided cover for the Almaz military program, as the name was used for both the military and civilian stations. Flights related to the secret military stations—as well as the stations themselves—were depicted as part of the readily publicized civilian Salyut program, and their true nature was obscured by stressing those elements of the two types of missions that were similar.

For example, details of the physical exercise necessary to the health of the crewmembers could be publicly released without risk that the crew's other activities would also be revealed. Similarly, mission managers could freely describe the maintenance routines or scientific work that would be performed in any space station, regardless of the details of any larger missions that might be conducted within the station.

In spite of the technical, political, and military challenges of adapting the military Almaz development to the civilian Salyut program, Soviet engineers and space program officials displayed great flexibility in solving the difficulties of creating the Salyut stations. The basic Soyuz spacecraft was modified to create a workable space station design, and the 18-day flight of Soyuz 9 in June 1970 provided the basic medical data to ensure the safety of future space station crews.

The Salyut program began with the nonmilitary Salyut 1, and cosmonauts Georgy Dobrovolsky, Vladislav Volkov, and Viktor Patsayev were given the honor of being the first human beings to live and work in space for an extended period. Sadly, at the end of their successful 23-day stay at Salyut 1, they were killed during their return to Earth, after a malfunction in their Soyuz 11 spacecraft.

That the Salyut program was able to continue after the loss of the first crew was a testament to both the capabilities of the engineers who modified the Soyuz spacecraft in the aftermath of the tragedy, and to the crew members themselves, as the achievement of their mission aboard the space station proved an irresistible glimpse of the potential for future long-duration crews, despite the grave risks involved.

Designed to test the concept of living and working in space, the first five Salyuts (and several additional failed attempts) were designed with a single docking port. They could not be resupplied or refueled, and were as a result limited to relatively short stays in orbit. Salyut 6 and Salyut 7 each had two docking ports, which allowed for refueling and the delivery of supplies via unmanned Progress spacecraft, which in turn led to longer duration missions that stretched the boundaries of human beings' ability to live and work in space.

Despite the secrecy of the Salyut program's hidden Almaz component, the tragic Soyuz 11 accident and the frequently harrowing mishaps that plagued many of the flights associated with the project, the Salyut series of space stations constituted a major, lasting success for the Soviet space program. The Salyut development effort gave the Soviets the opportunity to methodically expand their maturing capabilities, and the engineers and administrators of the Soviet program responded to the challenges inherent in the creation of an orbiting space station with great daring and admirable persistence.

The experience and expertise gained in the Salyut program served the Soviets well in the case of the subsequent Mir space station in particular, and in the future course of the Russian space program in general.

Salyut Space Stations:

Station	*Purpose*	*Launch Date*	*De-Orbited*
Salyut 1	civilian	April 19, 1971	October 11, 1971
Salyut 2	military	April 3, 1973	May 28, 1973
Salyut 3	military	June 24, 1974	January 24, 1975
Salyut 4	civilian	December 26, 1974	February 2, 1977
Salyut 5	military	June 22, 1976	August 8, 1977
Salyut 6	civilian	September 29, 1977	July 29, 1982
Salyut 7	civilian	April 19, 1982	February 7, 1991

Missions to Salyut Stations:

Salyut 1:	Soyuz 10, Soyuz 11
Salyut 2:	None.
Salyut 3:	Soyuz 14, Soyuz 15
Salyut 4:	Soyuz 17, Soyuz 18, Soyuz 20 (unmanned)
Salyut 5:	Soyuz 21, Soyuz 23, Soyuz 24
Salyut 6:	Soyuz 25, Soyuz 26, Soyuz 27, Soyuz 28, Soyuz 29, Soyuz 30, Soyuz 31, Soyuz 32, Soyuz 33, Soyuz 34 (unmanned), Soyuz T-1 (unmanned), Soyuz 35, Soyuz 36, Soyuz T-2, Soyuz 37, Soyuz 38, Soyuz T-3, Soyuz T-4, Soyuz 39, Soyuz 40
Salyut 7:	Soyuz T-5, Soyuz T-6, Soyuz T-7, Soyuz T-8, Soyuz T-9, Soyuz T-10–1, Soyuz T-10, Soyuz T-11, Soyuz T-12, Soyuz T-13, Soyuz T-14, Soyuz T-15

SALYUT 1

(April 19, 1971–October 11, 1971)

Soviet Space Station; First-ever Manned Space Station

Launched unmanned from the Baikonur Cosmodrome atop a Proton rocket on April 19, 1971, the civilian Salyut 1 was the world's first space station. The crew of Soyuz 10 was to have been the first to occupy Salyut 1, but a malfunction prevented them from entering the station. The Soyuz 11 crew—commander Georgy Dobrovolsky, flight engineer Vladislav Volkov, and research engineer Viktor Patsayev—had the honor of being the first human beings to live and work on an orbiting space station. Their daily activities were broadcast on Soviet television, and they were embraced by the Russian people for their heroic achievement and for their obvious good humor.

The joy of the remarkable Soyuz 11/Salyut 1 achievement turned to sorrow, however, when after 23 days in space a malfunction in their Soyuz spacecraft caused Dobrovolsky, Volkov, and Patsayev to be killed during re-entry.

A difficult period of introspection and slow recovery followed the Soyuz 11 accident, and although the fault that had killed the crew was confined to the Soyuz craft and was unrelated to the space station, the development of the Salyut program understandably slowed, along with the rest of the Soviet space program.

SALYUT 2

(April 3, 1973–May 28, 1973)

Soviet Military Space Station

Soviet hopes to outperform the American Skylab space station suffered a series of frustrating setbacks in the aftermath of the Soyuz 11 disaster, which resulted in the loss of three cosmonauts in October 1971.

In July 1972, the first attempted Salyut launch after Salyut 1 failed because of a rocket malfunction, and in April 1973 the first Almaz station—which was publicly referred to as Salyut 2—failed in orbit less than two weeks after it was launched.

Shortly after the Salyut 2 failure, an apparent third attempted Salyut launch also ended in frustration due to an error in the craft's stabilization system. The flight was referred to as Cosmos 557, and Soviet space officials quickly disowned it, denying even that it was a civilian Salyut that had been launched.

Because the stations were all launched unmanned, no crews were imperiled by the failures. At the same time, the rapid, successive loss of the space stations before crews could reach them raised serious questions about the program's future, and about the ability of future stations to support crews safely.

Gradually, the concerns eased, and development of the involved technologies and systems continued. In December 1973, the crew of Soyuz 13 reportedly carried out tests of the photographic equipment that had long been intended for use aboard the Almaz space stations, achieving a necessary step in the program's development despite the failure of Salyut 2.

SALYUT 3

(June 24, 1974–January 24, 1975)

Soviet Military Space Station

A series of successful unmanned test flights helped to mitigate the frustrations of failed Salyut launch attempts in 1972 and 1973, and on June 24, 1974, the Soviets launched Salyut 3—the first manned military surveillance outpost in space. On July 3, 1974, Soyuz 14 crew members Pavel Popovich and Yuri Artyukhin followed

Salyut 3 into space, and, during their stay aboard the station, they reportedly conducted the kind of photo reconnaissance first envisioned by the Soviets in the original 1964 proposal for the Almaz program of manned military spy satellites. Popovich and Artyukhin also performed a variety of nonclassified activities, which were widely publicized. They landed on July 19, 1974, having achieved the first entirely successful Soviet space station flight.

The Almaz/Salyut 3 station was also visited by the crew of Soyuz 15, but a malfunction prevented the Soyuz 15 crew from docking, and their scheduled stay had to be aborted.

SALYUT 4

(December 26, 1974–February 2, 1977)

Soviet Space Station

Launched unmanned on December 26, 1974, the civilian Salyut 4 space station successfully hosted the Soyuz 17 crew, which lived and worked on the station for 29 days, and the crew of Soyuz 18, which endured a difficult 63 days marred by faults in the Salyut's environmental system that gave rise to fogged windows and a green mold that grew on the walls of the station. Fortunately, the Soyuz 18 crew suffered no lasting ill effects to their health.

The Soyuz 18/Salyut 4 flight overlapped with the joint Soviet-American Apollo-Soyuz Test Project (ASTP) flight, and separate teams of Soviet controllers managed the two flights simultaneously, confirming the Russians' assurances to the Americans that they could handle the two flights at the same time without any difficulties or threat to the safety of the involved cosmonauts and astronauts.

SALYUT 5

(June 22, 1976–August 8, 1977)

Soviet Military Space Station

The Salyut 5 space station was launched on June 22, 1976. The last of the original Salyut design (featuring a single docking port) and also the last Almaz military station, Salyut 5 was used by the Soyuz 21 and Soyuz 24 crews, both of which were apparently focused on conducting military photo surveillance. The true purpose of the flights as military surveillance missions was indicated by the use of scrambled voice communications and by the return of the unmanned third section of Salyut 5—the Descent Module thought to be used for the return of reconnaissance film and data—following the end of the Soyuz 24 flight in February 1977.

SALYUT 6

(September 29, 1977–July 29, 1982)

Soviet Space Station

Representing a major step forward in space station design and capabilities, the Salyut 6 space station was launched on September 29, 1977. Outfitted with two docking ports, Salyut 6 (and the subsequent Salyut 7) could receive supplies and fuel ferried to it by unmanned Progress spacecraft. The Progress supply ships were based on the basic Soyuz design; the first Progress flight, to Soyuz 6, was launched on January 20, 1978.

The ability to resupply and refuel the later Salyuts meant that cosmonauts could stay aboard the stations for much longer periods of time, and could as a result conduct more complex and scientifically rewarding research.

The Soviets also took advantage of an opportunity that arose from the limitations of the Soyuz spacecraft. Because the electronic systems aboard the Soyuz spacecraft tended to deteriorate within a maximum of 90 days in space, the Soyuz craft docked at the space station during long-duration missions had to be replaced before its 90-day safety limit expired. Thus, frequent short "replacement" flights had to be flown, and the Soviets used such flights as opportunities for representatives of Soviet-bloc nations or other allies of the U.S.S.R. to make their country's first flight in space, during brief visits to Salyut 6 and Saylut 7. The visiting cosmonaut flights were collectively known as the manned portion of the Intercosmos program.

During its five years in orbit (1977–1982), the Salyut 6 space station hosted 17 crews, 5 of which made long-duration stays aboard the station. The long-duration missions incrementally extended the record for longest stay in space as mission planners became more confident in the station's ability to support cosmonauts for longer periods, and more comfortable with the vastly improved reliability of the Soyuz spacecraft.

In addition to its impressive records involving manned flights, Salyut 6 was also visited by 12 unmanned Progress spacecraft, which were controlled by mission handlers on the ground who directed their automated docking with the station. The Progress flights delivered 20 tons of supplies and equipment to the resident crews of Salyut 6, and refueled the station at regular intervals throughout its half-decade of operations.

At the end of its five-year mission, the Salyut 6 space station was also used in an automated docking test involving a large module known as Cosmos 1267. The test confirmed that a large module—such as those used in the later Mir and International Space Station (ISS)—could be automatically linked to an orbiting space station by controllers on the ground.

SALYUT 7

(April 19, 1982–February 7, 1991)

Soviet Space Station

On April 19, 1982, the Soviets launched the last of the Salyut space stations, Salyut 7. Over the course of the next four years, the final Salyut would be the orbital home of 10 groups of cosmonauts. Six crews made long-duration visits to the station, and the missions included a record-setting stay of 237 days by Soyuz T-10/Salyut 7 cosmonauts Leonid Kizim, Vladimir Solovyov, and Oleg Atkov.

The station was also visited by 13 Progress cargo spacecraft, and was used in test dockings with two unmanned experimental modules, Cosmos 1443 and Cosmos 1686.

Perhaps most indicative of the enormous advances that had been made in the Soviet space station program over the years, Salyut 7 was reactivated and reoccupied after the station failed in orbit in February 1985. A fault in the station's battery system had caused the failure, and at the time, Soviet space officials were apparently willing to accept the situation as the end of the station's useful life after nearly three years of operations.

But on June 6, 1985, Vladimir Dzhanibekov and Viktor Savinykh launched aboard Soyuz T-13 with the unique mission of trying to salvage the Salyut 7 space station. They succeeded in their remarkable orbital repair duties, and the station was again occupied until November 21, 1985.

In May 1986, Soyuz T-15 cosmonauts Leonid Kizim and Vladimir Solovyov returned to Salyut 7 in the midst of their stay aboard the new space station Mir. They spent nearly two months aboard the Salyut, and their visit seemed a fitting bridge between the last of the original Russian space stations and the forward-looking Mir.

After the visit from Kizim and Solovyov, Salyut 7 (and the attached Cosmos 1686) was boosted into a higher orbit. Salyut 7 then remained in orbit, unoccupied and linked with the Cosmos 1686 module, until 1991.

On February 7, 1991, the station and module fell from orbit and made a fiery return to Earth, raining debris over parts of Argentina.

SAUDI ARABIA: FIRST CITIZEN IN SPACE.

See Al-Saud, Salman Abdulaziz

SAVINYKH, VIKTOR P.

(1940–)

Soviet Cosmonaut

A veteran of three spaceflights, including the remarkable 1985 Soyuz T-13/Salyut 7 space station repair mission, Viktor Savinykh was born in Berezkiny, Russia, on March 7, 1940. He attended the Moscow State University of Geodesy and Cartography, where he graduated in 1969.

Savinykh was chosen for training as a cosmonaut as a civilian specialist candidate in December 1978. His initial duties as a cosmonaut included service as a member of the backup crew for Soyuz T-3, which launched in November 1980.

Savinykh began his first spaceflight on March 12, 1981, as flight engineer for the Soyuz T-4/Salyut 6 long-duration mission. Launched from the Baikonur Cosmodrome with Vladimir Kovalyonok, who served as commander for the flight, Savinykh spent two-and-a-half months in space while serving as a member of the fourth and final Salyut 6 long-duration crew.

During their stay, he and Kovalyonok carried out scientific experiments and performed the engineering and space station maintenance chores necessary to keep the station in good working order. They also welcomed two visiting crews that included guest cosmonauts from Mongolia and Romania.

The international visits were part of the manned Soviet Intercosmos Program, which provided spaceflight opportunities for the citizens of nations friendly to or aligned with the Soviet Union.

On March 22, Soviet cosmonaut Vladimir Dzhanibekov arrived with guest cosmonaut Jugderdemidiyn Gurragcha of Mongolia aboard Soyuz 39. In keeping with the protocol established for the Intercosmos flights, the four cosmonauts participated in ceremonies broadcast on Soviet television, and then, after a visit of about a week, the visitors left and returned to Earth.

Leonid Popov commanded the second visiting flight, Soyuz 40, which included Dumitru Prunariu, the first citizen of Romania to fly in space. After a week at Salyut 6 with Savinykh and Kovalyonok, Popov and Prunariu returned to Earth on May 22, followed by their hosts four days later, on May 26, 1981.

During his first trip into orbit, Savinykh spent more than 74 days in space. He and Kovalyonok had the distinction of being the last individuals to occupy Salyut 6, which was de-orbited over the Pacific Ocean on July 29, 1982.

Savinykh next served as a member of the backup crew for the Soyuz T-7, Soyuz T-8, Soyuz T-10/Salyut 7 and Soyuz T-12 missions.

He made his second flight into space on June 6, 1985 when he launched as flight engineer for the remarkable Soyuz T-13/Salyut 7 space station repair mission, with Vladimir Dzhanibekov serving as commander.

During a period in which it was unoccupied, the Salyut 7 station had suffered a sudden failure in orbit on February 5, 1985. Since its April 1982 launch, it had supported three long-duration crews and four visiting crews, but the sudden loss of its electrical system seemed to signal the end of the station's usefulness to the Soviet space program. Salyut 7 was to be the last of the Salyut series; it would be replaced by the more advanced Mir space station.

After much consideration, Soviet space officials decided that some attempt should be made to examine the station, to diagnose what had gone wrong and if possible, to repair whatever damage may have occurred. Savinykh and Dzhanibekov were given the difficult assignment, and they launched and traveled to the station in their Soyuz T-13 craft uncertain of what they might find when they arrived at the cocooned Salyut 7.

They were met with an eerie sight when they first encountered the dark station. Because its control systems were not in working order, Salyut 7 was slowly rolling, passing through orbit in a blind torpor. Despite the station's monotonous spin, Dzhanibekov was able to dock Soyuz T-13 with Salyut 7 manually, with Savinykh helping the process by operating a laser range-finder.

The two cosmonauts then made a cautious first trip into the interior of the cold, dark shell that had once served as an orbital home for previous crews of cosmonauts. They searched for signs of obvious damage, and after several forays into the cocooned station they narrowed the source of the problem to a fault in one of the batteries that stored the power generated by the station's solar arrays. By rewiring the connection between the arrays and the batteries, they were able to get the station's systems powered up again, and Salyut 7 gradually blinked back to life.

After completing their initial fix-it work, Savinykh and Dzhanibekov made a list of the tools and spare parts they would need to make repairs to all the equipment damaged by the four frigid months of the station's unintended hibernation. Mission controllers immediately began preparing an unmanned Progress cargo spacecraft for a flight to Salyut 7, loaded with the requested hardware.

Also on its way to the station by mid-July was the Cosmos 1669 module, which when docked with Salyut 7 would effectively double the station's available living and work space. Docked for a little more than a month before being jettisoned to make room for the arrival of Soyuz T-14, the Cosmos 1669 module proved a welcome addition for Savinykh and Dzhanibekov, who had suffered through intensely cold and unpleasant working conditions during their first days aboard the station.

Savinykh and Dzhanibekov continued their repair work during a five hour spacewalk on August 2, 1985, when they ventured outside of Salyut 7 to install additional panels on the station's solar arrays.

Then, following the arrival of Soyuz T-14 on September 17, Savinykh bade farewell to Dzhanibekov, who returned to Earth in Soyuz T-13, and welcomed Vladimir Vasyutin and Alexander Volkov, who would join him aboard Salyut 7 as the station's next long-duration crew.

Savinykh, Vasyutin and Volkov initially established a productive routine, performing scientific experiments and maintaining the station—and themselves, via a program of daily exercise—in good working order. As the flight progressed, however, they were enveloped in a dramatic test of their courage and compassion, when Vasyutin, the commander of the long-duration mission, became ill. He first showed signs of discomfort in October, and by mid-November he was too sick to carry out his daily activities. Worried mission controllers placed Savinykh in command, as medical personnel tried to establish a long-distance diagnosis of Vasyutin's condition, and to evaluate the risk his illness might pose to his fellow crew mates.

Although the mission had reportedly been planned to last six months, Soviet space officials decided to end the flight four months early. Unsure of the breadth of the crisis they faced, Savinykh and his crew mates boarded Soyuz T-14 and returned to Earth on November 21, 1985. They were immediately placed in isolation while doctors studied every aspect of Vasyutin's illness, and then, when he was found to have an infection not related in any way to his experience in space, Savinykh and Volkov were released from the hospital. Vasyutin recovered fully after a month-long treatment.

During his remarkable second flight, in which he and Dzhanibekov had effected the near-miraculous revival of Salyut 7 by equal parts courage and sheer hard work, and which continued through the uncertainty and strain of supporting a comrade through an unprecedented health crisis in the cramped confines of an orbiting space station, Savinykh amassed more than 168 days in space.

Savinykh was awarded a candidate of technical sciences degree in 1985. His next spaceflight assignment was as a member of the backup crew for the Soyuz TM-3 Intercosmos flight, which took place in July 1987. The manned Intercosmos Program provided spaceflight opportunities for citizens of nations aligned with or friendly to the Soviet Union.

Savinykh then commanded the Soyuz TM-5 Intercosmos mission—his third flight in space. Launched on June 7, 1988, Savinykh, flight engineer Anatoly Solovyov and guest cosmonaut Aleksandr Aleksandrov of Bulgaria traveled to the Mir space station, where they visited with the station's third long-duration crew, Vladimir Titov and Musa Manarov, for about a week. They returned to Earth on June 17, 1988, aboard Soyuz TM-5.

Viktor Savinykh spent more than 252 days in space, including five hours in EVA, during his career as a cosmonaut.

In 1990, he earned a doctorate of technical sciences degree. After his retirement from the cosmonaut corps, he served as rector of his alma mater, the Moscow State University of Geodesy and Cartography, and in 2006 he became a correspondent member of the Russian Academy of Sciences.

SAVITSKAYA, SVETLANA Y.

(1948–)

Soviet Cosmonaut and First Woman to Conduct a Spacewalk

The first woman to participate in a spacewalk and the second woman to fly in space, Svetlana Savitskaya was born in Moscow, Russia on August 8, 1948. She attended the Moscow Aviation Institute, where she graduated in 1992 after having received a candidate of technical sciences degree in 1986. She also joined the Soviet Air Force, and served as a parachutist, test pilot and aeronautical engineer prior to her selection as a cosmonaut.

Savitskaya was chosen for training as a member of the second female group of cosmonauts, who were selected in July 1980. She was initially chosen to serve as commander of a female Soyuz crew, but that assignment was not developed beyond its initial planning stages, and was later abandoned.

On August 19, 1982, Savitskaya began her first trip into space as a research engineer for the Soyuz T-7 flight, which launched from the Baikonur Cosmodrome with commander Leonid Popov and flight engineer Aleksandr Serebrov. With her first spaceflight, Savitskaya became the second woman to fly in space; Valentina Tereshkova had been first, during Vostok 6 in June 1963.

Soyuz T-7 was a Soyuz switching flight, intended to replace the Soyuz T-5 vehicle that had been docked at the Salyut 7 space station before it reached its safe operations limit.

At Salyut 7, Savitskaya, Popov and Serebrov visited for about a week with the station's resident crew members Anatoli Berezovoi and Valentin Lebedev, who were in the midst of a record setting stay as the station's first long-duration crew. Savitskaya and her crew mates returned to Earth aboard Soyuz T-5 on August 27, 1982, leaving their Soyuz T-7 at the station for Berezovoi and Lebedev.

During her first trip into orbit, Savitskaya spent 7 days, 21 hours and 52 minutes in space.

She flew again in space during the Soyuz T-12 flight, in July 1984. Launched on July 17, 1984 as flight engineer for the historic flight, with commander Vladimir Dzhanibekov and research engineer Igor Volk, Savitskaya's mission profile for the Soyuz T-12 called for her to become the first woman to walk in space.

At Salyut 7, she and her crew mates were met by the station's resident crew members Leonid Kizim, Vladimir Solovyov, and Oleg Atkov. The six cosmonauts carried out joint operations for about a week, and then, on July 25, 1984, Savitskaya and Dzhanibekov ventured outside of Salyut 7 to test an electron beam hand tool that she had helped to develop during her long pre-flight training on Earth.

Using the Universalny Rabochy Instrument (URI), a multipurpose hand tool, she carried out tests on a variety of sample materials to demonstrate the tool's capabilities for cutting, welding, soldering and brazing.

Dzhanibekov also took a turn using the URI, and together they performed a thorough test of the device during an EVA of 3 hours and 35 minutes. They also retrieved experiment packages from the exterior of the station, and installed new ones in their place.

Savitskaya, Dzhanibekov and Volk returned to Earth in Soyuz T-12 on July 29.

Svetlana Savitskaya spent more than 19 days in space, including 3 hours and 35 minutes in EVA, during her remarkable career as a cosmonaut.

Savitskaya continued to serve the Soviet space program in the years following her milestone EVA flight, as an employee of NPO Energia. She later embarked on a career in politics, and in 1989 she was elected to the Russian Parliament, the Duma.

SCHIRRA, WALTER M., JR.

(1923–2007)

U.S. Astronaut

A pioneering American space hero and the only astronaut to fly in space during the Mercury, Gemini and Apollo programs, Walter "Wally" Schirra was born in Hackensack, New Jersey on March 12, 1923. He attended Newark College of Engineering in 1941, and then joined the United States Navy and attended the United States Naval Academy, where he received a bachelor of science degree in 1945.

During his outstanding military career, Schirra received the U.S. Navy Distinguished Service Medal, and he was awarded the Navy's Distinguished Flying Cross and Air Medals on three separate occasions. He graduated from the U.S. Navy Test Pilot School in 1958.

NASA chose him for training as an astronaut as one of its original selection of astronauts, the Mercury Seven, on April 9, 1959. He was a lieutenant commander at the time of his selection by the space agency; he would rise to the rank of captain in the United States Navy by the time of his retirement from the service.

Pioneering astronaut Walter "Wally" Schirra, the only astronaut to fly in space during the Mercury, Gemini, and Apollo programs. [NASA/courtesy of nasaimages.org]

Schirra served as spacecraft communicator and chase plane pilot during the first flights of the U.S. Mercury program, supporting his fellow Mercury Seven astronauts as they carried out the first four manned missions of the U.S. space program.

Then, after a rigorous training and evaluation period followed by extensive flight simulations and mission-specific training, Schirra first flew in space during the Sigma 7 Mercury flight on October 3, 1962. He was the fifth U.S. astronaut—and the ninth individual overall—to fly in space.

Schirra named his Mercury spacecraft "Sigma 7" (sigma being the engineering symbol for summation, and the numerical suffix signifying the unity of the original seven Mercury astronauts) in honor of the effort that NASA engineers had put into the design and development of the vehicle, its mission plan, and the systems and equipment that would accompany its pilot into space.

The plan for the fifth manned Mercury flight called for doubling the duration of each of the previous orbital missions, which had each lasted three orbits and about four hours. The flight began with a 7:15 A.M. lift-off on October 3, 1962, and, due to a minor error in the timing of the cut-off of the launch vehicle sustainer engine, Schirra entered an orbit of 175.8 by 100 miles—the highest orbit of any Mercury flight. The slight error would also result in his speed in orbit, 17,557 miles per hour, being the fastest achieved during the Mercury program.

During his first space mission, Schirra successfully carried out a wide array of navigation and flight tests in Sigma 7, carefully conserving fuel and executing each planned action in his mission profile with precision. He also flew the spacecraft in powered-down drifting mode, and then tested the capabilities of its systems and equipment when he powered them up again.

The Sigma 7 flight continued flawlessly through six orbits, and Schirra then oriented the vehicle for re-entry into the Earth's atmosphere. The return to Earth went equally as well as the rest of the flight, and Sigma 7 splashed down less than five miles from its targeted landing site in the Pacific Ocean on October 3, 1962, after a total flight of 143,983 miles, in 9 hours, 13 minutes and 11 seconds.

After his recovery, Schirra emerged from the Sigma 7 capsule tired but otherwise demonstrating no ill effects of his six orbits in space. NASA officials later characterized his superb Mercury mission as a "textbook flight."

Schirra continued his career as an astronaut as the pioneering Mercury program came to a close and NASA embarked on Project Gemini. He served as the backup command pilot for Gemini 3—the first manned Gemini flight, which took place on March 23, 1965.

His second flight in space began on December 15, 1965, when he lifted off as commander of the Gemini 6-A mission, with Pilot Tom Stafford. A key flight in the development of rendezvous procedures crucial to the progress of the U.S. space program, Gemini 6-A was initially hampered by a series of frustrating delays, including the explosion in orbit of the unmanned Agena target vehicle that Schirra and Stafford were to have used for their rendezvous test during their originally scheduled launch on October 25. Although NASA officials briefly considered moving the planned rendezvous test to a later flight, the goal of mastering rendezvous was ultimately deemed too important to delay, and it was decided that Schirra and Stafford would carry out their rendezvous attempt with a new target: the manned Gemini 7.

Designed as the first extended-duration flight of the U.S. space program, Gemini 7 launched on December 4, 1965 with commander Frank Borman and pilot James Lovell. Immediately after Borman and Lovell lifted off, launch pad technicians at Cape Canaveral began preparations for the launch of Schirra and Stafford in Gemini 6-A.

As they prepared for their launch a second time, after the scrubbed launch of October 25, Schirra and Stafford sat in their Gemini 6-A spacecraft (renamed 6-A to distinguish it from the earlier launch attempt, which had been Gemini 6) and followed the countdown to lift-off all the way to zero. The instruments within the capsule indicated that they had lifted off the launch pad, but a minor fault in the Titan II launch vehicle caused the rocket's engines to shut down. If the instrument readings had been correct and the launch vehicle had indeed left the launch pad, the resulting settling back onto the pad could have set off a catastrophic explosion.

In the brief instant they had to decide whether or not the instruments were correct and to make a choice between remaining within the spacecraft or activating a risky abort procedure that would cause them to be violently ejected from the vehicle, veteran pilots Schirra and Stafford instantly chose to ignore the readings they were seeing on the craft's instrument panel. Their intuition was correct; the readings were wrong, and the Gemini 6-A capsule remained safely in place atop the Titan II launcher, which had in fact remained on the launch pad.

Where a decision to eject might have led to one or both astronauts being injured, the decision to remain within the spacecraft resulted in their being safely evacuated and ready for another attempt at launch—which was ultimately successful—three days later.

Once they were finally launched on their rendezvous mission, Schirra and Stafford entered orbit at a distance of 1,992 kilometers behind Borman and Lovell in Gemini 7. Five hours and 16 minutes later, during their fourth orbit in Gemini 6-A, Schirra and Stafford caught up with their rendezvous target.

At 2:33 P.M. on December 15, 1965, Schirra successfully achieved the first-ever constant rendezvous of two manned spacecraft, when he brought Gemini 6-A to a halt at steady distance of about 40 meters from Gemini 7.

The two spacecraft then achieved "stationkeeping" (keeping the two spacecraft at a constant chosen distance) for more than five hours, at distances ranging from 1 to 295 feet. They flew in a variety of positions during three orbits, and Schirra found the exercise to be pleasant duty, later expressing joy at the ease with which he was able to maneuver his spacecraft.

In keeping with the oft-expressed good humor for which he and his fellow astronauts had become known, Schirra indulged in an improvised rendition of "Jingle Bells" before returning to Earth, while Stafford reported seeing Santa Claus traveling in polar orbit. In understandably high spirits after having finally achieved their long-sought rendezvous goals, Schirra and Stafford splashed down on December 16, 1965, after 16 orbits and a flight of 1 day, 1 hour, 15 minutes and 58 seconds.

With the successful completion of Project Gemini and the advent of the Apollo lunar landing program, Schirra continued his astronaut career as one of just three of the original Mercury Seven astronauts on active duty as the first Apollo flight approached. Although Alan Shepard and Donald "Deke" Slayton would later rebound from medical problems to fly in space and John Glenn would ultimately return to orbit during the space shuttle era, only Schirra, Virgil "Gus" Grissom and Gordon Cooper were available for assignment during the formulation of plans for the first Apollo missions.

Schirra's close friend Grissom was given the honor of commanding what was scheduled to be the first manned Apollo flight, with veteran Edward White and Group Three astronaut Roger Chaffee as his crew. Sadly, the flight was not to be: Grissom, White and Chaffee were killed in a fire during a test of their Apollo spacecraft on January 27, 1967.

After a difficult period of recovery following the tragic fire, NASA officials assigned Schirra as the commander of the next manned Apollo mission, Apollo 7. As the first manned flight after the Apollo 1 fire, and given its mission profile, which called for the crew to test the spacecraft's maneuvering, rendezvous and docking abilities, Apollo 7 was crucially important to the future of the U.S. space program. In choosing Schirra to command the flight, NASA officials anticipated the flawless performance that Schirra had displayed in his two earlier spaceflights, and for which he had become rightly celebrated.

From the start of his training for the mission, in which he and his crew mates, Command Module pilot Donn Eisele and lunar module pilot Walter Cunningham, spent 600 hours simulating every facet of the Apollo 7 flight on the ground prior to launch, to the spacecraft's splashdown at the end of the flight, Schirra delivered the command and leadership necessary to place NASA back on course in the space agency's efforts to make a lunar landing before the end of the 1960s.

Schirra, Eisele and Cunningham launched on October 11, 1968, and quickly set to work on the major goals of their mission profile. They successfully achieved rendezvous with the burnt-out second stage of their Saturn IB launch vehicle, and ultimately conducted eight successful tests of the service module engine. They flawlessly achieved the engineering objectives and systems and equipment tests that they had been assigned, and they validated the Apollo spacecraft—which had undergone an extensive redesign in the aftermath of the Apollo 1 fire—as a safe and responsive vehicle for future manned flights.

They were also the first crew to provide extensive live television coverage of a spaceflight in progress, and in doing so, demonstrated the positive aspects of their flight to a national audience.

The few difficulties that did arise during the flight had little to do with substantive issues or with the mission's objectives. Schirra and his crew mates shared a minor illness—apparently a common cold—and the crew became increasingly agitated with what it viewed as unreasonable requests for them to perform activities not explicitly covered in the original mission plan. The stress of the mission's importance and the lingering emotional effects of the Apollo 1 tragedy might well have added to the difficulty of the circumstances; in any case, an unusual tension developed between the crew and NASA officials on the ground.

Despite his discomfort, Schirra ensured that all of the flight's major goals were achieved during the 163 orbits that he, Eisele and Cunningham were in space. They splashed down on October 22, 1968, after a flight of 10 days and 20 hours, and were recovered within 45 minutes.

The success of Apollo 7 helped to ensure the forward progress of the Apollo program, and played a key role in the development of the hardware, systems and procedures that would later culminate in the successful Moon landings of the program's later years.

During his remarkable career as an astronaut, as the only person to fly in space during the Mercury, Gemini and Apollo programs, Wally Schirra accumulated more than 295 hours in space.

Even before the launch of Apollo 7, Schirra had announced that the flight would be his last space mission. He became a member of the board of directors of the Imperial American oil company in 1967, and served in a similar capacity for a wide variety of corporations during the 1970s and 1980s, including the J. D. Jewel chicken company; the First National Bank of Englewood, Colorado; Rocky Mountain Airlines; Carlsberg Oil and Gas; Advertising Unlimited; Electromedics; the Finalco leasing company; Cherokee Data Systems; Net Air International; Kimberly-Clark; and Zero Plus Telecommunications.

He also served as a trustee of the Detroit Institute of Technology and National College in South Dakota, and as an adviser to Colorado State University.

In 1969 he became president of Regency Investors, and the following year he founded the Environmental Control Company (ECCO). He subsequently became chairman of the Sernco corporation, and in 1975 he joined the Johns-Manville Sales Corporation in Denver, Colorado as vice president. He also served as president of Prometheus Systems in the early 1980s.

From 1979, he operated Schirra Enterprises as an independent consultant.

Among the many honors he received for his achievements as an astronaut, Schirra was awarded the Robert J. Collier Trophy, the Clifford B. Harmon International Trophy, and the Ivan C. Kincheloe Award of the Society of Experimental

Test Pilots, and he twice received the Haley Astronautics Award of the American Institute of Aeronautics and Astronautics (AIAA).

In 1988 he authored the book *Schirra's Space*, with Richard Billings.

Wally Schirra died of a heart attack on May 3, 2007, at the age of 84. He is survived by his wife and two grown children.

SCHLEGEL, HANS

(1951–)

European Space Agency Astronaut

A veteran of two spaceflights, Hans Schlegel was born in Überlingen, Germany on August 3, 1951. In 1968, as a student in the American Field Service exchange program, he attended Lewis Central High School in Council Bluffs, Iowa, from which he graduated. He also graduated from Hansa Gymnasium in Cologne, Germany in 1970, and then attended the University of Aachen in Germany, where he graduated with a diploma in physics in 1979.

Schlegel served as a paratrooper in Germany's Federal Armed Forces for two years before leaving the service in 1972 with the rank of second lieutenant. In 1979 he became a staff member at the University of Aachen, where he performed research in semiconductors and solid state physics until 1986. He then worked in private industry until he was selected for training as an astronaut candidate in 1988.

In preparation for his career as an astronaut, Schlegel received training at the German Aerospace Center (DLR), at the Gagarin Cosmonaut Training Center (GCTC) in Russia, and at NASA's Johnson Space Center (JSC) in Houston, Texas. Among his initial assignments as an astronaut, he served as a communications coordinator for the Soyuz TM-25/Mir space station scientific project conducted by Reinhold Ewald in 1997.

Schlegel became a member of the European Space Agency (ESA) astronaut corps in 1998, and in May, 2005 he was designated the agency's lead astronaut at JSC.

He first flew in space as a payload specialist aboard the STS-55 flight of the space shuttle Columbia, which lifted off on April 26, 1993. The second Spacelab mission devoted to German scientific research, STS-55 numbered two medical doctors among its crew, including Schlegel's fellow German astronaut Ulrich Walter. The crew worked 24 hour days in two shifts, and completed 88 experiments during the flight, which lasted just under 10 days.

Schlegel was a subject of one milestone experiment in the life sciences portion of the program, when he was among those crew members who were intravenously administered a saline solution to study the I.V. method for replacing fluids lost during the body's adaptation to weightlessness. The experiment marked the first time the intravenous method was used in space.

The STS-55 mission ended when Columbia landed at Edwards Air Force Base in California on May 6, 1993.

In February, 2008—nearly 15 years after his first spaceflight—he launched on a second space mission, aboard the STS-122 flight of the space shuttle Atlantis. The STS-122 crew delivered the ESA Columbus Laboratory to the International Space Station (ISS), and Schlegel participated in one of the three spacewalks devoted to installing the laboratory module.

He was originally scheduled to participate in the first EVA of the STS-122 flight, but a medical issue led mission controllers to delay his involvement until the second spacewalk instead. When he ventured outside the shuttle with fellow STS-122 mission specialist Rex Walheim on February 13, 2008, he was able to complete his assigned duties, and in the process he accumulated nearly seven hours of EVA.

During the period of docked operations at the space station, Schlegel's remarkable return to space was celebrated in a congratulatory call from Angela Merkel, the chancellor of Germany.

Atlantis returned to Earth on February 20, 2008.

In his two spaceflights, Hans Schlegel spent more than 22 days in space.

SCHMITT, HARRISON H.

(1935–)

U.S. Astronaut; Twelfth Human Being to Walk on the Surface of the Moon

The twelfth human being to walk on the Moon, Harrison "Jack" Schmitt was born in Santa Rita, New Mexico on July 3, 1935. He graduated from Western High School in Silver City, New Mexico, and then attended the California Institute of Technology, where he earned a bachelor of science degree in science in 1957.

An outstanding scholar, Schmitt received a Fulbright Fellowship to study at the University of Oslo in Norway during the 1957–1958 academic year. While in Norway he put his skills to use in helping to gather data for the Norwegian Geological Survey.

Upon his return to the United States, he attended Harvard University, where he earned a doctorate in geology in 1964. During his association with Harvard, he was a recipient of the Kennecott Fellowship in geology, the Harvard Fellowship, the Parker Traveling Fellowship, and the National Science Postdoctoral Fellowship. He served as a teaching fellow at the university in 1961.

He also worked on the United States Geological Survey, in New Mexico and Montana, and worked as a geologist in Alaska for two summers.

His association with NASA began when he was an employee of the U.S. Geological Survey Astrogeology Center in Flagstaff, Arizona, where he served as chief of the center's program for the development of methods of geological research for use on the lunar surface.

Widely recognized for his expertise as a geologist, Schmitt is a fellow of the American Geophysical Union, the American Association of Petroleum Geologists, the American Association for the Advancement of Science, and the

Harrison Schmitt gathers lunar sample materials on the surface of the Moon during Apollo 17, December 12, 1972. [NASA/courtesy of nasaimages.org]

American Institute of Aeronautics and Astronautics. He is also an honorary fellow of the Geological Society of America and the Geological Society of London, and an honorary member of the American Institute of Mining, Metallurgical and Petroleum Engineers, the New Mexico Geological Society and the Norwegian Geographical Society.

NASA chose him for training as an astronaut as a member of its Group Four selection of scientist astronauts in June, 1965. He underwent an intensive training period that included flight training at Williams Air Force Base in Arizona, and was qualified for assignment as a lunar module pilot for Apollo lunar landing missions.

As a pilot, he has accumulated over 2,100 hours of flying time, including 1,600 hours in jet aircraft.

His initial assignments at the space agency included service as the backup lunar module pilot for the Apollo 15 mission, which launched in July, 1971, and he also advised his fellow astronauts about the geological features of the lunar surface and the scientific value of various types of sample materials that were subsequently collected during lunar landing flights.

On December 7, 1972, Schmitt lifted off from the Kennedy Space Center (KSC) as lunar module pilot during the Apollo 17 mission, with commander Eugene Cernan and command module pilot Ronald Evans, in the sole night-time launch of the Apollo program. Once they entered lunar orbit, Schmitt and Cernan left Evans in the Apollo 17 command module America and traveled to the lunar surface in the lunar module Challenger. They touched down in the Taurus-Littrow area of the Moon, just southeast of Mare Serenitatis, on December 11, 1972.

Four hours after they landed, Cernan and Schmitt began their EVA work in earnest. Cernan made the ceremonial first step onto the lunar surface and as he followed the veteran commander out of the Challenger lunar module, Schmitt jokingly took notice of Cernan's footprints in the moondust and wondered aloud, "Who's been tracking up my lunar surface?"

The first portion of their first Moonwalk was taken up with the removal of the Lunar Roving Vehicle (LRV) from its storage spot on the lunar module, and the deployment of experiments and the planting of the American flag. They also gathered samples of the lunar crust for later study.

In the first major excursion of their planned geological survey of the lunar surface, Schmitt and Cernan used the LRV—the electric "Moon car"—to travel to the Steno Crater, where they collected sample materials. They had originally planned to explore Emory Crater, but the distance was deemed too great for the amount of time remaining in their first scheduled EVA. On their way to the site of their first exploration, they set explosive charges that were later detonated, with the result recorded on the instruments they had set up at the start of the EVA.

During their ride they lost the fender of the LRV, which resulted in a plume of dust following them throughout their travels.

The first EVA proved unexpectedly difficult for both astronauts. The accumulating dust caused Schmitt to suffer a bout of hayfever, and Cernan was badly bruised after struggling with the drill used to collect samples from beneath the lunar crust. Fortunately, they were able to return safely to the Challenger lunar module after a total first EVA of 7 hours and 12 minutes.

At the start of their second Moonwalk, on December 12, Schmitt and Cernan repaired the LRV. They then used the vehicle to travel to Nansen Crater, where they collected samples from the South Massif landslide.

They also planted more explosives, and explored Shorty Crater—where Schmitt made a colorful discovery: he found orange soil. The find initially set off a wave of excitement among NASA scientists, who pondered the possibility that the soil might have been an indicator of recent volcanic activity. Once they'd had a chance to study Schmitt's carefully collected samples, however, the scientific consensus was that the oddly tinted soil was volcanic glass that had resulted from the explosive volcanic activity that had formed the crater a million years earlier.

The longest EVA of the Apollo program, the second lunar surface excursion by Schmitt and Cernan lasted 7 hours and 37 minutes. They traveled more than 11 miles (19 kilometers) in the LRV before returning to the lunar module.

Schmitt and Cernan made the final Apollo exploration of the lunar surface on December 13, 1972, when they explored the North Massif area. Then, before returning to their Challenger lunar module, they unveiled a plaque on the spacecraft's spindly-legged lower stage, which would remain on the surface after they lifted off in the upper stage. Commemorating the conclusion of the final flight of the first manned program to explore the Moon, the plaque read "Here man completed his first exploration of the Moon, December 1972 A.D. May the spirit of peace in which we came be reflected in the lives of all mankind." The sentiment seemed particularly fitting in light of the remarkable achievements embodied by the activities of the Apollo 17 crew, which included new records for the longest manned lunar flight, the most EVA time on the lunar surface of any of the Moon landing missions, the largest amount of samples returned, and the longest stay in lunar orbit.

Schmitt and Cernan left the lunar surface on December 14, and then rejoined Evans in the America command module for the return trip to Earth. They had gathered approximately 250 pounds of carefully selected rocks and soil samples, and took 2,120 photographs during their time on the lunar surface.

The three Apollo 17 crew mates splashed down in the Pacific Ocean on December 19, 1972.

During his remarkable Apollo 17 flight, Harrison Schmitt spent 12 days, 13 hours and 52 minutes in space, including 22 hours and five minutes in EVA on the surface of the Moon.

He resumed his academic career in 1973, while still working for NASA, when he became a Sherman Fairchild Fellow at the California Institute of Technology. In February, 1974 he became chief of NASA's scientist astronauts, and later that same year, he was made assistant administrator for energy programs at the space agency.

Schmitt left NASA in August 1975 to launch a political career. He was elected to represent New Mexico in the United States Senate in November 1976, and subsequently served as the ranking Republican member of the Select Committee on Ethics; the Science, Technology and Space subcommittee of the Commerce, Science and Transportation committee; and the consumer subcommittee of the Banking, Housing and Urban Affairs committee. He was defeated in his re-election campaign of 1982.

In 1994 he became chairman and president of the Annapolis Center for Environmental Quality, and he has also served as an adjunct professor of engineering at the University of Wisconsin and worked as a consultant, writer and public speaker specializing in issues related to geology, public policy, science, space and technology.

Among the many honors he has received for his achievements as an astronaut and as a scientist, Schmitt has received the G.K. Gilbert Award and the Penrose Medal of the Geological Society of America, and the 1981 Engineer of the Year Award of the National Society of Professional Engineers.

SCHWEICKART, RUSSELL L.

(1935–)

U.S. Astronaut

A member of the third group of astronauts selected by NASA, who flew in space during the agency's epic Apollo program, Russell "Rusty" Schweickart was born in Neptune, New Jersey on October 25, 1935. He graduated from Manasquan High School in New Jersey, and then attended the Massachusetts Institute of Technology (MIT), where he earned a bachelor of science in aeronautical engineering and a master of science in aeronautics and astronautics.

During his time at MIT, he worked in the institute's Experimental Astronomy Laboratory as a research scientist. His research included study of the physics of the upper atmosphere, and stabilization of stellar images.

He also served in the United States Air Force and in the Air National Guard from 1956 to 1963, and accumulated 3,900 hours of flying time as a pilot, including 3,500 hours in jet aircraft.

NASA chose him for training as an astronaut as a member of its Group Three selection of pilot astronauts in October 1963.

On March 3, 1969, Schweickart lifted off as lunar module pilot during the Apollo 9 mission, with commander James McDivitt and command module pilot David Scott. Five days later, he and McDivitt entered the Apollo 9 lunar module Spider and began the first solo flight of the strange looking vehicle that would, in later versions, ferry astronauts to the surface of the Moon.

Schweickart and McDivitt fired the lunar module's engine during the test flight to propel the craft into a higher orbit than the command module Gumdrop, which was piloted by Scott. Then, after separating from the lunar module's descent stage, they achieved a flawless rendezvous and docking with the command module—a key test of the systems and procedures that would be used in lunar orbit during later missions.

Despite suffering motion sickness during the early portion of the flight, Schweickart also conducted a spacewalk designed to test the Apollo A7L Extravehicular Mobility Unit (EMU) spacesuit and Portable Life Support System (PLSS) backpack that were later worn by the astronauts who walked on the Moon during the Apollo 11, Apollo 12, and Apollo 14 missions.

Originally planned to last about two hours, the spacesuit test was trimmed to less than half its original length in consideration of the motion sickness that afflicted both Schweickart and his EVA partner David Scott during their first day in space.

On March 6, 1969, Schweickart tried out the 185 pound (85 kilogram) suit while inside the lunar module with the hatch door open; the hatch on the Gumdrop command module was also open during the EVA. He also tested the PLSS, which provided air to the astronaut and water to cool the spacesuit's temperature control system. Once he was fully suited up and the hatch was opened, he felt well enough to venture outside of the lunar module, and conducted a portion of the EVA while he was secured in the foot restraint on the narrow porch of the lunar module.

Schweickart also took photographs of Scott, who conducted his portion of the EVA from the command module as part of a test of the ability of command and lunar module crew to transfer from one vehicle to another in the event of an emergency, and also briefly worked his way around the exterior of the lunar module to retrieve experiment samples from the outside of the Spider lunar module. Each astronaut then returned to his respective craft at the end of the EVA, and after docking, Schweickart and McDivitt rejoined Scott in the command module. The three crew mates returned to Earth on March 13, 1969.

Rusty Schweickart spent 10 days and 1 hour in space, including 46 minutes in EVA, during the Apollo 9 flight.

He subsequently served as the backup commander for the first manned Skylab mission, and played a key role in devising repairs to the space station after it was damaged during launch. He also helped to develop the primary astronomy instrument used during Skylab, the Apollo Telescope Mount (ATM), and helped to plan the spacewalks conducted by Skylab crews.

In 1974 he transferred from the NASA Astronaut Office at the Johnson Space Center (JSC) in Houston to the agency's Office of Applications at NASA headquarters in Washington, D.C., where he served as director of user affairs.

SCOBEE, FRANCIS R.

(1939–1986)

U.S. Astronaut

During his distinguished career as an astronaut, Francis "Dick" Scobee was chosen as a crew member for two space shuttle missions, and as the commander of the ill-fated final mission of the space shuttle Challenger.

Born in Cle Elum, Washington on May 19, 1939, Scobee graduated from Auburn Senior High School in Auburn, Washington in 1957 and then enlisted in the United States Air Force. After receiving training as a reciprocating engine mechanic, he was assigned to Kelly Air Force Base in Texas.

Continuing his education concurrently with his military service, Scobee attended night school and accumulated two years of college credits. He was selected for the Air Force Airman's Education and Commissioning Program, and attended the University of Arizona, where he received a bachelor of science in aerospace engineering in 1965.

He was commissioned that same year, and received his wings in 1966. He was then deployed overseas, serving a combat tour during the Vietnam War.

After his return to the United States, he attended the Air Force Aerospace Research Pilot School at Edwards Air Force Base in California, and then served as a test pilot flying a variety of aircraft, including the Boeing 747, the X-24B, F-111 and C-5 aircraft.

During his outstanding military career, Scobee accumulated more than 6,500 hours of flying time in 45 different types of aircraft. He was a recipient of the Air Force Distinguished Flying Cross and the Air Medal, and was a member of the Society of Experimental Test Pilots, Tau Beta Pi, the Air Force Association, and the Experimental Aircraft Association.

NASA selected him for astronaut training in 1978, and he became an astronaut the following year. His duties at the space agency included service as an instructor pilot on the Boeing 747 shuttle carrier airplane.

Scobee first flew in space as pilot of the space shuttle Challenger during STS-41C, which launched on April 6, 1984. The flight was commanded by Robert Crippen, and included fellow crew members George Nelson, James van Hoften and Terry Hart. The STS-41C crew achieved a landmark on-orbit salvage mission, as they retrieved and repaired the Solar Maximum satellite (which was whimsically nicknamed "Solar Max").

The STS-41C launch featured the first direct ascent trajectory of the shuttle program—meaning that the shuttle Challenger achieved orbit by a longer-than-usual firing of its Space Shuttle Main Engines (SSMEs), and just one (rather than the usual two) firings of its Orbital Maneuvering System (OMS) engines. In the typical sequence, one OMS burn maneuvers the shuttle into an elliptical orbit, and a second OMS firing places the vehicle in an almost-circular orbit of the Earth. In the case of STS-41C, the direct ascent trajectory was utilized to propel Challenger directly into a higher-than-usual orbit (nearly 300 miles from Earth) so the shuttle could rendezvous with the Solar Max satellite.

After the Challenger crew tracked down and achieved rendezvous with the satellite, Nelson and van Hoften retrieved the Solar Max craft by using the nitrogen-propelled Manned Maneuvering Units (MMUs) to sidle up to the spacecraft and wrestle it into the shuttle's cargo bay. The task proved more difficult than anticipated, and mission controllers extended the flight by one day to make sure that the astronauts would have enough time to make the planned repairs.

Once they'd managed to fasten Solar Max firmly to Challenger's cargo bay, Nelson and van Hoften replaced the satellite's malfunctioning altitude control system and its coronagraph/polarimeter, and then returned the craft to separate flight. A remarkable achievement, the repair mission was the first of its kind—the first-ever planned fix to be made on a spacecraft while it was in orbit.

The crew also successfully achieved its other primary objective for STS-41C with the deployment of the Long Duration Exposure Facility (LDEF), an orbiting cylinder containing 57 experiments designed to test the impact of long stays in space on a variety of items, including various materials and seeds.

STS-41C crew members also used Cinema 360 and IMAX cameras to record their activities, and oversaw a student experiment studying the impact of the microgravity environment on the honeycomb-building activities of bees.

Challenger returned to Earth on April 13, 1984, after a flight of 6 days, 23 hours, 40 minutes and 7 seconds, and 108 orbits.

Scobee next served as commander of STS-51L, which launched on January 28, 1986 at the Kennedy Space Center (KSC) in Florida.

Tragically, commander Scobee and his fellow Challenger crew mates were killed when a fault in one of the shuttle's huge solid rocket boosters caused a fuel leak that ignited and caused a massive explosion just 73 seconds after liftoff. He was 46 at the time of his death.

Dick Scobee is survived by his wife, June, and two children. He was posthumously awarded the Congressional Space Medal of Honor.

SCOTT, DAVID R.

(1932–)

U.S. Astronaut; Seventh Human Being to Walk on the Surface of the Moon

The seventh person to walk on the surface of the Moon and a veteran of three spaceflights, David Scott was born in San Antonio, Texas on June 6, 1932. He graduated from Western High School in Washington, D.C., and then attended the United States Military Academy at West Point, New York, where he received a bachelor of science degree while graduating fifth in his class of 633 students.

After receiving his initial training as a pilot at Webb Air Force Base in Texas and receiving gunnery training at Laughlin Air Force Base in Texas and Luke Air Force Base in Arizona, Scott served from 1956 to 1960 at Soesterberg Air Base in the Netherlands while assigned to the 32nd Tactical Fighter Squadron.

In 1960 he returned to the United States to attend the Massachusetts Institute of Technology (MIT), where he subsequently earned a master of science degree in aeronautics and astronautics, and an engineer degree in aeronautics and astronautics.

He also graduated from the Air Force Experimental Test Pilots School and the Air Force Aerospace Research Pilot School at Edwards Air Force Base in California.

During his exceptional military career, Scott accumulated more than 5,600 hours of flying time. He rose to the rank of colonel in the United States Air Force by the time of his retirement from the service in March 1975.

NASA chose Scott for training as an astronaut as a member of its Group Three selection of pilot astronauts in October 1963.

He first flew in space as pilot of the Gemini 8 mission, with Neil Armstrong, who served as commander of the flight. Lifting off on March 16, 1966, Scott and

Apollo 15 commander David Scott salutes the American flag on the surface of the Moon, August 1, 1971. [NASA/courtesy of nasaimages.org]

Armstrong were given the task of attempting to rendezvous and dock with an unmanned Agena rocket that was launched the same day to act as a target vehicle for their Gemini 8 craft.

At the beginning of the flight, Scott and Armstrong entered an elliptical orbit of 160 by 272 kilometers, and were 1,963 kilometers distant from the Agena. During their first two trips around the Earth, the Gemini 8 crew members began their pursuit of the target vehicle with several firings of their spacecraft's thrusters. At the end of each of the first two firings, Armstrong noticed that the thrusters were not cleanly shutting off at the end of each burn. The problem posed no immediate danger, but made it difficult for the astronauts to determine whether or not the maneuvers were achieving their intended effect.

Additional difficulties cropped up with the balky thrusters as Gemini 8 drew nearer to the Agena target vehicle, and Armstrong was forced to make an additional adjustment to the spacecraft's orbital position to fully close the distance between the Gemini spacecraft and the Agena target. Then, with Scott calling out radar readings to help him align the two vehicles, Armstrong slowed Gemini 8 to a stop alongside its target, successfully achieving the second rendezvous of the U.S. space program. A period of "stationkeeping" (keeping the two vehicles at a constant chosen distance) followed, for about 36 minutes.

After confirming the successful rendezvous, Scott and Armstrong were cleared to attempt a milestone achievement: the first docking of the U.S. space program. Armstrong expertly maneuvered the Gemini craft toward the Agena target vehicle, and docked without difficulty—giving rise to a crescendo of cheers in Mission Control.

The easy docking was a pleasant surprise to the engineers and mission managers who were following the flight's progress from Earth. It fulfilled the primary goal of the Gemini 8 mission profile, and added an important skill to NASA's portfolio of skills that were crucial for subsequent flights.

For Scott and Armstrong, the docking exercise was just the first of a series of maneuvers they intended to carry out. As they prepared for their next activity, in which they would use the Agena's attitude control system to alter the orbital position of the linked Agena and Gemini vehicles, they received disturbing news from their fellow astronaut James Lovell, who was serving as spacecraft communicator ("Capcom") at Mission Control. Lovell informed them that the target vehicle's attitude control system had apparently suffered a malfunction. He advised them to shut the system down if it appeared to be acting erratically, and use Gemini 8 to control the linked flight instead.

Lovell's warning was the last interaction Scott and Armstrong had with the ground before they passed out of communications range as they passed over China. With the possibility of control difficulties foremost on his mind, Scott began the first docked maneuver, testing the Agena's ability to move the two docked spacecraft. The Agena seemed at first to behave as expected, but Scott noticed that the movement he initiated cut off more quickly than planned. Checking off the next item on his list, he remotely activated the Agena's tape recorder, but as he did, he noticed that the docked vehicles had begun to rotate slightly.

As the instruments within their spacecraft began to register ominous readings about the rapidly developing predicament, indicating that they were rolling at an angle of 30 degrees, Armstrong tried to counter the rolling motion by firing the Gemini craft's maneuvering rockets. At first, the procedure appeared to work; but after a short period of level flight, the linked Gemini and Agena began to revolve again.

Taking Lovell's advice, Scott tried shutting down the target vehicle's attitude control system, which briefly resulted in a period of level flight. He tried the procedure again when the vehicles resumed their spinning motion, and then tried to document the increasingly alarming situation by photographing the Agena's behavior through the window of Gemini 8.

With the spinning motion increasing as the docked flight progressed, both Gemini 8 crew members recognized the dangers posed by their circumstances. In addition to the possibility that the rapid spinning might cause damage to their spacecraft, the rotating motion could also have affected their ability to maintain control of the vehicle, and could lead directly to their becoming ill or incapacitated. They quickly agreed that the best course of action would be to disengage their Gemini 8 craft from the target vehicle.

Scott transferred control of the Agena to flight controllers on the ground, and Armstrong then undocked their vehicle from the target. Rather than solving the problem, however, the separation from the Agena resulted in Gemini 8 spinning even more rapidly. Scott radioed the astronauts' concern to Mission Control: "We have serious problems here . . . We're tumbling end over end up here."

The bad news from orbit stunned mission managers, who tried to react to the situation even as it drastically changed on a virtually minute-by minute basis. Without the luxury of much time to try to develop an approach to the problem, NASA's engineering teams struggled to come up with immediate answers to the unexpected sudden problems, sorting through the large amount of data available from the spacecraft's telemetry system to do whatever they could to help Scott and Armstrong as quickly as possible.

Equipped with years of expertise gained from their long careers as pilots and test pilots, and drawing upon their intimate knowledge of the Gemini 8 spacecraft, Scott and Armstrong were able to quickly diagnose their problem as being the result of a fault in their spacecraft's thruster system. They tried to turn off the vehicle's maneuvering rockets, but found that they couldn't.

Spinning at a rate of one full end-over-end revolution each second, both Scott and Armstrong experienced a sickening dizziness and blurred vision. They conferred briefly, and fully understanding both the gravity of their circumstances and the potential consequences of the solution they'd each hit upon, they decided to try to counter the revolving motion by using Gemini 8's re-entry control system. The procedure involved great risk, as the firing of the re-entry rockets threatened to prevent them from properly orienting their spacecraft for its return to Earth; at the same time, they needed to find some way of immediately ending the spinning motion or they would not survive long enough to worry about returning to Earth.

The first two attempts to counter the spinning by using the re-entry control system failed. Armstrong tried the procedure first, with no effect; and then Scott tried, and again the system failed to engage. They then ran through the prescribed re-entry procedures—maintaining an almost unimaginable presence of mind and courageous professionalism despite the imminent threat of their circumstances—and were able to get the re-entry system working. Within seconds Gemini 8 stopped spinning, and they were able to analyze the malfunction that had set the vehicle in motion. As they had suspected, one of the spacecraft's maneuvering thrusters had become stuck in the "open" position and pushed the craft continuously around in a circular motion.

The early use of the spacecraft's re-entry system dictated an immediate end to the mission, and mission controllers instructed Scott and Armstrong to begin preparations for an early return to Earth shortly after the astronauts had solved the thruster problem. They splashed down in the Pacific Ocean (their return had originally been planned for the Atlantic) on March 17, 1966, after a flight of 10 hours, 41 minutes and 26 seconds.

Scott began his second space mission on March 3, 1969, as command module pilot of Apollo 9. James McDivitt served as commander for the flight, and Russell "Rusty" Schweickart was lunar module pilot during the first test of the complete Apollo system—with the Apollo command, service and lunar modules put through a series of exercises designed to prove their capabilities prior to the planned lunar landing missions.

Shortly after separation from their launch vehicle, Scott piloted the Apollo 9 "Gumdrop" command module through the process of docking with the lunar module, which had been given the call sign "Spider," and the joined spacecraft then flew together in orbit for the first time.

McDivitt and Schweickart entered the lunar module several days later and undocked from the command module to perform a series of tests. They first moved into a higher orbit, and then ran through a simulation of the process that would be used by later crews during the lunar landings.

Although Scott and Schweickart experienced symptoms of motion sickness during the early portion of the flight, they gamely carried out an important test of the Apollo A7L Extravehicular Mobility Unit (EMU) spacesuit and Portable Life Support System (PLSS) backpack that were later used during the first lunar landing missions. Their EVA was originally scheduled to last about two hours in open space, but McDivitt and officials in Mission Control decided it would be best to limit the exercise to less than an hour, and for Scott and Schweickart to conduct their activities from the inside of their command and lunar modules, with the hatch of each vehicle opened to allow them to work in the appropriate conditions of weightlessness.

The EVA involved a test of the crew members' ability to transfer from one vehicle to another in the event of an emergency, and Schweickart also tested the 185 pound (85 kilogram) spacesuit and PLSS during the 46 minute spacewalk.

The three crew members then reunited in the Gumdrop command module, jettisoned the Spider lunar module, and returned to Earth on March 13, 1969, after a flight of 10 days and 1 hour.

After his first two spaceflights, Scott served as backup commander for the second lunar landing mission, Apollo 12.

On July 26, 1971, he launched on a remarkable third spaceflight, as commander of Apollo 15. The first of three lunar landing flights designed to conduct a more extensive exploration of the lunar surface, Apollo 15 was also the first mission to feature the use of a lunar roving vehicle (LRV), the electric Moon car designed by the Boeing Corporation.

Scott's crew for Apollo 15 included James Irwin, who served as pilot of the Falcon lunar module, and Alfred Worden, who piloted the command module Endeavor. Together, they traveled to the Moon, where Worden completed 74 lunar orbits in Endeavor over the course of 145 hours, and Scott and Irwin spent 66 hours and 54 minutes on the lunar surface and made three trips outside the Falcon lunar module to explore the Hadley Rille valley near the Apennine Mountains of the Moon.

Although it was not part of the three primary EVAs, Scott made the first partial, tentative excursion of the lunar stay when he ventured halfway out of the lunar module's hatch on July 30 to examine the area in which he and Irwin had set down on the surface.

Because Apollo 15 was the first of the lunar landing missions to set astronauts down in a region removed from the Moon's equator, and its mission profile called for longer and more extensive exploration of the geologically rich area they were to explore, Scott took advantage of the early opportunity to survey the terrain that he and Irwin would cross the following day. Surveying the route they would take during the first scheduled EVA, he got his first good look at Mount Hadley, which towered to an elevation of 16,500 feet (5,000 meters) nearby the landing site. Then, satisfied with his survey of the area, he closed the hatch after 33 minutes.

Scott and Irwin made their first "official" Moonwalk the following day, July 31, 1971. Irwin collected the "contingency sample" of lunar surface materials—the sample the astronauts would spirit away in the event that some unforeseen emergency forced them to leave the surface earlier than expected—and Scott set up a television camera that would record their movements. They then pulled the LRV from its storage space on the lunar module and prepared the vehicle for its first use on the lunar surface. The task took longer than it had in simulations on Earth, and the rover's front wheel steering mechanism was not working, but Scott was able to steer with the rear wheels to navigate across the lunar surface. He and Irwin traveled several miles south of their landing site, along the Elbow Crater to the St. George Crater, where they collected samples of surface materials. When they completed their first geological exploration, they returned to the lunar module, where they set up the Apollo Lunar Surface Experiments Package (ALSEP).

Several difficulties arose during the first EVA. A frightening moment occurred when Irwin's life support system backpack unit suddenly flashed a failure warning; the reading turned out to be false, however, and was the result of the unit having been tilted to one side when it was recharged in the lunar module prior to the start of the EVA. Because Falcon had landed with one footpad in a shallow crater, the lunar module was slanted about 30 degrees, which had led to the problem with Irwin's backpack.

Scott also suffered a difficulty with his life support system during the first EVA; he used more oxygen than expected, which led mission controllers to cut the exercise short by about 30 minutes, to a total of 6 hours and 34 minutes.

The backpack recharging process also delayed the start of the second EVA by 30 minutes.

During their second trip outside the lunar module, on August 1, Scott and Irwin were surprised to find the front wheel steering of their rover somehow miraculously restored, and they made a seven-and-a-half mile (12.5 kilometers) tour of the area between their landing site and Mount Hadley, driving to an elevation of 300 feet above their Falcon lunar module.

Scott and Irwin collected several remarkable samples during their second excursion, including the "Genesis Rock," which provided scientists with clues to how the Earth and Moon had been formed.

They also planted the American flag in the lunar soil before the end of the second EVA, which lasted 7 hours and 13 minutes.

On August 2 they made their final exploration of the Moon, adding 4 hours and 20 minutes to their EVA total for the mission. Another scary moment preceded the third EVA, when medical personnel monitoring their health from Earth found that both astronauts were registering irregular heartbeats. After an extra rest period of an hour and 45 minutes, they were cleared to carry on as planned. Their momentary health difficulties were later diagnosed as the result of a potassium deficiency.

When they emerged for their final trek across the lunar landscape, Scott set a then-record with the fifth EVA of his career, a mark that was not eclipsed until 1984.

He and Irwin got a closer look at the Hadley Rille during their third trip outside the lunar module. They also carried out several activities that were intended to demonstrate the less serious aspects of scientific adventure on the lunar surface—and which in one case resulted in controversy after the conclusion of the mission.

Connecting for a moment the Earthly duties of the postman and the otherworldly adventure of the lunar explorer, the astronauts hand-canceled a U.S. postage stamp—creating a sought-after collectible item. Along with several other similarly prepared souvenirs, the item was later sold, and the entire incident resulted in a great deal of criticism of the astronauts and of NASA.

A far more positive reaction greeted an innovative demonstration of the theory of gravity first postulated by Galileo Galilei as "the law of uniformly accelerated motion toward the Earth," which Scott corroborated by simultaneously dropping a hammer and a feather, which each fell to the surface in the light lunar gravity and landed at the same time.

Scott is also well remembered for two other aspects of his participation in the Apollo 15 mission. In the first, he and Irwin left a small plaque on the lunar surface that bore the names of the astronauts and cosmonauts who had perished during the long, intense competition between the United States and the Soviet Union to be the first to land on the Moon. Next to the commemorative plaque, the astronauts left a tiny figure, symbolic of their fallen comrades on both sides of the ideological divide who, by their inclusion in the tiny memorial, reached the lunar surface at least in memory, if not in fact.

The Apollo 15 commander also eloquently placed his lunar odyssey in the larger context of the higher spiritual goals of space exploration. Speaking from the lunar surface, he expressed his wonder at the experience: ". . . I realize a fundamental truth about our nature. Man must explore. And this is exploration at its greatest."

At the conclusion of their third traverse across the lunar surface, Scott and Irwin returned to the Falcon lunar module and closed out their EVA activities. They then rejoined Worden in the Endeavor command module, bringing with them 169 pounds (76.6 kilograms) of carefully selected lunar surface samples. The three astronauts splashed down in the Pacific Ocean on August 7, 1971, after a flight of 12 days, 17 hours and 12 minutes.

David Scott spent more than 23 days in space, including more than 19 hours in EVA during five spacewalks, during his career as an astronaut.

SCULLY-POWER, PAUL D.

(1944–)

U.S. Astronaut; First Native of Australia to Fly in Space

Paul Scully-Power was born in Sydney, Australia on May 28, 1944. He was educated in Sydney and in London, England, and attended the University of Sydney, where he graduated with honors, earning a bachelor of science degree and a post graduate diploma of education in 1966.

In 1967 he founded the first oceanographic unit of the Royal Australian Navy, which he led for the next five years, simultaneously serving as a scientific officer. He was assigned to the U.S. Naval Underwater Systems Center in New London, Connecticut and to the Office of Naval Research in Washington, D.C. in July 1972, as an Australian Navy exchange scientist working with the United States Navy.

Scully-Power began his association with NASA during his time as an exchange scientist, when the agency asked him to participate in the Skylab space station program as a member of the Skylab Earth Observations team.

Following his return to Australia in March 1974, he organized the first-ever oceanographic and acoustic measurement of an ocean eddy, the Anzus Eddy project, which involved scientists from Australia, New Zealand, and the United States.

In 1976, he resumed his involvement with space activities as the principal non-U.S. investigator for the Heat Capacity Mapping remote sensing satellite.

Scully-Power took up full-time residence in the United States the following year, to work as a senior scientist and technical specialist at the U.S. Naval Underwater Systems Center. He became an American citizen on September 17, 1982.

Widely regarded for his academic research and oceanographic expertise, Scully-Power has served as chief scientist during 13 scientific cruises, and has been a part of 24 oceanographic expeditions. He has published 60 articles in scientific journals. A qualified Navy diver, he has received 18 special achievement awards from the United States Navy.

On October 5, 1984, Scully-Power became the first native of Australia to fly in space, when he lifted off as a payload specialist aboard the space shuttle Challenger during STS-41G.

Robert Crippen commanded the STS-41G mission, and the crew also included pilot Jon McBride, mission specialists Kathryn Sullivan, Sally Ride and David Leestma, and Scully-Power's fellow payload specialist Marc Garneau, who with STS-41G became the first citizen of Canada to fly in space.

Scully-Power conducted oceanographic research for the U.S. Naval Research Laboratory during the flight.

The crew also deployed the Earth Radiation Budget Satellite (ERBS) and conducted three experiments for NASA's Office of Space and Terrestrial Applications (OSTA-3); Garneau conducted a series of experiments dubbed "CANEX" in honor of their having been designed by scientists in Canada; and on October 11 Sullivan became the first American woman to perform a spacewalk, when she and Leestma spent 3 hours and 29 minutes outside the shuttle to conduct a test of the Orbital Refueling System (ORS) that proved that it is possible to refuel satellites in orbit.

Challenger returned to Earth on October 13, 1984, landing at the Kennedy Space Center (KSC) in Florida.

Paul Scully-Power spent 8 days, 5 hours, 23 minutes and 38 seconds in space during the STS-41G mission, traveling more than 3.4 million miles and making 132 orbits of the Earth.

SEARFOSS, RICHARD A.

(1956–)

U.S. Astronaut

Richard Searfoss was born in Mount Clemens, Michigan on June 5, 1956. In 1974 he graduated from Portsmouth Senior High School in Portsmouth, New Hampshire, and then attended the United States Air Force Academy, where he received a bachelor of science degree in aeronautical engineering in 1978. At graduation from the Air Force Academy, he was honored with the Harmon, Fairchild, Price and Tober Awards as the top overall graduate, and most outstanding academic, engineering and aeronautical engineering graduate.

After graduating from the Air Force Academy, he received a National Science Foundation scholarship to attend the California Institute of Technology, where he earned a master of science degree in aeronautics in 1979.

He also attended the Air Force Squadron Officer School, where he received the Commandant's Trophy as the top graduate in his class, the Air Command and Staff College, and the Air War College, and he was recognized as a distinguished graduate of the Air Force Fighter Weapons School. He has also been honored with an award for excellence in turbine engine design by the Air Force Aero Propulsion Laboratory.

Searfoss received his initial pilot training at Williams Air Force Base in Arizona, and was then assigned to Royal Air Force Base Lakenheath, in England, where he flew the F-111F aircraft. Upon his return to the United States, he served as an F-111A instructor pilot and weapons officer at Mountain Home Air Force Base in Idaho. In 1985 he received the Tactical Air Command F-111 Instructor Pilot of the Year award.

In 1988 he attended the U.S. Naval Test Pilot School at Patuxent River, Maryland as an Air Force exchange officer, and he was subsequently assigned to the U.S. Air Force Test Pilot School at Edwards Air Force Base in California, where he served as a flight instructor.

During his military career, Searfoss accumulated more than 5,000 hours of flying time in 56 different types of aircraft. He holds FAA Airline Transport Pilot, glider and flight instructor ratings. He retired from the United States Air Force in 1998 as a colonel.

NASA selected him for training as an astronaut in 1990.

Searfoss first flew in space as pilot of the space shuttle Columbia during STS-58, in 1993—the second Spacelab flight dedicated to life sciences. Launched on October 18, 1993, STS-58 was the longest flight of the shuttle program up to that time, at just over 14 days. Fourteen Spacelab experiments were conducted during the mission, in the areas of regulatory physiology, the cardiovascular and cardiopulmonary systems, the musculoskeletal system, and neuroscience. The crew also conducted 16 engineering tests and 20 extended duration medical experiments before Columbia returned to Earth at Edwards Air Force Base in California on November 1, 1993, after 225 orbits.

During his second spaceflight, Searfoss piloted the shuttle Atlantis for STS-76, in March 1996. The third docking mission between a shuttle and the Mir space station, STS-76 marked the beginning of a continuous U.S. presence aboard the Russian space station, with the transfer of NASA astronaut Shannon Lucid from Atlantis to Mir to begin a four-and-a-half month stay. The U.S.-Russian program of cooperative living in space was part of the two nations' initial arrangements for their subsequent cooperation in the International Space Station (ISS) program.

The STS-76 crew also made the first use of the SPACEHAB pressurized module to support the shuttle-Mir docking. During five days of linked flight, the Atlantis crew and Mir 21 cosmonauts Yuri Onufrienko and Yury Usachev transferred 2 tons of equipment and supplies and 1,500 pounds of water from the shuttle to the station. At the conclusion of STS-76, Atlantis returned to Earth at Edwards Air Force Base on March 31, 1996.

On April 17, 1998, Searfoss lifted off on a remarkable third space mission, as commander of STS-90—the Spacelab Neurolab flight. The twenty-fifth and final Spacelab flight, STS-90 featured 26 experiments designed to gauge the effects of weightlessness on the brain and nervous system, in tests designed by scientists at a variety of organizations, including NASA, the Canadian Space Agency (CSA), the French space agency Centre National d'Études Spatiales (CNES), the German space agency Deutsches Zentrum für Luft- und Raumfahrt e.V. (DARA), the European Space Agency (ESA) and the National Space Development Agency (NASDA) of Japan.

The 16-day mission—the twenty-fifth flight of the space shuttle Columbia—came to a close on May 3, 1998, with the shuttle's landing at the Kennedy Space Center (KSC) in Florida, after 256 orbits.

Richard Searfoss accumulated more than 939 hours in space during his three space shuttle missions. He left NASA in 1998 to pursue a private sector career, and subsequently worked at the NASA Dryden Flight Research Center (DFRC) at Edwards Air Force Base as a research test pilot until February 2003, when he returned to private business interests.

SEDDON, MARGARET RHEA

(1947–)

U.S. Astronaut

Rhea Seddon was born in Murfreesboro, Tennessee on November 8, 1947. She attended the University of California, Berkeley, where she received a bachelor of arts in physiology in 1970. In 1973 she received a doctorate of medicine from the University of Tennessee College of Medicine.

She completed a surgical internship and began her medical career in Memphis, Tennessee, where she served for three years of general surgery residency. She also worked as an emergency room physician in hospitals in Mississippi, Tennessee and Texas. She has taken a special interest in the role that nutrition plays in the health and recovery abilities of surgery patients, and has also conducted clinical research into the impact that radiation therapy has on the nutrition of cancer patients.

Seddon became an astronaut in 1979. She first flew in space as a mission specialist during STS-51D in April 1985, aboard the space shuttle Discovery. During that mission, she operated the shuttle's robotic arm (the Remote Manipulator System, or RMS) as part of an effort with fellow mission specialists David Griggs and Jeffrey Hoffman to salvage the deployment of the SYNCOM IV-3 satellite.

In June 1991 she served aboard Columbia as one of three medical doctors assigned to the STS-40 Spacelab Life Sciences (SLS-1) mission, participating in a milestone effort that returned an unprecedented amount of medical data.

Seddon was payload commander for the second Spacelab Life Sciences mission, STS-58, aboard the shuttle Columbia in 1993. During the 14-day flight, the crew performed a wide array of medical experiments that added significantly to aerospace medicine and biomedical research.

During her three space missions, Rhea Seddon spent more than 722 hours in space.

In 1996 NASA assigned her to Vanderbilt University Medical School in Nashville, Tennessee, where she helped to design cardiovascular experiments that flew aboard the shuttle Columbia as part of the Neurolab experiments during the final Spacelab mission, STS-90, in 1998.

Seddon retired from NASA in 1997, and has since served as assistant Chief Medical Officer of the Vanderbilt Medical Group in Nashville. She is married to former NASA astronaut Robert L. Gibson; they have three children.

SEE, ELLIOT M., JR.

(1927–1966)

U.S. Astronaut

Elliot See was born in Dallas, Texas on July 23, 1927. He attended the U.S. Merchant Marine Academy, where he received a bachelor of science degree.

He began his career at the General Electric Company in 1949, where he worked as a flight test engineer, group leader and experimental test pilot. In 1953 he was designated a Naval Aviator, and in 1956 he returned to General Electric. His work there included service as a project pilot developing the J79-8 engine for the F4H aircraft, and he also conducted tests of a variety of other engines, while flying in F4H, F11F-1F, F-86, F-104, RB-66, T-38 and XF4D aircraft.

See also attended the University of California at Los Angeles, where he earned a master of science degree in engineering.

During his outstanding career as a pilot and test pilot, See accumulated more than 3,700 hours of flying time, including 3,200 hours in jet aircraft.

NASA selected him for training as an astronaut in September 1962, as a member of its Group Two selection of pilot astronauts. His fellow Group Two astronauts included Neil Armstrong, Frank Borman, Charles "Pete" Conrad, James Lovell, James McDivitt, Thomas Stafford, Edward White, and John Young.

When he joined the astronaut corps, See was given responsibilities focusing on the design and development of guidance and navigation systems. He served as a member of the backup crew for the Gemini 5 mission, and on November 8, 1965, he received word that he had been chosen to command the Gemini 9 mission in 1966, on which he was scheduled to fly with Group Three astronaut Charles Bassett.

But it was not to be. On February 28, 1966, while flying in a T-38 jet airplane to the McDonnell Aircraft plant in St. Louis, See and Bassett were killed when their plane encountered dense fog and crashed.

Their shocking deaths served as a poignant reminder of the deep devotion of the early astronauts, whose long hours of training and sacrifices of time and travel and study were largely lost to public attention that was, understandably, focused on the space missions themselves.

Elliot See was survived by his wife Marilyn and their three children.

SEREBROV, ALEKSANDR A.

(1944–)

Russian Cosmonaut

A veteran of 4 space missions, including 2 long-duration flights and 10 spacewalks, Aleksandr Serebrov was born in Moscow, Russia on February 15, 1944.

He attended the Institute of Physics and Technology in Moscow, where he graduated in 1967, and in 1974 he earned the degree of candidate of technical sciences. He worked as an engineer before his selection as a cosmonaut.

Serebrov was chosen for training as a cosmonaut in 1978, as a member of the civilian specialist Group Six selection of cosmonaut candidates.

He made his first flight in space as flight engineer aboard Soyuz T-7, with commander Leonid Popov and research engineer Svetlana Savitskaya—who with the Soyuz T-7 flight became the second woman to fly in space (Valentina Tereshkova had been first, during Vostok 6 in June 1963).

Launched on August 19, 1982, Soyuz T-7 was a spacecraft switching flight, in which a Soyuz that had been docked at a space station for a long period would be replaced by a fresh spacecraft before the expiration of the first vehicle's 90-day safe operations limit.

Serebrov, Popov, and Savitskaya traveled to the Salyut 7 space station, where they visited with resident crew members Anatoli Berezovoi and Valentin Lebedev for about a week. They then left their Soyuz T-7 craft docked at the station and returned to Earth in Soyuz T-5, after nearly eight days in space.

During his second space mission, as flight engineer during Soyuz T-8, Serebrov was scheduled to serve as part of a long-duration crew aboard Salyut 7 with commander Vladimir Titov and fellow flight engineer Gennady Strekalov. When they lifted off from the Baikonur Cosmodrome on April 20, 1983, however, the radar antenna on their spacecraft was torn loose. When they arrived at Salyut 7, it quickly became apparent that they would not be able to make an automatic docking as planned, and after several manual attempts at rendezvous and docking failed, the crew was forced to abandon the mission and return to Earth after a frustrating flight of just over two days in space.

Serebrov was next assigned with Titov to the Soyuz TM-2/Mir 2 long-duration mission, but a medical problem forced him to give up the assignment, and he and Titov were ultimately replaced by their backup crew, Yuri Romanenko and Aleksander Laveykin. He also served as a member of the backup crew for the Soyuz TM-7/Mir 4 long-duration crew, which launched in November 1988.

His third spaceflight began on September 6, 1989, when he served as flight engineer with commander Aleksandr Viktorenko during Soyuz TM-8/Mir 5. During a stay of more than five months aboard the Mir space station, Serebrov and Viktorenko carried out a busy, productive mission in which they performed scientific experiments, marked the expansion of the space station with the addition of the Mir Kvant 2 module, received supplies from the first of the new Progress "M" series of cargo resupply spacecraft, and made five remarkable spacewalks, during which Serebrov tested the Sredstvo Peredvizheniy Kosmonavtov (SPK; translates to "Cosmonaut Maneuvering Equipment" in English) EVA backpack in space.

The Kvant 2 module was launched on November 26, 1989, and was to have been automatically docked with Mir on December 2, but a computer problem prevented that first docking attempt. Serebrov and Viktorenko helped mission controllers work out the problem, and Kvant 2 was successfully added to Mir on December 6.

The cosmonauts made their first spacewalk on January 8, 1990. They retrieved Mir Environmental Effects Payloads (MEEPs) experiment packages from the outside of the station and installed equipment during that first excursion, and then made a second, similar trek on January 11.

On January 26 they attached EVA hardware to the outside of the station, in preparation for the later MMU tests.

Then, on February 1, 1990, Serebrov and Viktorenko conducted the first-ever test of the SPK backpack. The Soviet equivalent of the U.S. Manned Maneuvering Unit (MMU), the SPK was designed to allow the wearer to propel himself freely through space, without necessarily requiring the use of a tether (although Serebrov was tethered during the test, as a safety precaution).

Serebrov had trained for several years on Earth before embarking on the SPK test in orbit, and his long preparation was amply demonstrated in the successful test. He put the SPK through a series of planned maneuvers, and thoroughly demonstrated the capabilities of the EVA backpack while Viktorenko documented the exercise on videotape during an EVA of about five hours. The cosmonauts carried out a similar exercise during a 3 hour, 45 minute EVA on February 5, and then, with their long, busy mission complete, they returned to Earth on February 19, 1990, in Soyuz TM-8.

During his third spaceflight, Serebrov spent more than 166 days in space, and accumulated 17 hours and 36 minutes in EVA during five spacewalks.

He next served as a member of the backup crew for Soyuz TM-16/Mir 13, which launched in January 1993.

On July 1, 1993, Serebrov began a remarkable fourth space mission when he launched aboard Soyuz TM-17 with Vasili Tsibliyev and visiting cosmonaut Jean-Pierre Haigneré of France. He again traveled to Mir, where he and Tsibliyev became the station's fourteenth long-duration crew, while Haigneré spent several weeks at Mir and then returned to Earth with the Mir 13 crew.

During the Mir 14 long-duration flight, Serebrov made another five spacewalks, while also carrying out scientific work and performing the engineering and maintenance tasks necessary to keeping the station in good working order.

The first two EVAs he and Tsibliyev made during their Mir 14 stay, on September 16 and September 20, 1993, were devoted to the installation of a truss segment, Rapana. The work was thought to be related to plans for a second Mir space station, a larger, more technically advanced version that would have been able to accommodate visits from the Russian space shuttle Buran. Although both the Mir-2 station and the shuttle program were ultimately abandoned, Serebrov and Tsibliyev were able to attach the 57.2 pound Rapana Truss to the station as planned over the course of their first two EVAs.

They made their third spacewalk on September 28, to begin the Panorama survey—a detailed inspection of the outside of Mir, which was intended to collect evidence that the station was in good enough condition to be used for future cooperative missions with NASA and the European Space Agency (ESA). During the first attempt to complete the survey, Tsibliyev's spacesuit developed a malfunction, and the cosmonauts were forced to cut their EVA short after 1 hour and 52 minutes. Serebrov was able to capture video and still images of a good portion of the station's exterior, but the originally planned four hour spacewalk was curtailed before he could complete the work.

A second attempt to finish the Panorama survey also had to be given up, when Serebrov suffered a malfunction in the oxygen supply of his Orlan-DMA spacesuit on October 22. Later investigation revealed that the suit had been used on 13 previous spacewalks, and should have been discarded.

Serebrov and Tsibliyev were finally able to complete the survey of Mir's exterior on October 29, during an EVA of 4 hours and 12 minutes. With his fifth Mir 14 spacewalk, Serebrov set a new record for the most career EVAs up to that time, at 10.

While they worked outside of Mir during the October 29 EVA, the cosmonauts experienced an eerie encounter with a piece of metal, which inexplicably drifted past them. They could not account for the sudden appearance of the debris.

They had a more pleasant time at the end of their final spacewalk, when they engaged in a bit of mischievous good humor while discarding the spacesuit that had failed Serebrov earlier in the flight. They strapped the arm of the suit to its neckpiece, to create the illusion that the empty spacesuit was engaged in a salute, and then released the empty, worn-out suit from the open hatch of the Mir Kvant 2 module.

Then, with their long mission finally concluded on January 14, 1994, Serebrov and Tsibliyev returned to Earth in Soyuz TM-17 after a remarkable flight of more than 196 days, including a total of 14 hours and 13 minutes in EVA.

Aleksandr Serebrov accumulated a total of more than 372 days in space during his career as a cosmonaut, including more than 31 hours in EVA during 10 spacewalks.

He left the cosmonaut corps in 1995 but continued to serve the Russian space program as an adviser after leaving active duty.

SEVASTYANOV, VITALI I.

(1935–)

Soviet Cosmonaut

A veteran of two long-duration space missions, Vitali Sevastyanov was born in Krasnouralsk, Russia on July 8, 1935. He attended the Moscow Aviation Institute, where he graduated in 1959, and worked as an engineer at the Korolev Design Bureau, where he helped to develop the Vostok spacecraft. In 1964 he received the degree of candidate of technical sciences.

Sevastyanov was chosen for training as a cosmonaut in 1967. His initial assignments as a cosmonaut included training for the planned Soviet lunar landing program, which was later abandoned when the Soviet space program changed its focus to the development of space stations in Earth orbit. He also served as a member of the backup crew for Soyuz 8, which flew in October 1969.

On June 1, 1970, Sevastyanov lifted off from the Baikonur Cosmodrome as flight engineer aboard Soyuz 9, in the first night-time launch of the manned Soviet space program. Traveling into space with Soyuz 9 commander Andrian Nikolayev, Sevastyanov would spend more than 17 days in space, while he and Nikolayev helped to gather medical data about the impact that longer duration stays in space might have on human beings.

The flight took on a particular urgency in light of the Soviet Union's plans to launch the first Salyut space station, which would routinely accommodate long-duration crews, the following year.

Carrying out a demanding routine of rigorous medical tests that included careful monitoring, collection of sample materials and strenuous exercise, Sevastyanov and Nikolayev worked 16 hour days during their time in orbit, and collected valuable data for study by Soviet medical personnel and mission planners. They also carried out psychological tests and navigation exercises, and completed a variety of photography assignments.

During their time in orbit, the Soyuz 9 spacecraft was placed in a slow roll to keep the vehicle's solar array panels pointed toward the Sun. Known as a "spin-stabilized" position and frequently used for unmanned spacecraft, the maneuver resulted in unpleasant sensations for the crew members and was thought to have contributed to the fatigue and weakness that both cosmonauts experienced.

Sevastyanov and Nikolayev took a day off from their taxing routine when the flight reached its mid-way point. They spent their break reading and playing chess—a particular interest of Sevastyanov's; he later became president of the Soviet Chess Federation. They returned to their intense work schedule the following day and kept up their rigorous routine from that point on, until their return to Earth on June 19.

In the initial period of post-flight physical examinations and debriefings, both cosmonauts displayed pronounced fatigue and unexpectedly severe weakness, but neither suffered any long-term damage to their health. They were released from medical isolation 10 days after their return from space.

During the Soyuz 9 mission, Sevastyanov and Nikolayev spent 17 days, 16 hours and 59 minutes in space—a new record for the longest spaceflight up to that time. Their Soyuz 9 flight provided key medical data that helped to support the planning of future missions, and contributed to the safe operation of the Salyut series of space stations, the first of which was launched in April 1971.

Sevastyanov next served as a member of the backup crew for the April 5, 1975 launch of what came to be known as the Soyuz 18-1 mission, in which Vasili Lazarev and Oleg Makarov endured the first manned launch abort in the history of world spaceflight.

He then launched on May 24, 1975 as flight engineer for the Soyuz 18/Salyut 4 long-duration mission, which was commanded by Pyotr Klimuk. Sevastyanov and Klimuk spent more than two months in space while serving as the station's second long-duration crew. They carried out a diverse program of scientific work during their long stay aboard Salyut 4, including astronomical observations, photographic and spectrographic studies of the Sun, and extensive photography of the sprawling land mass that constituted the Soviet Union at that time.

In a less well-known aspect of their long flight, Sevastyanov and his crew mate endured a frightening health scare when an unidentified green mold began to grow on the walls of the Salyut 4 space station. The situation became dire enough at one point for mission controllers to consider the possibility of cutting the cosmonauts' stay short; but that option was dismissed because the Soyuz 18/Salyut 4 flight overlapped with the joint Soviet-American Apollo-Soyuz Test Project flight, and the Soviets had assured their American partners that the two missions would proceed simultaneously without any difficulty. Mission controllers worried that news of the strange mold within Salyut 4 would become public knowledge if Sevastyanov and Klimuk left the station ahead of schedule.

Despite the increasingly unpleasant circumstances aboard Salyut 4, the cosmonauts were able to complete their planned stay. Sevastyanov and Klimuk returned to Earth on July 26, 1975, after a flight of more than 62 days. Post-flight examinations revealed no lasting damage to either crew member.

Sevastyanov continued to serve the Soviet space program after his second space station flight; he later received training for assignment to a flight aboard the Buran space shuttle, but ultimately the Soviet shuttle program was abandoned in the economic disarray that followed the dissolution of the Soviet Union.

Vitali Sevastyanov amassed a total of more than 80 days in space during his two long-duration space missions.

He left the cosmonaut corps in 1993 to embark on a political career and in 1994 was elected to the Russian Parliament, the Duma.

SHARIPOV, SALIZHAN S.

(1964–)

Russian Cosmonaut

Salizhan Sharipov was born in Uzgen, in the Oshsk region of Kirghizia, on August 24, 1964. In 1987 he graduated from the Air Force Pilot School, and then served in the Air Force as a pilot-instructor, flying MIG-21 and L-39 aircraft. He has accumulated more than 950 flying hours during his career as a pilot.

He was selected for cosmonaut training in 1990. In addition to his extensive space training, he also continued his education, graduating from Moscow State University in 1994 with a degree in cartography.

He first flew in space as a mission specialist during the STS-89 flight of the space shuttle Endeavour, which took part in the eighth docking of a shuttle and the Mir space station. Endeavour was the second shuttle to be modified to dock with Mir (Atlantis was used for the first seven shuttle dockings with the space station).

The STS-89 flight began with the launch of Endeavour on January 22, 1998. At Mir, NASA astronaut Andrew Thomas joined the space station crew, replacing David Wolf. Wolf returned to Earth aboard Endeavour after having served at Mir for 119 days.

The STS-89 crew also delivered more than 8,000 pounds of equipment and supplies to Mir. The flight came to a close with the Endeavour's landing at the Kennedy Space Center (KSC) in Florida on January 31, 1998.

Sharipov's next space mission brought him to the International Space Station (ISS) as a member of the long-duration ISS Expedition 10 crew.

Sharipov and Expedition 10 commander Leroy Chiao, launched from the Baikonur Cosmodrome in Kazakhstan on October 13, 2004, joined aboard Soyuz TMA-5 with cosmonaut Yuri Shargin of the Russian Space Forces. Shargin returned to Earth with the Expedition 9 crew on Soyuz TMA-4 on October 23. Sharipov and Chiao lived and worked aboard the ISS for the next six months.

During their Expedition 10 stay on the ISS, Sharipov and Chiao made two spacewalks, each lasting more than five hours. In the first, on January 26, they worked on the space station module Zvezda. Then, during their second EVA, on March 28, they prepared the station for docking with European space vehicles and launched a small Russian satellite.

At the end of their successful mission, Sharipov and Chiao returned to Earth on April 24, 2005 aboard Soyuz TMA-5. For their return trip, they were accompanied by European Space Agency (ESA) astronaut Roberto Vittori of Italy, who had traveled to the station on April 14 with the Expedition Eleven crew.

During two spaceflights, Salizhan Sharipov has spent more than 201 days in space, including 9 hours and 58 minutes in EVA.

SHATALOV, VLADIMIR A.

(1927–)

Soviet Cosmonaut

A veteran of three space missions, Vladimir Shatalov was born in Petropavlovsk, Kazakhstan on December 8, 1927. He received his initial education at school number four in what was then known as Leningrad, which has since been renamed Saint Petersburg. In 1943 he completed his education in Petropavlovsk, and then joined the Soviet Air Force.

He attended the Air Force sixth Voronezh special school in Karaganda and Lipetzk, and in 1949 he graduated from the Kacha A. F. Miasnikov Red Banner Higher Aviation School of Pilots. After graduation, he remained at the school to serve as an instructor pilot, attached to the 706 Air Regiment, and in June 1951 he was elevated to the position of instructor of flying technique. Later that same year he became an instructor pilot in the 707 Air Regiment.

Shatalov next attended the Air Force Red Banner Higher Military Academy in Monino, where he graduated with honors in 1956, and was subsequently assigned as deputy commander and then commander of an air squadron. In May, 1960 he was named deputy commander of an Air Force regiment, and the following year he was assigned to the combat training department of the 48 Air Army in the Odessa military district, as a senior inspector pilot. He also received the rating of first class military pilot in 1960.

He was chosen for training as a cosmonaut as a member of the Air Force Group Two selection of cosmonaut candidates in January 1963. After a period of intensive training, he became a cosmonaut in 1965. His initial assignments as a cosmonaut included service as a member of the standby backup crew for the Voskhod 3 mission, which was subsequently canceled. He also served as backup to Georgi Beregovoi for the Soyuz 3 flight in October 1968.

Shatalov began his first spaceflight on January 14, 1969, when he launched from the Baikonur Cosmodrome aboard Soyuz 4 as part of the first-ever docking and crew transfer between two manned spacecraft. Boris Volynov, Aleksei Yeliseyev and Yevgeni Khrunov launched the following day aboard Soyuz 5, and on January 16, Shatalov achieved a rendezvous and docking with the Soyuz 5 spacecraft, which acted as a passive target for the operation.

Although the two Soyuz craft were docked, they were not equipped for a direct crew transfer, so Yeliseyev and Khrunov made a 37 minute spacewalk to move from Soyuz 5 to Soyuz 4.

The successful flight fulfilled the primary goal of the early years of the Soyuz program; the docking had been the mission profile for the planned Soyuz 1/Soyuz 2 missions of April 1967, which had ended in tragedy when Vladimir Komarov's Soyuz 1 malfunctioned, leading to his death.

With Soyuz 4 and Soyuz 5, the Soviets made a major advance that greatly increased the expertise and confidence of the engineers responsible for the nation's space program. Having completed the major goal of his first flight, Shatalov returned to Earth in Soyuz 4 with Yeliseyev and Khrunov on January 17, 1969.

Shatalov's next flight in space was as commander of Soyuz 8, with flight engineer Aleksei Yeliseyev, as part of a remarkable group flight of three Soyuz spacecraft. The Soyuz 8 launch on October 13, 1969 followed Soyuz 6 on October 11 and Soyuz 7 on October 12; the three vehicles were in space simultaneously until Soyuz 6 landed on October 16.

The Soyuz 8 mission profile officially called for Shatalov and Yeliseyev to make scientific investigations into the reflection of sunlight in the Earth's atmosphere, but Shatalov and Yeliseyev were apparently also scheduled to test a modified docking system by docking with Soyuz 7. Although the docking was not announced as part of the mission, it is thought that the operation could not be carried out because of a failure of the docking mechanism.

Whatever the difficulties with the docking plan, the launch and simultaneous flight of three separate spacecraft was an impressive achievement for the Soviet space program. Seven cosmonauts were involved in the group flight: Georgi Shonin and Valeri Kubasov aboard Soyuz 6; Anatoli Filipchenko, Vladislav Volkov, and Victor Gorbatko on Soyuz 7; and Shatalov and Yeliseyev aboard Soyuz 8.

The impressive simultaneous flights utilized many of the skills necessary for the next phase of the Soviet program, which included the Salyut space stations of the 1970s and 1980s.

Shatalov and Yeliseyev were able to rendezvous with Soyuz 7 on October 15, while the crew of Soyuz 6 looked on. Then, as the last of the three crews to launch, they were also the last to land, when they returned to Earth on October 18, 1969.

On April 22, 1971, Shatalov lifted off as commander of Soyuz 10, which was intended to deliver him, with flight engineer Aleksei Yeliseyev and research engineer Nikolai Rukavishnikov, to the Salyut 1 space station to serve as the station's first long-duration crew. It was Shatalov's third space mission with Yeliseyev; but unlike their first two successful flights, the Soyuz 10 mission would end in frustration.

The first space station in history, Salyut 1 was launched unmanned on April 19, 1971, two days before Soyuz 10 lifted off with Shatalov, Yeliseyev and Rukavishnikov. The cosmonauts arrived at the station and were able to rendezvous and dock without difficulty; but they were then unable to transfer from their Soyuz spacecraft into Salyut 1. Working with engineers on the ground, Shatalov and his crew mates tried to diagnose the difficulty, and made several attempts to solve the problem by undocking and trying again to dock and enter—with similarly disappointing results.

When all their attempts to enter the station ultimately failed, Shatalov, Yeliseyev and Rukavishnikov were ordered to cut their flight short and return to Earth. They landed on April 24, 1971, after a flight of less than two days. They were the first Soviet crew to make a night-time landing, and a malfunction in their Descent Module caused a life-threatening vapor leak. Fortunately, they emerged from the harrowing ordeal without permanent injury.

Shatalov was given the qualification of first class cosmonaut in 1971. The following year, he earned a candidate of technical sciences degree.

During his career as a cosmonaut, Vladimir Shatalov spent more than nine days in space.

He served as chief of the Gagarin Cosmonaut Training Center (GCTC) in Star City from 1987 to 1991, and left the cosmonaut corps in 1998 to become assistant for cosmonaut training and space mission provision to the commander-in-chief of the Russian Air Force. He has served as a lieutenant general of aviation in the reserve since 1992.

Among many awards and honors for his achievements as a cosmonaut, Shatalov was honored with two Gold Star Hero of the Soviet Union medals, three Orders of Lenin, the Order of Merits to the Country of the IV class, and the State Prize of the Soviet Union for the Intercosmos program. He is also claimed as an honorary citizen of eight cities, including Houston, Texas, and a crater on the Moon has been named in his honor.

In 1978 Shatalov authored the book *The Hard Road to Space.*

SHAW, BREWSTER H., JR.

(1945–)

U.S. Astronaut

A veteran of three space shuttle missions, Brewster Shaw was born in Cass City, Michigan on May 16, 1945. In 1963 he graduated from Cass City High School, and then attended the University of Wisconsin at Madison, where he received a bachelor of science in engineering mechanics in 1968, and a master of science in engineering mechanics in 1969.

Following his graduation form the University of Wisconsin, Shaw attended Officer Training School, where he was recognized as a distinguished graduate. He received his initial pilot training at Craig Air Force Base in Alabama, and received his wings in 1970. An outstanding pilot, he was honored upon completion of his undergraduate pilot training with the Commander's Trophy, Outstanding Flying Trophy, and Outstanding Academic Trophy, and also received awards as the Best T-38 Pilot and Top Formation Pilot.

Shaw served at Luke Air Force Base in Arizona while assigned to the F-100 Replacement Training Unit, and was honored with the F-100 Barry Goldwater Top Gun Award for his abilities as an F-100 pilot. He was then assigned to the 352nd Tactical Fighter Squadron at Phan Rang Air Base in the Republic of Vietnam in March 1971. He served as an F-100 combat fighter pilot until August 1971, when he returned to the United States for training in the F-4 aircraft, as a member of the F-4 Replacement Training Unit at George Air Force Base in California.

He served a second tour of duty during the Vietnam War while assigned to the 25th Tactical Fighter Squadron at Ubon Royal Thai Air Force Base in Thailand, where he flew combat missions as an F-4 fighter pilot.

In April 1973 he returned to George Air Force Base to serve as an F-4 instructor, as a member of the 20th Tactical Fighter Training Squadron. He then attended the U.S. Air Force Test Pilots School at Edwards Air Force Base in California, where he was recognized as a distinguished graduate. He remained at Edwards after graduation, initially as a test pilot assigned to the 6512th Test Squadron and then as an instructor at the Test Pilot School.

During his outstanding military career as a pilot, combat pilot, flight instructor and test pilot, Shaw accumulated over 5,000 hours of flying time in more than 30 different types of aircraft, including 644 hours of combat in F-100 and F-4 aircraft during his two tours of duty in Vietnam. He retired from the United States Air Force in 1996 with the rank of colonel.

NASA selected Shaw for training as an astronaut in 1978. He first flew in space as pilot of the space shuttle Columbia during STS-9, which launched on November 28, 1983. Commanded by Gemini and Apollo veteran John Young, the STS-9 crew also included mission specialists Owen Garriott and Robert Parker, and payload specialists Byron Lichtenberg and European Space Agency astronaut Ulf Merbold, the first ESA astronaut to fly on the space shuttle. STS-9 was the first flight of the ESA Spacelab orbital laboratory.

Jointly developed by NASA and ESA, the Spacelab laboratory and observation platform enabled 73 scientific investigations across a variety of disciplines including astronomy, physics, atmospheric physics, Earth observation, life sciences, materials sciences, space plasma physics, and technology.

A series of disturbing difficulties cropped up near the end of the otherwise successful mission when several instruments—including two of the spacecraft's on-board computers—failed. Landing was delayed for about 8 hours, extending the total duration of the mission to 10 days, 7 hours, 47 minutes and 24 seconds. When the shuttle did land, on December 8, 1983 at Edwards Air Force Base in California, two of its three auxiliary power units caught fire. Fortunately, no harm came to the crew, and no serious damage was done to the spacecraft.

For his participation in the STS-9 mission, Shaw was awarded the 1984 National Space Award of the Veterans of Foreign Wars.

He served as commander of his second spaceflight, STS-61B, which launched aboard the shuttle Atlantis on the evening of November 26, 1985. The mission marked the first flight in space by a Mexican citizen, Dr. Rodolfo Neri Vela, and the STS-61B crew deployed three communications satellites, the Mexican MORELOS-B, the Australian AUSSAT-2 and RCA Americom's SATCOM KU-2.

Mission specialists Jerry Ross and Sherwood Spring made two EVAs during STS-61B, to conduct exercises designed to test assembly procedures for building structures in space, the Experimental Assembly of Structures in Extravehicular Activity (EASE) and the Assembly Concept for Construction of Erectable Space Structure (ACCESS).

payload specialist Charles Walker of McDonnell Douglas operated the company's Continuous Flow Electrophoresis (CRFES) experiment during the STS-61B flight, and the crew conducted a Getaway Special experiment for Telesat of Canada and tested the shuttle's digital autopilot. Atlantis landed on December 3, 1985 at Edwards Air Force Base.

Shaw played a key role in the investigation of the Challenger accident of January 1986, supporting the efforts of the Rogers Commission to determine the cause of the accident and methods of correcting the errors that had led to it. He also led the return-to-flight team responsible for ensuring the safety of future shuttle missions.

On August 8, 1989, he lifted off on a third spaceflight, as commander of STS-28—the fourth shuttle mission devoted to the classified activities of the U.S. Department of Defense—aboard the space shuttle Columbia. His crew for the flight included pilot Richard Richards and mission specialists James Adamson, Mark Brown and David Leestma. Columbia returned to Earth at Edwards Air Force Base in California on August 13, 1989.

Brewster Shaw spent more than 22 days in space during his career as an astronaut.

He continued to serve NASA after his third shuttle flight, initially as deputy director of Space Shuttle Operations at the Kennedy Space Center (KSC) in Florida, responsible for all operational aspects of the shuttle program, and as chair of the Mission Management Team, responsible for the ultimate decision to launch or not launch any given shuttle mission.

Subsequently named deputy program manager for the Space Shuttle, Shaw served as a NASA headquarters employee located at KSC; he then became director of Space Shuttle Operations. He ultimately served as Space Shuttle Program Manager, and led an effort to make the shuttle program more efficient and cost-effective.

Shaw left NASA in 1996, and put his extensive expertise to use as an employee of the Rockwell company, which was acquired by Boeing that same year.

Over the course of the next eight years he served in positions of successively increasing responsibility as he played a major role in the development of the systems and equipment Boeing provided to NASA for the International Space Station (ISS). Boeing is the space agency's prime contractor and supplier of all U.S. hardware and software for the ISS program.

In 2003 Shaw became chief operating officer of the United Space Alliance, NASA's prime contractor for the space shuttle program, responsible for the activities of 10,000 employees located in the United States and in Russia.

In January 2006, he was named vice president and general manager, Space Exploration, for Integrated Defense Systems at the Boeing Company.

SHENZHOU V

(October 15, 2003)

People's Republic of China Manned Spaceflight

On October 15, 2003 the People's Republic of China became the third nation on Earth to launch one of its own citizens into space, with the flight of Shenzhou V.

Lifting off from launch pad CZ-2F at the Jiuquan Satellite Launching Center in the first manned Shenzhou spacecraft, taikonaut Liwei Yang (Yang Liwei in China, where naming conventions prescribe that the surname be listed first) became the first citizen of China to fly in space. The major national and international significance of the event was signified by the presence of China's president, Hu Jintao, at the Jiuquan facility for the launch.

The Shenzhou V flight was part of an ambitious, $2 billion long-term development program of manned flights that are seen as critically important to China's national development.

By all accounts, Yang had a smooth flight. He was in space for nearly a full day and made 14 orbits while fulfilling the two most important objectives of the inaugural flight: safely launching, and safely landing.

Launched by the workhorse Chinese "Long March" rocket (known by the prefix CZ–Chang Zheng, which means "Long March" in Chinese), the Shenzhou spacecraft is adapted from the Russian Soyuz design, and can accommodate crews of three taikonauts.

Yang was the single occupant of Shenzhou V for the first manned Chinese flight.

The successful completion of his historic flight placed Yang in the company of space pioneers Yuri Gagarin, the Soviet cosmonaut who made the first manned spaceflight in history aboard Vostok 1 on April 12, 1961, and the first American in space, Alan Shepard, who flew aboard Freedom 7 on May 5, 1961. While other nations have sent their citizens into space, only Russia, the United States and China have done so using their own launch vehicles and spacecraft from their own launch facilities.

During his historic flight, Yang spent a total of 21 hours and 23 minutes in space. He returned to Earth on October 15, 2003, landing safely in Shenzhou V in Dorbod Xi, in the Gobi Desert in central inner Mongolia.

SHENZHOU VI

(October 12–16, 2005)

People's Republic of China Manned Spaceflight

The second manned space launch of the People's Republic of China, Shenzhou VI lifted off on October 12, 2005 from the Jiuquan Satellite Launching Center in the Gobi Desert.

Junlong Fei (Fei Junlong in China, where naming conventions prescribe that the surname be listed first) was commander of Shenzhou VI; he was joined on the flight by Haisheng Nie (Nie Haisheng), who served as flight engineer. Fei and Nie were reportedly chosen for the flight from a pool of six taikonauts after a long period of training.

Media coverage of the Shenzhou VI mission was unprecedented, given the tight control of press coverage that the Chinese government normally imposes on the dissemination of information about the nation's space program. Representatives of the official Chinese press were allowed to witness the launch at the Jiuquan Satellite Launching Center, and the government allowed the launch to be televised live (in contrast with the first manned Chinese flight, Shenzhou V, in October 2003), but with a slight delay so the broadcast could be interrupted in the event of a problem.

Fortunately, the launch went well, and Chinese newspapers carried photos of Shenzhou VI commander Junlong Fei floating in the space capsule as it passed through orbit during the flight. The government-controlled news agency also relayed greetings and good wishes from Fei and Nie to their fellow Chinese, including citizens of Hong Kong, Macao, and Taiwan.

The Shenzhou VI mission was designed as part of an ambitious, $2 billion long-term program of manned flights seen as critically important to China's national development. Chinese space officials have developed the Shenzhou spacecraft (the term "Shenzhou" is translated into English as "Divine Vessel"), systems, and equipment from Russian Soyuz technology, making extensive modifications to the basic Russian components and manufacturing within China all the elements of the program that were flown in space.

The Shenzhou spacecraft are launched by the homegrown Chinese "Long March" rocket (known by the prefix CZ–Chang Zheng, which means "Long March" in Chinese).

During the nearly five days they spent in orbit during the Shenzhou VI flight, Fei and Nie reportedly conducted a series of scientific experiments, and made observations of the atmosphere and of patterns of pollution in the Earth's oceans.

After just one previous mission of less than one day, the Chinese space program took a major step forward with the successful Shenzhou VI flight. Fei and Nie completed 76 orbits during their flight, and spent 4 days, 19 hours and 32 minutes in space. They returned to Earth on October 16, 2005, landing in Siziwang Banner in the Gobi Desert in Mongolia.

SHENZHOU VII

(September 25–28, 2008)

People's Republic of China Manned Spaceflight; First Spacewalk of Chinese Space Program

China's third manned spaceflight began on September 25, 2008, when commander Zhigang Zhai (Zhai Zhigang in China, where naming conventions prescribe that the surname be listed first) and flight engineers Buoming Liu (Liu Buoming) and Haipeng Jing (Jing Haipeng) lifted off from the Jiuquan Satellite Launching Center on a mission that called for the nation's first-ever spacewalk.

After little more than 24 hours in orbit, Zhai and Liu began the task of assembling their complex EVA spacesuits. Manufactured in China, the Feitian EVA suits ("Feitian" is the name of a Buddhist goddess, and translates to "flying in the sky" in English) represented a particular point of pride for the Chinese space program, and a substantial investment of technology, with a per-suit cost estimated at $4 million.

Carefully following a step-by-step protocol methodically worked out during the 10 years of training for the Shenzhou VII crew (each of the three Shenzhou VII taikonauts had been selected for the Chinese space program in 1998), Zhai and Liu gathered the various components of the Feitian suits from storage spots around the spacecraft's cabin and assembled the pieces to form the finished suits, which weigh 120 kilograms. The laborious process required more than 20 major steps, and took about 12 hours.

At approximately 4:40 P.M. local time on September 27, 2008, Zhigang Zhai floated out of the Shenzhou VII spacecraft to begin the first spacewalk of the Chinese space program. He was supported by Buoming Liu, who performed a stand-up EVA in the capsule's hatchway, while Haipeng Jing monitored his crew mates' activities from within Shenzhou VII.

Zhai began his historic EVA with a wave to a camera mounted on the outside of the spacecraft, signaling to those following his exploits from Earth that he was well and progressing as planned. The milestone spacewalk was televised live to an audience estimated by the Chinese national news agency Xinhua as numbering in the tens of millions.

Xinhua also reported that Chinese President Hu Jintao watched the spacewalk with a contingent of other government leaders at the Beijing Aerospace Control Center.

While tethered to the spacecraft, Zhai carefully maneuvered himself along the vehicle's outer edge until he was within reach of an experiment package mounted on the exterior. He retrieved the package, therein completing the primary practical activity of the EVA.

Estimates of the exact length of Zhai's historic spacewalk differ, but Western sources generally conclude that the EVA took about 18 minutes to complete.

In the course of the rest of their flight, the Shenzhou VII crew deployed the small BanXing satellite, and conducted several satellite communications experiments.

The Shenzhou VII crew landed in inner Mongolia on September 28, 2008, after a flight of 2 days, 20 hours and 28 minutes, and 45 orbits of the Earth.

SHEPARD, ALAN B., JR.

(1923–1998)

U.S. Astronaut; First American to Fly in Space; Fifth Human Being to Walk on the Moon

The first American to fly in space, and the fifth human being to walk on the surface of the Moon, Alan Shepard was born in East Derry, New Hampshire on November 18, 1923. He received his initial education in East Derry and Derry, New Hampshire, and then attended the United States Naval Academy, where he received a bachelor of science degree in 1944.

During World War II Shepard served in the Pacific Ocean aboard the destroyer Cogswell. He received his initial training as a pilot in Corpus Christi, Texas and in Pensacola, Florida, and received his naval aviator's wings in 1947. He then completed several tours of duty aboard aircraft carriers in the Mediterranean Sea while attached to Fighter Squadron 42, and was chosen to attend the United States Naval Test Pilot School at Patuxent River, Maryland, where he graduated in 1951.

He served as a test pilot at the school after graduating, where his work included flight tests designed to study how light interacts with air masses at various altitudes across North America. He then served at Moffett Field in California in a night fighter unit flying Banshee jet aircraft while assigned to Fighter Squadron 193, where he served as operations officer. In that capacity, he was twice deployed in the Pacific Ocean aboard the aircraft carrier Oriskany.

Shepard then returned to the U.S. Naval Test Pilot School, where he helped to test a variety of aircraft, including the F8U Crusader, the F3H Demon, the F11F Tigercat and the F4D Skyray. He served as project test pilot for the F5D Skylancer, and as a flight instructor.

In 1957 he graduated from the U.S. Naval War College in Newport, Rhode Island, and then served as aircraft readiness officer on the staff of the commander in chief of the Atlantic Fleet.

During his outstanding military career, Shepard accumulated more than 8,000 hours of flying time, including 3,700 hours in jet aircraft, and rose to the rank of rear admiral in the United States Navy. He retired from the service in August 1974.

NASA selected Shepard as a member of the Mercury Seven—the first group of astronauts chosen for the U.S. space program—in April, 1959. With his fellow

Alan Shepard explores the Moon as commander of Apollo 14, February 5, 1971. [NASA/courtesy of nasaimages.org]

Mercury astronauts Virgil "Gus" Grissom and John Glenn, Shepard entered intensive training for one of the first three Mercury missions, and he was ultimately given the honor of being assigned to the first Mercury flight.

On May 5, 1961, Alan Shepard became the first American and second human being to fly in space. Only Soviet cosmonaut Yuri Gagarin had flown in space before Shepard, in April 1961.

A worldwide television audience was introduced to the novel technicalities of the space age with Shepard's launch, carefully following the pre-flight arrival of the astronaut at launch pad LC-5 at Cape Canaveral in Florida, his trip up the gantry apparatus alongside the rocket he would ride into orbit, his subsequent entrance into the Mercury capsule, which Shepard had given the name Freedom 7, and the countdown to launch, when the launch vehicle—a Redstone rocket—would lift off.

The Freedom 7 flight was designed to be a sub-orbital mission, in which the spacecraft would reach an altitude at which it would be in space, but would not enter into orbit around the Earth.

Shepard endured a series of planned and unplanned delays while awaiting the start of his trip into space, and wound up sitting in his contoured couch within Freedom 7 for 4 hours and 14 minutes before the countdown was finally completed and the launch began.

He lifted off at 9:34 A.M. on May 5, 1961, inaugurating a new era of space exploration for the United States and elevating the nation into a long competition for supremacy in space achievements with its superpower rival, the Soviet Union.

The first phases of the short flight proceeded as planned, with Shepard enjoying a smooth ride for about 45 seconds and then experiencing an expected turbulence until the Redstone rocket shut down 2 minutes and 22 seconds into the flight. Freedom 7 was automatically flipped from a top-forward position to a bottom-first orientation at the three minute mark, and it would remain in that position, with its heatshield leading the way forward, for the remainder of the flight.

Shepard remained comfortable within his spacesuit, which was equipped with a temperature control system, even when the temperature within the spacecraft shot up to 91 degrees Fahrenheit.

Once in orbit, he switched control of his Mercury capsule from automatic flight mode to manual operation, and tried out the spacecraft's controls, which governed the craft's pitch—the vehicle's up and down movements; yaw, which directed the craft's motion to the left and right; and roll, which controlled the spacecraft's ability to revolve.

Despite the novelty of the vehicle, the veteran pilot was able to direct Freedom 7 through a variety of maneuvers, and at the end of the flight, he successfully placed the craft into the necessary position for firing its retrorockets before re-entering the Earth's atmosphere.

He was also able to invoke the "fly-by-wire" flying technique successfully, firing the capsule's hydrogen peroxide jets to direct the vehicle's flight at two separate points in the mission.

In addition to expertly piloting his Mercury capsule, Shepard also communicated clearly and effectively with communicators tracking his flight from Earth, putting to rest worries about whether or not an astronaut could perform useful work in space. Because the flight was so early in the history of space exploration, and little was known about the impact that exposure to the space environment might have on a pilot, NASA medical personnel worried that an astronaut might become insensible, or suffer unexpected physical trauma brought on by the stress of the experience. Shepard's positive experience cleared the way for future flights.

He also carried out a limited program of scientific experiments, including Earth observations in which he located and described various landmarks. He had inadvertently made his observation work in space more difficult during his long wait on the launch pad; to pass the time during the long delays in the countdown, he had tried out the spacecraft's periscope with a filter that later hindered his view. His first attempt to remove the filter brought him into close contact with the control that would be used to invoke an abort, and he decided as a result that it would be best to leave the filter in place for the rest of the flight.

Shepard splashed down in Freedom 7 in the Atlantic Ocean on May 5, 1961, 15 minutes and 28 seconds after launch. He had traveled a total distance of 303 miles, and reached an altitude of 116.5 statute miles during his historic first spaceflight. Recovery forces reached Freedom 7 just minutes after it hit the water, and delivered Shepard to the primary recovery ship, the aircraft carrier USS *Lake Champlain*.

Widely celebrated for his courage during the flight and the good-natured way he handled the intense public interest that accompanied his sudden emergence as the premier American space hero, Shepard was received by President John F. Kennedy at the White House after the mission. Kennedy celebrated Shepard's success by setting an amazing agenda for the future of the U.S. space program, when he announced on May 25 his desire for Congress to fund the research and development necessary to land an American astronaut on the Moon before the end of the 1960s.

The single most important decision of the first decades of humanity's efforts to explore space, Kennedy's lunar landing challenge was the ultimate expression of the optimism and courage that Shepard generated among his fellow citizens with his successful Freedom 7 mission.

Shepard was named chief of NASA's Astronaut Office in 1963. In that capacity, he oversaw astronaut training procedures, ensured that the astronauts' expertise and experience was reflected in the agency's overall spaceflight operations and the specific details of each flight, and served as the primary representative of the astronaut corps in its dealings with NASA officials.

He was next chosen to command the first manned flight of the Gemini program, Gemini 3, in which he was scheduled to fly with Thomas Stafford in March 1965. A medical problem intervened, however, when an inner-ear condition caused him to experience dizzy spells and severe losses of balance. As a result, NASA medical personnel removed Shepard from active duty status as an astronaut, and he and Stafford were forced to turn over the Gemini 3 assignment to backup crew members Virgil "Gus" Grissom and John Young, who successfully completed the mission on March 23, 1965.

Although he continued to serve NASA as chief of the Astronaut Office, Shepard was frustrated by his removal from active flight status, and diligently sought a means to correct the condition that prevented him from flying. He eventually underwent surgery that restored his balance and led to his being returned to active duty in May 1969.

He made his second flight in space as commander of Apollo 14, the third U.S. flight to land astronauts on the lunar surface. Following the harrowing ordeal of Apollo 13, in which James Lovell, Fred Haise and Jack Swigert had barely survived the impact of an explosion in an oxygen tank aboard their spacecraft, the success of the Apollo 14 flight was important to the continuance of the Apollo program, which included plans for three subsequent lunar landing flights that were to feature longer stays on the Moon and more extensive scientific goals.

Apollo 14 launched on January 31, 1971, with Shepard as commander, accompanied by command module pilot Stuart Roosa and lunar module pilot Edgar Mitchell. Roosa overcame a problem that could have foiled the planned Moon landing when he was able to dock the Apollo 14 command module Kitty Hawk with the lunar module Antares after several attempts had failed. His expertise and persistence paid off, and salvaged the mission profile's goal of having Shepard and Mitchell land on the lunar surface.

Shepard and Mitchell faced a new problem after they reached lunar orbit, boarded the Antares lunar module, and began their descent to the surface. A short circuit in Antares's abort circuitry nearly caused the vehicle's on-board systems to invoke the abort procedure. The fault caused a warning light on the spacecraft's instrument panel to light, and whether as a result of their frustration or merely by accident, Shepard and Mitchell reacted to the warning by striking the panel—which caused the warning light to stop glowing.

Suspecting that the easily dimmed light might indicate an electrical problem rather than a more serious difficulty that would force the crew to give up the planned landing, engineers tracking the flight from Mission Control looked for alternate sources for the problem, and were able to trace the near-abort to a malfunctioning switch. They devised a work-around that involved reprogramming Antares's on-board computer—a delicate, exacting task that required Mitchell to quickly and accurately enter a long, intricate sequence of software code without making a single error, despite the tenseness of the situation.

Mitchell performed his impromptu assignment flawlessly, and the planned lunar landing was once again back on track.

A further challenge arose as a result of the reprogramming, however. Although the new software code bypassed the malfunctioning hardware aboard the lunar module, it also rendered part of the spacecraft's guidance system inoperative. As a result, Shepard was forced to guide Antares down to the lunar surface with virtually no margin for error.

Just as his crew mates had risen to the earlier challenges, Shepard achieved a near-perfect landing in the Moon's Fra Mauro Highlands on February 5, 1971, expertly guiding the lunar module to the surface at a point just 175 feet (53 meters) distant from the planned landing site. Antares perched in the rocky area at a slight angle, which caused Shepard and Mitchell some uneasiness during their later rest period, but the tilt had no impact on the safety of the landing, or on their later ascent from the surface.

While the astronauts readied themselves for their first Moonwalk, their support teams at Mission Control worked to troubleshoot problems in their communications systems, which delayed the first EVA by 49 minutes. When the communications problem was surmounted, Shepard signaled the crew's readiness to venture out onto the lunar surface in his inimitable good humor, radioing to Houston that he and Mitchell were ready "to go out and play in the snow."

Then, at the start of the 4 hour, 49 minute first EVA of the Apollo 14 mission, on February 5, 1971, Alan Shepard stepped out of the Antares lunar module, climbed down the spacecraft's ladder, and took his first step onto the lunar surface—the fifth human being to walk on the Moon.

He had arrived at the spot after years of remarkable, unprecedented achievements, and crushing setbacks. Chosen as one of the original seven U.S. astronauts, making his country's first-ever flight in space, being celebrated as a national and international hero; and then suffering a debilitating medical condition, searching for years for a suitable treatment, enduring in his astronaut career through the Gemini years when he was unable to fly in space, experiencing the loss of his friends and colleagues Gus Grissom, Ed White and Roger Chaffee in the tragic Apollo 1 fire . . . the weight of all those experiences, good and bad, added an emotional gravity to Shepard's first few moments on the Moon that make his Apollo 14 odyssey unique even among the unprecedented lunar landing missions.

Mitchell joined Shepard on the surface a few minutes later, and they set to work on the initial chores planned for the EVA. They first gathered a "contingency sample"—a rough assortment of sample material that they would spirit away in the lunar module in the event that some unforeseen circumstance forced them to leave the surface earlier than planned. Then they set up the American flag, a television camera and a dish antenna for S-band communications, and deployed the Apollo Lunar Surface Experiments Package (ALSEP) and a laser reflector that would be used in later experiments.

Mission controllers allowed Shepard and Mitchell an extra half hour on the surface, to some degree as compensation for the delay that had preceded the start of the exercise. The astronauts traveled a total of 1,815 feet (550 meters) on the lunar surface during their first trip outside the lunar module, and then returned to Antares to rest and eat.

While Roosa circled overhead in lunar orbit in the Apollo 14 command module Kitty Hawk, Shepard and Mitchell attempted to sleep for a few hours within Antares on the lunar surface. They found it difficult to rest, and were pleased when mission managers following their progress from Earth expressed a willingness to cut the scheduled rest period short.

Shepard and Mitchell emerged from the lunar module again on February 6, about two-and-a-half hours before they had originally been scheduled to do so, to begin their second exploration of the Moon. Their objective for the second EVA was to hike to the Cone Crater, a 1,000 foot (300 meter) wide hole in the lunar surface that was thought to have been created by the long-ago impact of a meteor.

They were equipped for their exploration chores with a two-wheeled cart, the modular equipment transporter (MET), on which they carried tools, photographic equipment and containers for storing samples of the materials they would encounter along the way and at the site.

Just getting to the area where they were to carry out their geological survey took up a good portion of the EVA; they arrived at the base of Cone Crater and began climbing up toward the rim about 90 minutes after they'd begun their second EVA.

Scientists were most interested in the material at the rim of the crater because they assumed that samples from that area would be the oldest. The slope around the rim was a treacherous mess of loose dust dotted with rocks and boulders in a staggering array of sizes and shapes, some with sharp edges or unseen points. Shepard and Mitchell were forced to pick their way up the 2,800-foot (850 meter) debris blanket very carefully, as they struggled toward the rim.

Fighting their way along the oddly sloping terrain, which limited their view of the crater's rim as they worked their way forward, the astronauts were covered with dust that was kicked up off the surface by their movements. They became increasingly weary as the struggle continued, and stopped twice to rest along the way.

Mission controllers extended the EVA by a half hour to allow the crew more time to try to reach the crater's rim, but finally decided that the effort was placing too much strain on the astronauts. Capcom Fred Haise—who would have made the arduous trek himself, with James Lovell, if Apollo 13 had been able to land on the Moon as planned—radioed the word that Shepard and Mitchell were cleared to stop their climb, and should gather what samples they could where they were, and then prepare to conclude the EVA.

Shepard appeared relieved at the prospect of ending the difficult climb, but Mitchell initially favored going a bit further up the crater slope. A three-way dialogue ensued, with Shepard and Haise trying to convince Mitchell of the wisdom of cutting the climb short, but Mitchell's reply carried the day—albeit temporarily—when he protested, "I think you're finks."

As a result, the astronauts' trek went on a bit further, until ultimately everyone involved agreed that the exhausted Moonwalkers had hiked as far as they should. Shepard and Mitchell turned around at that point without having reached the crater's rim, and began their long hike back to Antares.

Along the way, Shepard stopped to engage in a bit of mischievous fun that would long be remembered as a pop culture highlight of the Apollo program. Using the device he'd earlier employed to collect the contingency sample, to which he attached the head of a golf club (a six iron), he took a one-handed swing at a golf ball he'd pulled from a pocket of his spacesuit. The ball skipped off across the surface, bouncing oddly along the undulating Moonscape in the weak throes of lunar gravity.

Dissatisfied with the results of his first shot, he produced a second golf ball, which he hit more squarely. As the tiny ball soared upward and then flew off into the distance, Shepard narrated its flight, "There it goes! Miles and miles . . ."

Resplendent in his bulky spacesuit, which was adorned with red stripes for easy identification during post-flight examination of the photos taken during the mission, Shepard displayed both the irreverent good humor and casual courage that was characteristic of a large majority of the early astronauts. His lunar golf experiment would over time become nearly as widely known and cherished a memory as any incident of the first era of space exploration.

At the conclusion of their second EVA, Shepard and Mitchell returned to the interior of Antares after spending 4 hours and 46 minutes in EVA. They blasted off in their lunar module after a total stay of 33 hours and 30 minutes on the surface, and returned to lunar orbit, where they docked with the Kitty Hawk command module and rejoined Roosa. The three crewmembers then returned to Earth, splashing down on February 9, 1971 with 94.6 pounds (slightly more than 43 kilograms) of Moon rocks and lunar surface sample materials.

Alan Shepard spent a total 9 days, 57 minutes and 28 seconds in space, including 9 hours and 35 minutes in EVA on the surface of the Moon, during his two epic spaceflights.

President Richard Nixon appointed Shepard as a delegate to the 26th General Assembly of the United Nations in 1971, and the veteran astronaut represented the United States in that capacity throughout the assembly, from September to December 1971.

Shepard also continued to serve NASA after his second space mission, continuing as chief of the agency's Astronaut Office until his retirement from NASA and the U.S. Navy on August 1, 1974. He then pursued a private sector career, serving as president of the nonprofit Mercury Seven Foundation, which provides science scholarships for deserving students.

Alan Shepard died on July 21, 1998. He was survived by his wife, Louise, who died on August 25, 1998; and by daughters Julie, Laura and Alice, and six grandchildren.

SHEPHERD, WILLIAM M.

(1949–)

U.S. Astronaut

A veteran of four space missions, William Shepherd was born in Oak Ridge, Tennessee on July 26, 1949. In 1967 he graduated from Arcadia high School in Scottsdale, Arizona, and then attended the United States Naval Academy, where he received a bachelor of science degree in aerospace engineering in 1971.

His military service has included assignments with the Navy's Underwater Demolition Team Eleven, Special Boat Unit Twenty, and Navy SEAL Teams One and Two, and he has risen to the rank of captain in the United States Navy.

Continuing his education concurrently with his military career, he attended the Massachusetts Institute of Technology (MIT), where in 1978 he earned the degrees of ocean engineer and master of science in mechanical engineering.

NASA selected Shepherd for astronaut training in 1984. He first flew in space as a mission specialist during STS-27, aboard the space shuttle Atlantis in December 1988. The STS-27 mission was the second flight following the Challenger accident in January 1986, and the third shuttle flight devoted to the classified activities of the U.S. Department of Defense. During the launch on December 2, insulating material from one of the shuttle's SRBs came loose and hit Atlantis about a minute-and-a-half after the shuttle left the launch pad; similar damage years later led to the loss of the STS-107 crew and the shuttle Columbia, in 2003. Despite the mishap, Atlantis and her crew returned safely after a little more than four days in space, landing at Edwards Air Force Base on December 6, 1988.

Shepherd's second space mission was STS-41, in October 1990. As a mission specialist aboard the space shuttle Discovery, he participated in the deployment of the Ulysses solar polar orbiter for the European Space Agency (ESA).

Ulysses had evolved out of a plan that originally envisioned NASA and ESA collaborating on two spacecraft designed to study the sun. NASA withdrew from the project in 1981, but ESA continued development of its portion of the mission, and the re-named Ulysses was launched with a mission profile calling for intense study of the sun's polar areas.

The deployment of the Ulysses spacecraft featured the first combination of an Inertial Upper Stage (IUS) and a Payload Assist Module-S (PAM-S) motor, which together placed the probe on the appropriate trajectory for its journey to the Sun.

Shepherd made his third flight in space as a mission specialist aboard the shuttle Columbia during STS-52, which launched on October 22, 1992. A landmark in international cooperation in space science, STS-52 carried a diverse array of international scientific payloads for sponsors that included the Italian space agency Agenzia Spaziale Italiana (ASI), the Canadian Space Agency (CSA), and the European Space Agency (ESA).

Jointly sponsored by NASA and the Italian space agency Agenzia Spaziale Italiana (ASI), the Laser Geodynamic Satellite (LAGEOS) II satellite was propelled into orbit by the Italian Research Interim Stage (IRIS), which was used for the first time during STS-52.

The international flavor of the mission was also reflected in the Canadian Experiment-2 (CANEX-2), whose experiments included the Space Vision System (SVS)—which entailed the deployment, on the ninth day of the mission, of the Canadian Target Assembly (CTA) satellite.

The ESA was represented by the Attitude Sensor Package in the shuttle's cargo bay. Mounted on a "Hitchhiker" plate, the sensor package included a modular star sensor, a yaw Earth sensor and a low altitude conical Earth sensor.

The STS-52 crew also conducted a series of experiments that were bundled together as the U.S. Microgravity Payload-1 (USMP-1), which was carried in Columbia's cargo bay.

After his third spaceflight, Shepherd served in management capacities in NASA's space station program.

On October 31, 2000, he began a fourth space mission, lifting off from the Baikonur Cosmodrome in Kazakhstan aboard Soyuz TM-31 as commander of the first long-duration crew aboard the International Space Station (ISS).

Shepherd and his fellow ISS Expedition 1 crew members Sergei Krikalyov and Yuri Gidzenko of Rosaviakosmos lived and worked aboard the station for five-and-a-half months. They returned to Earth aboard the space shuttle Discovery during STS-102, landing on March 21, 2001. From launch to landing, the ISS Expedition 1 crew spent 140 days, 23 hours and 40 minutes in space.

Although his stay at the ISS occurred before the Internet phenomenon of blogging was widespread, Shepherd introduced a forward-looking innovation when he kept a log of his thoughts and activities that was made available via NASA's Web site.

William Shepherd spent more than 159 days in space during his four spaceflights.

SHRIVER, LOREN J.

(1944–)

U.S. Astronaut

Loren Shriver was born in Jefferson, Iowa on September 23, 1944. He attended the United States Air Force Academy, where he received a bachelor of science degree in aeronautical engineering in 1967. He was commissioned in the Air Force after graduation. He continued his education concurrently with his military service, and in 1968 he earned a master of science degree in astronautical engineering from Purdue University.

He served at Vance Air Force Base in Oklahoma from 1969 to 1973, as a T-38 academic instructor pilot. Later, in 1973, he received F-4 combat crew training at Homestead Air Force Base in Florida, and he then made an overseas tour in Thailand, where he served until October 1974.

Upon his return to the United States, he attended the U.S. Air Force Test Pilots School at Edwards Air Force Base in California, and subsequently worked in the test and evaluation of the T-38 lead-in fighter while assigned to the 6512th Test Squadron at Edwards. He also served as a test pilot for the F-15 Joint Test Force during his time at the base.

During his exceptional career as a pilot and test pilot, Shriver accumulated more than 6,200 hours of flying time in jet aircraft. He holds commercial pilot and private glider ratings. He rose to the rank of colonel in the United States Air Force prior to his retirement from the service.

NASA chose Shriver for astronaut training in January 1978. In September 1982 he received his initial flight assignment, as pilot for STS-10. Scheduled as the first mission dedicated to the classified activities of the U.S. Department of Defense (DOD), STS-10 was subsequently canceled.

Shriver did fly on the first shuttle flight devoted to the DOD, when he launched from the Kennedy Space Center (KSC) aboard the shuttle Discovery at the start of STS-51C, on January 24, 1985. The fifteenth mission of the space shuttle program, STS-51C was originally planned to be flown aboard the shuttle Challenger, but problems with Challenger's thermal tiles made it necessary to shift the flight to Discovery. The flight was also delayed for a day by freezing weather conditions.

The shift from Challenger to Discovery and concerns about the cold weather became significant in retrospect a year later, when freezing weather conditions contributed to the accident that caused the loss of the STS-51L crew in the Challenger explosion, on January 28, 1986.

For STS-51C, Apollo 16 and STS-4 veteran Thomas Mattingly served as commander, accompanied by Shriver, mission specialists Ellison Onizuka and James Buchli, and payload specialist Gary E. Payton. Except for the veteran Mattingly, the entire crew was making its first flight in space.

Although the shuttle missions devoted to the Department of Defense were considered classified and little information is known about their payloads, NASA did reveal that the STS-51C crew successfully deployed a U.S. Air Force Inertial Upper Stage booster during the flight. Discovery landed at KSC on January 23, 1985, after a flight of slightly more than three days.

On April 24, 1990 Shriver launched aboard the shuttle Discovery as commander of STS-31—the deployment of the Hubble Space Telescope (HST). The thirty-fifth flight of the space shuttle program, STS-31 brought Discovery to a much-higher-than-usual orbit of 380 miles to facilitate the launch of the HST. The crew also used a hand-held IMAX camera to record their activities, and conducted experiments in protein crystal growth, polymer membrane processing, and the impact of microgravity and magnetic fields on an ion arc.

Discovery returned to Earth on April 29, 1990, landing at Edwards Air Force Base in California after 76 orbits. The STS-31 landing marked the first use of the new carbon brake system (as opposed to earlier versions which used beryllium) that had been integrated into the shuttle after a major re-design. Discovery had made a harsh landing at the end of STS-51D in April 1985, suffering a blown tire and extensive brake damage and causing NASA officials to transfer subsequent landings to Edwards until the shuttle's steering system could be overhauled.

Shriver's role in the HST deployment was widely recognized for its contributions to space science; he was honored with the 1990 Flight Achievement Award of the American Astronautical Society, and he also received the American Institute of Aeronautics and Astronautics' Haley Space Flight Award for 1990.

He also served as commander during his third flight in space, aboard the shuttle Atlantis during STS-46 in 1992.

Dedicated to international flight experiments, STS-46 featured the first flight in space of an Italian citizen, payload specialist Franco Malerba, and the first flight in space of a citizen of Switzerland, Claude Nicollier. Nicollier deployed one of the flight's two primary payloads, the European Retrievable Carrier (EURECA) of the European Space Agency (ESA), which would remain in orbit until its retrieval during STS-57 in 1993.

The flight also featured the first major test of tethered spaceflight, in a joint project of NASA and the Italian space agency Agenzia Spaziale Italiana (ASI). Prior to STS-46, teams of engineers from NASA and ASI had collaborated for more than a decade to design the equipment, the plan, and the procedures that were to have been used during the Tethered Satellite System (TSS-1) test aboard Atlantis.

Despite all the careful preparations, however, a mechanical hitch foiled the test, when the tether jammed after it had been unwound to a distance of just 840 feet. The crew attempted to release the stuck line for several days, but the experiment finally had to be written off as a failure. The TSS satellite was retrieved and returned to Earth. Atlantis landed at KSC on August 8, 1992.

Loren Shriver has accumulated more than 386 hours in space during his three shuttle flights.

He continued to serve NASA after his third space mission, first as deputy chief of the Astronaut Office, and subsequently as Space Shuttle program manager for launch integration, and then as deputy director for launch and payload processing at KSC.

He retired from the United States Air Force in 1993, with the rank of colonel.

In 2000 Shriver joined United Space Alliance, NASA's prime contractor for the space shuttle program, as deputy program manager. He has since risen to the position of vice president for engineering and integration, and has been designated the company's Chief Technology Officer.

SHUKOR, SHEIKH MUSZAPHAR

(1972–)

First citizen of Malaysia to fly in space

A medical doctor and the first citizen of Malaysia to fly in space, Dr. Sheikh Muszaphar Shukor was born in Kuala Lumpur, Malaysia on July 27, 1972. He attended Kasturba Medical College in Manipal, India, where he graduated with a bachelor of medicine and surgery degree, and began his medical career in 1998 at Hospital Seremban in Malaysia. The following year he worked at Kuala Lumpur General Hospital, and he became a staff member at Hospital Selayang in 2000. He currently serves as an orthopedic surgeon and university medical officer at the University Kebangsaan in Malaysia, where he was pursuing his masters degree in orthopedic surgery at the time of his selection for spaceflight training.

Shukor was chosen for training as a representative of the Malaysian National Space Agency Angkasa under an agreement between the governments of Malaysia and Russia which resulted in the Malaysia Angkasawan ("Astronaut") Program of manned spaceflight. In its commitment to the agreement, Russia pledged to train several Malaysian cosmonaut candidates, one of whom would be selected to fly to the International Space Station (ISS) aboard a Russian Soyuz spacecraft. In return, Malaysia agreed to purchase 18 Russian fighter jets, as one provision of a national defense agreement the two countries signed in 2003.

With his backup, Faiz Khaleed, Shukor spent 18 months at the Gagarin Cosmonaut Training Center in Russia while training to prepare for his flight aboard Soyuz TMA-11.

On October 10, 2007, Shukor lifted off in Soyuz TMA-11 with Russian cosmonaut Yuri Malenchenko and NASA astronaut Peggy Whitson. Launched from the Baikonur Cosmodrome, the Soyuz TMA-11 launch was viewed by a large Malaysian television audience that included Malaysian Prime Minister Abdullah Badawi, who followed Shukor's progress along with hundreds of school children from the Kuala Lumpur Convention Center.

During his stay at the ISS, Shukor carried out several health related experiments and a protein crystal experiment.

A devout Muslim, he also celebrated Ramadan during his time at the ISS. In anticipation of the unique circumstances of his stay at the station, Islamic religious authorities created a guidebook to assist Shukor and future Muslims who fly in space in properly performing Islamic rites in the space environment.

At the end of his stay aboard the ISS, Shukor returned to Earth in Soyuz TMA-10 with ISS Expedition 15 crew members Fyodor Yurchikhin and Oleg Kotov, on October 21, 2007. The Soyuz TMA-10 crew endured a rough landing after making a ballistic re-entry, with the spacecraft spinning rapidly under the influence of gravity. They landed more than 300 kilometers from their intended landing site in Kazakhstan, but emerged from the ordeal unharmed.

During his historic Soyuz TMA-11 flight to the International Space Station, Dr. Sheikh Muszaphar Shukor spent 10 days, 21 hours and 14 minutes in space, and made 171 orbits of the Earth.

SKYLAB PROGRAM

(May, 1973–February, 1974)

U.S. Manned Space Station Program

Plans for what would eventually become the U.S. Skylab space station program began as early as 1964, when Wernher von Braun spearheaded the program's development as one of his first duties as director of the Marshall Space Flight Center (MSFC) in Huntsville, Alabama.

In its original incarnation, as envisioned by von Braun and other NASA officials, the scope of the space station project extended far beyond the vehicle that would eventually fly as Skylab. In the mid- to late-1960s, as public interest in the U.S. space program grew and anticipation of the nation's first lunar landing attempt increased, far-thinking NASA mission planners envisioned a permanent orbiting laboratory, the "Space Base," with a staff of 100. As a follow-up to the Apollo Moon landing program, the first U.S. space station was to have been completed by 1975, and the planning also involved the development of a reusable spacecraft, which the agency first referred to in 1968 as a "space shuttle." The crews flying to and from the space station aboard the space shuttle would eventually be continuing their trip beyond the station to a base on the Moon.

None of these grandiose plans took into account the sharp reduction in U.S. space plans that followed the end of the Apollo program—indeed, even before the end of the Apollo flights, as public attention focused elsewhere and the projected final three lunar landing flights were canceled. The ensuing changes to the plans for the first U.S. space station included deletion of the lunar base, the space shuttle, and the large, permanent "Space Base" with its large staff.

Re-named Skylab and greatly scaled down, the revised space station program, wherever possible, made use of components, equipment and systems left over from the Apollo program. The Skylab station itself was created from the third stage of one of the Saturn V rockets that had been designed for Apollo, and the three Skylab crews traveled to the station aboard Apollo spacecraft that had

Damaged at launch, the Skylab space station was repaired by its first crew. [NASA/courtesy of nasaimages.org]

originally been intended for use during trips to the Moon. The Skylab station was lifted into orbit by a Saturn V launcher; the three Skylab crews were launched aboard Saturn 1B rockets.

While U.S. plans for a post-Apollo space station gradually evolved into the Skylab program, the Soviet Union launched the world's first space station, Salyut 1, on April 19, 1971. The crew of Soyuz 11—commander Georgy Dobrovolsky, flight engineer Vladislav Volkov, and research engineer Viktor Patsayev—spent

23 days aboard Salyut 1, becoming the first human beings to live and work on an orbiting space station. Sadly, they were killed during re-entry because of a malfunction in their Soyuz spacecraft.

In addition to establishing a U.S. presence in orbit and maintaining the progress of the U.S. space program during the period following Apollo and preceding the first launch of the space shuttle—which had by the early 1970s become a priority on its own, apart from any space station, lunar base or other destination—the Skylab program also sought to gather medical data about the impact of long-duration stays in space, and to expand scientific knowledge through extensive study of the Sun and of Earth resources.

Skylab was launched unmanned on May 14, 1973. The station was 36 meters long and 6.7 meters in diameter, and weighed 91 metric tons. It was placed in an orbit of 268.1 x 269.5 miles, with an orbital period of about 93 minutes.

An excess of vibration at lift-off led to the loss of a shield that had been designed to protect the station from meteorite impacts, and when the shield broke off it carried away one of the lab's two solar panels. The accident also left debris wrapped around the station's second solar panel, essentially rendering it useless.

The accident had a paradoxical result: because the meteor shield was also designed to protect the space station's primary work area from direct exposure to the Sun, its loss left Skylab vulnerable to extreme heat; at the same time, the effective loss of the solar panels prevented them from converting the Sun's energy into the electricity necessary to operate Skylab's crucial systems. Thus, the station would have more exposure to the Sun than it needed, and yet would be unable to make appropriate use of the exposure it received.

With a determined, skillful effort that rivaled the most impressive of any of the agency's previous engineering feats, NASA set out to save the damaged space station, and thereby salvage the Skylab program. Pooling its expertise with that of its vast network of contractors and sub-contractors and members of the scientific community, the space agency devised a series of exacting repairs to counter the damage done by the faulty launch, and to implement the modifications necessary to make the station operational.

Their flight delayed for 10 days, the first Skylab crew—commander Charles "Pete" Conrad, Paul Weitz and Joseph Kerwin—traveled to the station on May 25, 1973, as the Skylab 2 crew (the launch of the station itself had been designated Skylab 1). They painstakingly implemented the repairs that had been carefully worked out in simulations on Earth, and gradually brought the station back to an acceptable operating level. They also achieved much of the scientific work that had originally been planned for the mission, and returned to Earth after 28 days and 49 minutes.

Alan Bean, Jack Lousma and Owen Garriott served as the Skylab 3 crew. Lifting off on July 28, 1973, they spent 59 days, 11 hours and 9 minutes in space, and performed well above expectations in completing their program of scientific experiments and astronomical observations. They dedicated themselves to working long hours and carefully completing their assigned tasks, as well as taking on additional work as the mission progressed, definitively proving that it is possible to do productive work in space for extended periods of time.

For the final Skylab mission, Skylab 4, Gerald Carr, William Pogue and Edward Gibson all made their first flight in space, and in doing so they set a new long-duration record of 84 days, 1 hour and 16 minutes—the longest stay in orbit up to that time. They achieved impressive scientific results, recording the first complete photographic record of a solar flare, and making extensive observations and photographs of the comet Kohoutek.

Like the first five Soviet Salyut space stations, Skylab was not designed to be resupplied or refueled. When the final crew left the station in February 1974, NASA shut the station's systems off and left Skylab to drift in orbit while the U.S. space program shifted its resources fully into development of the shuttle transportation system.

An unexpectedly heavy period of solar activity hastened the dormant station's re-entry. In a mirror image of the work they'd done to keep the damaged station aloft and useful after its launch in 1973, NASA engineers worked throughout the Fall of 1978 to determine the most likely path Skylab would take during its return to Earth, and to try to figure out how much of the structure might survive re-entry and potentially cause damage, if the debris should strike land.

The agency suffered a great deal of criticism in the months leading up to Skylab's fall from orbit, with media outlets around the world wondering aloud about NASA's ability to steer the station clear of a catastrophic impact. Suffering both serious criticism and ridicule that ranged from gentle jibes to lacerating sarcasm, NASA officials remained focused on the job of safely landing whatever might be left of Skylab after the station's fiery trip through Earth's atmosphere.

On July 12, 1979, Skylab made a partially controlled descent from orbit, largely burning up as it re-entered. Although some pieces of the spacecraft survived and impacted land areas, the debris mostly fell into the Indian Ocean, with some fragments landing in isolated regions of Western Australia, where they caused no injuries. The safe return was largely lost, however, in the wake of the public outcry that had preceded it. Even worse, in succeeding years the substantial achievements of the Skylab program have sometimes been given less attention in media accounts of the station than the criticism that preceded its de-orbiting.

In the three manned Skylab flights, the U.S. space program accumulated a cumulative total of more than 170 days in which its astronauts lived and worked in space. The three Skylab crews made a combined 2,203 orbits, and spent a total of more than 824 hours on medical observations and experiments, over 941 hours studying the Sun, and more than 569 hours studying Earth resources. They conducted a total of about 300 experiments, and successfully carried out extensive solar astronomy observations and photography.

They also conducted a total of 10 spacewalks, and accumulated a total of 41 hours and 46 minutes in EVA. The total amount of time the crews spent in space and the total EVA time were each greater than the combined totals of all previous stays and all previous EVAs up to that time.

Perhaps most importantly, the Skylab astronauts also definitively achieved the primary goals that had been set for the program, proving that human beings could safely live and work in space for extended periods.

SKYLAB 1

(May 14, 1973–July 12, 1979)

U.S. Space Station, Launched Unmanned and Occupied by Three Long-Duration Crews

America's first space station, Skylab, lifted off unmanned from launch pad 19 at the Kennedy Space Center (KSC) in Florida at 1:30 P.M. on May 14, 1973. Developed from the third stage of a Saturn V rocket, the cylindrical laboratory and living space was 36 meters long and 6.7 meters in diameter, and weighed 91 metric tons. The station was lifted into orbit by a Saturn V; subsequent manned Skylab flights utilized Saturn 1B rockets as launch vehicles.

The Skylab program was designed to measure the impact that long-duration stays in space might have on the astronauts who would occupy the station, and to enable extensive studies of the Sun.

For the latter task, the station was equipped with a mini-observatory known as the Apollo Telescope Mount (ATM), which contained two X-ray telescopes; a camera capable of capturing X-ray and extreme ultraviolet images; an ultraviolet spectroheliometer; an extreme ultraviolet spectroheliograph; an ultraviolet spectroheliograph; a white light coronoagraph; and two hydrogen-alpha telescopes. The Skylab observatory program was designed and directed by NASA engineers at the Marshall Space Flight Center (MSFC) in Huntsville, Alabama, and the instruments were developed at a variety of advanced research facilities, including the MSFC, the U.S. Naval Research Laboratory, the Harvard College Observatory, American Science and Engineering, and the High Altitude Observatory.

Unexpectedly high vibrations during Skylab's launch caused a meteoroid shield, which was to have shaded the station's workshop area in orbit, to deploy prematurely. As the Saturn V sped toward orbit, the shield broke loose, and yanked one of the station's two solar arrays from its storage area. The solar array was subsequently knocked loose when the rocket's second stage fired; and a piece of debris from the shield also became entangled on the second solar array, jamming it in place.

Despite the ugliness of the launch, the station did manage to achieve its intended orbit, at 268.1 x 269.5 miles, at an altitude of 270 miles with an orbital period of about 93 minutes. Mission managers used the solar panels of the ATM, which were separate from the main arrays that had been damaged, to funnel as much electric power as possible to the ailing station, but the grim reality of the launch accident was that the station was perilously close to being lost completely, just hours after it had reached orbit.

Even as Skylab was starved of electrical power because of the damage to its solar arrays, it also suffered extremely high temperatures—as high as 126 degrees

Fahrenheit (52 degrees Celsius) in the astronauts' workshop area—because of the position it had to maintain in order to generate even the small amount of electricity it gained from the ATM panels.

NASA officials responded to the grave crisis confronting the station with an all-out rescue effort that drew upon all of the agency's resources and personnel, and the contractors, sub-contractors and the scientific and academic communities that had come to support the U.S. space program during the Mercury, Gemini and Apollo projects.

Through extensive study of the data being collected from Skylab's systems by engineers at Mission Control and by detailed simulations of proposed solutions to the station's problems, NASA officials were able to create a detailed list of repairs to be attempted by the first crew.

For their part, the Skylab 2 crew (the station itself had been designated Skylab 1) of Charles "Pete" Conrad, Paul Weitz and Joseph Kerwin quickly shifted their attention from the careful mission plan for which they had trained to the hurriedly prepared repair tasks. Their launch was postponed for 10 days during the planning for the salvage mission; they lifted off on May 25, 1973, and thanks to the careful and detailed preparations—as well as the outstanding performance of the crew—they were able to implement the proposed repairs. They installed a shield to lower the temperatures inside of Skylab, and made a spacewalk to pull the remaining solar array free.

Although the results were less than optimal, with the station operating for some time on less than half of its intended electrical power and the large shield giving the orbiting laboratory a decidedly "patched" look, the repairs of its first crew saved Skylab, and made it habitable for its two subsequent crews.

Despite its shaky start, the station supported the work of long-duration crews thru February 1974, and remained in orbit until July 11, 1979.

In the Fall of 1978, Skylab presented NASA with another crisis, as its fall from orbit drew near. Fears that pieces of the station might survive re-entry and land in populated areas of the Earth gave rise to harsh criticism of the agency. The criticism proved no serious distraction for the engineers who were given the job of safely directing the station's de-orbiting, however, and although some pieces of debris did survive the return trip, they fell into the Indian Ocean and in isolated regions of Western Australia, where they caused no injuries.

SKYLAB 2

(May 25, 1973–June 22, 1973)

First U.S. Space Station Crew

Charles "Pete" Conrad, who had walked on the Moon during Apollo 12 and also flew in space aboard Gemini 5 and Gemini 11, served as commander of the first crew to live and work aboard Skylab—the first U.S. space station. He was joined on the mission by pilot Paul Weitz and scientist astronaut Joseph Kerwin.

Adapted from the third stage of a Saturn V rocket, the 36-meter long, 91-metric ton Skylab was launched unmanned on May 14, 1973. An unexpectedly high degree of vibration during the launch caused the loss of the station's micrometeorite shield and one of its two solar arrays, and the second solar array was also jammed in place during the mishap.

As a result, the originally scheduled launch of the first manned flight, Skylab 2 (the station itself had been designated Skylab 1), was postponed for 10 days while NASA officials put all their available resources to work to devise solutions to the many problems caused by the accident.

After long months of training for their planned long-duration stay aboard Skylab, Conrad, Weitz and Kerwin were forced to quickly absorb an entirely new plan for the early portion of their mission, as they carefully worked through the detailed list of repairs that had been devised during NASA's all-out effort to save the ailing space station.

Outfitted with tools and equipment that included an innovative parasol-like shield that they would unfurl over the station's overheated work area, the Skylab 2 crew lifted off from launch pad 19 at the Kennedy Space Center (KSC) aboard a Saturn 1B rocket on May 25, 1973. At the space station, the astronauts found the damage as grave as mission managers had feared; the meteorite shield was indeed missing, as was one of the solar arrays, and the second array was jammed tight.

Paul Weitz made the first spacewalk of the Skylab program, standing in the open hatch of the Skylab 2 command module for 40 minutes while unsuccessfully trying to force the stuck solar array to deploy.

The solar array damage left the station starved for electricity, while at the same time the small amount of electric power generated by a secondary set of solar panels mounted in the Apollo Telescope Mount (ATM) observatory made it necessary to orient the station in such a way that the Sun shone directly into the astronauts' work area—subjecting the crew to unbearably high temperatures whenever they ventured into the orbiting laboratory.

Using the solution that the engineering teams had come up with during the extensive pre-flight simulations, the crew installed the "parasol" over the station's workshop and achieved the intended result: the temperature dropped from 126 degrees Fahrenheit (52 degrees Celsius) to 78 degrees Fahrenheit (26 degrees Celsius).

On June 7, Conrad and Kerwin set to work on the solar array problem, during a three hour, 25 minute spacewalk. They cut away the debris that pinned the array in place, and then pried the solar panels partially free, giving Skylab a chance to soak up and convert enough sunlight to bring the station's electrical power up to about half of the originally planned level.

As the on-site representatives of the huge NASA effort to save the Skylab station and program, Conrad, Weitz and Kerwin achieved a remarkable feat with their long hours of difficult repair work, and they ensured the station's viability for its two subsequent crews.

In addition to their salvage work, the Skylab 2 crew also achieved the majority of the medical experiments, solar observations and Earth resources studies that had originally been assigned to them.

On June 19, 1973, Conrad and Weitz ventured outside the station for a spacewalk of 1 hour and 36 minutes to repair a power module and to change the film cartridge in the station's ATM observatory.

Having successfully achieved their remarkable repair chores and most of their experiments, Conrad, Kerwin and Weitz returned to Earth on June 22, 1973, splashing down in the Pacific Ocean after 404 orbits and a flight of 28 days and 50 minutes.

SKYLAB 3

(July 28, 1973–September 25, 1973)

Second U.S. Space Station Crew

Alan Bean, who had walked on the Moon during Apollo 12, was commander of Skylab 3. He and his crew mates, pilot Jack Lousma and scientist astronaut Owen Garriott, lifted off atop a Saturn 1B rocket on July 28, 1973.

Skylab 3 was the second manned flight to the first American space station; the station itself, which launched unmanned on May 14, 1973, had been designated Skylab 1.

As the follow-up to the Apollo lunar landing program, Skylab featured substantial use of Apollo vehicles and equipment. The lab itself had been fashioned from the third stage of a Saturn V rocket, and Skylab crews used Apollo command modules to travel to and from the station. The use of reliable, flight-proven systems proved a positive during the Skylab 3 flight, when a fuel leak in the crew's command module was quickly diagnosed and solved.

Throwing themselves into the substantial program of experiments they had been assigned, Bean, Lousma and Garriott settled into a productive routine, often working 12 hour days to complete their work. During their two month flight, they accumulated more than 310 man hours in medical investigations, over 305 hours on solar observations, more than 223 hours on studies of Earth resources, and an additional 240-plus hours on other experiments. Their exceptional work ethic produced excellent results, and by the end of the flight, they had exceeded both the number of experiments and the amount of data collection that NASA scientists had originally anticipated.

The Skylab 3 astronauts also participated in regular television broadcasts, and in regularly scheduled exercise designed to combat the losses in blood plasma volume and bone density that routinely occur during long-duration space missions.

On August 6, 1973, Garriott and Lousma conducted a 6 hour, 31 minute spacewalk to install a tarp-like shade over the station's workshop, to shield the area from exposure to the Sun and lower the temperature in the station. The sunshade replaced an earlier version that had been installed by the first Skylab crew; the repairs had been necessitated by damages that the station had sustained during its shaky launch in May, 1973. In addition to putting up the new shade, Garriott and Lousma also replaced the film cartridge in the station's Apollo Telescope Mount (ATM) observatory.

Garriott and Lousma made a second spacewalk to change the ATM film cartridge on August 24, and they also repaired one of the ATM instrument's gyroscopes, while spending 4 hours and 31 minutes outside the space station.

On September 22, Bean and Garriott ventured outside the station to change the ATM film, adding another 2 hours and 41 minutes to the mission's total EVA time.

The Skylab 3 crew returned to Earth on September 25, 1973, after 858 orbits and a flight of 59 days, 11 hours and 9 minutes. In their three spacewalks, they accumulated a total of 13 hours and 43 minutes in EVA.

SKYLAB 4

(November 16, 1973–February 8, 1974)

Third U.S. Space Station Crew

The final mission to the Skylab space station, which set a record for the longest space mission up to that time, Skylab 4 lifted off atop a Saturn 1B rocket at 9:01 A.M. on November 16, 1973, at the Kennedy Space Center (KSC) in Florida. Commanded by Gerald Carr, the Skylab 4 crew also included pilot William Pogue and scientist astronaut Edward Gibson. Skylab 4 was the first flight in space for each of the crew members, and the third manned Skylab flight; the space station itself had been designated Skylab 1, when it was launched unmanned on May 14, 1973.

After an initial period of adjustment, Carr, Pogue and Gibson settled into a productive routine of scientific work during their stay of nearly three months aboard the space station. The Skylab 4 crew accomplished some of the most striking scientific objectives of the Skylab program, particularly in the area of solar observation and photography—a task to which they devoted a total of 519 man hours.

Just a week after their arrival at the station, on November 22, 1973, Gibson and Pogue made the first spacewalk of the flight, accumulating 6 hours and 33 minutes in EVA while repairing an antenna and taking photographs.

The crew also devoted a good deal of their time to medical observations and experiments, and exercised regularly to combat the losses in blood plasma volume and bone density that regularly occur during long stays in space. The collection of medical data and information about the ability of human beings to live and work in space were primary goals of the Skylab program.

On December 25, 1973, Carr and Pogue ventured outside of the space station to change the film cartridge in the Skylab observatory, the Apollo Telescope Mount (ATM), and also captured dramatic images of the Comet Kohoutek, which had been discovered in March 1973. They spent seven hours and one minute in EVA during their holiday excursion.

The crew would make a thorough study of Kohoutek throughout the mission, taking ultraviolet photos of the comet with an advanced camera designed by scientists at the U.S. Naval Observatory. Carr and Gibson continued the Kohoutek photography during a three hour, 29 minute spacewalk on December 29.

In another scientific "first," Gibson captured a remarkable series of images that were the first to document a solar flare from beginning to end.

Over the course of their long stay, Carr, Gibson and Pogue carried out a scientific program of 56 experiments, 26 science demonstrations and 13 investigations designed by students, while traveling a total of more than 34 million miles in 1,214 orbits. Their work included observations of stellar objects, studies of Earth resources, and physiology experiments.

On February 3, 1974, Carr and Gibson made the final spacewalk of the flight—and the Skylab program—when they removed the ATM film cartridge. During the EVA, they added 5 hours and 19 minutes to the total spacewalking time for the mission.

As the end of their stay drew near, the Skylab 4 crew maneuvered the station into a higher orbit, where it would remain in a dormant, powered-down state for several years.

Carr, Gibson and Pogue returned to Earth on February 8, 1974. During their remarkable Skylab 4 mission, they established a new record for the longest stay in space up to that time, at 84 days, 1 hour and 16 minutes. Their long-duration record was a world-best for five years, until Soviet cosmonauts Yuri Romanenko and Georgi Grechko spent 96 days in space during the Soyuz 26/Salyut 6 mission, from December 10, 1977 to March 16, 1978.

In their 4 spacewalks, the final Skylab crew spent 22 hours and 22 minutes in EVA.

Some NASA officials hoped that the space shuttle, which was still in the early stages of its development when the Skylab program ended, might one day be able to visit the station; but as things turned out, Carr, Gibson and Pogue were the last astronauts to occupy Skylab. The station remained in orbit until July 1979, when it re-entered the Earth's atmosphere, and largely burnt up during its descent.

SLAYTON, DONALD K.

(1924–1993)

U.S. Astronaut

A pioneering astronaut who was chosen as one of NASA's Mercury Seven first selection of astronauts and who later flew in space during the historic joint U.S.-Soviet Apollo-Soyuz Test Project, Donald "Deke" Slayton was born in Sparta, Wisconsin on March 1, 1924. He graduated from Sparta High School, and in 1943, during World War II, he joined the United States Army Air Force.

He received his initial training as a pilot at Vernon, Texas and at Waco, Texas, and was then assigned to the 340th Bombardment Group in Europe, where he flew a B-25 aircraft during 56 combat missions.

After his return to the United States in 1944, Slayton served as a flight instructor in the B-25 and subsequently helped to conduct tests of the A-26 aircraft.

In April 1945, he was transferred to the 319th Bombardment Group in Okinawa, where he flew seven combat missions over Japan.

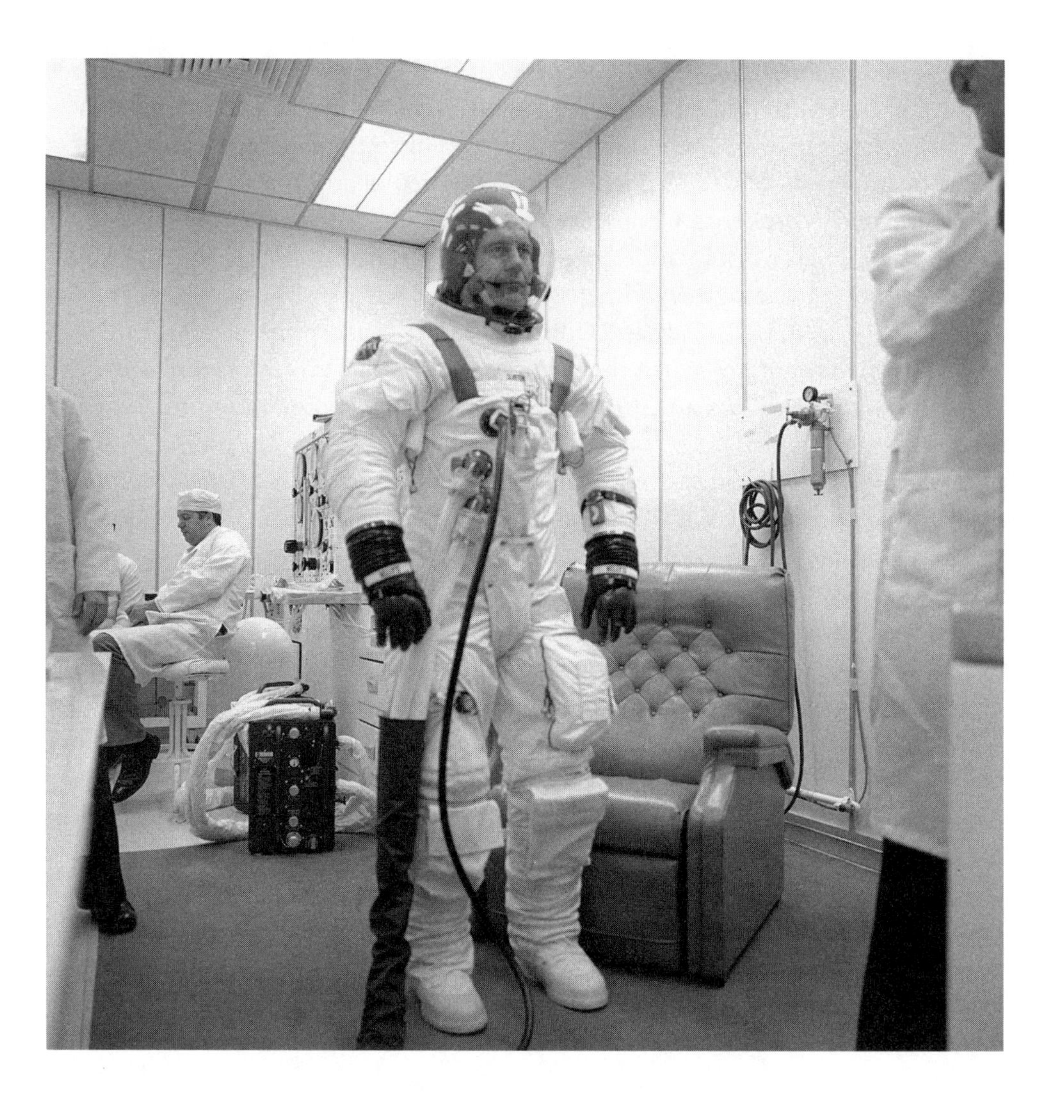

Donald "Deke" Slayton during training for the Apollo-Soyuz Test Project (ASTP). [NASA/courtesy of nasaimages.org]

Slayton continued his Army Air Force service for one year after the end of World War II, again serving as a flight instructor until he left the service to attend the University of Minnesota, where he earned a bachelor of science degree in aeronautical engineering in 1949.

He began his civilian career with the Boeing Aircraft Corporation. He worked as an aeronautical engineer at Boeing until 1951, when he was recalled to active duty with the Minnesota Air National Guard.

During his tour of duty with the Minnesota Air National Guard, Slayton initially served as maintenance flight test officer with an F-51 squadron in Minneapolis, and was then transferred to the headquarters of the Twelfth Air Force, where he served as a technical inspector. He also served overseas with the 36th Fighter Day Wing in Bitburg, Germany, as a fighter pilot and maintenance officer.

Upon his return to the United States in 1955, he was chosen to attend the U.S. Air Force Test Pilot School at Edwards Air Force Base in California. After graduation, he served as a test pilot at the school from January 1956 until his selection by NASA as one of the space agency's first group of astronauts.

During his outstanding military career, as a pilot, combat pilot, instructor pilot and test pilot, Slayton accumulated over 6,600 hours of flying time, including 5,100 hours in jet aircraft. He rose to the rank of major in the United States Air Force before leaving the military to become NASA's director of Flight Crew Operations in November 1963.

NASA chose Slayton as one of the Mercury Seven—the first U.S. astronauts—in April 1959. He underwent extensive training in preparation for a flight during the pioneering U.S. Mercury program, but in August 1959 NASA medical personnel found evidence that he had developed a heart condition, and removed him from active duty as an astronaut.

A crushing blow to his hopes of flying in space, the medical difficulty led to his assuming an administrative post within NASA that ultimately proved to be a key position with great influence on the future of the U.S. space program. After serving as coordinator of astronaut activities for a little over a year, Slayton resigned his commission in the Air Force in November 1963 to become NASA's Director of Flight Crew Operations—a position in which he would be responsible for overseeing virtually all phases of astronaut activities as crew members were chosen, trained and assigned to particular flights—for more than a decade, throughout the Gemini, Apollo and Skylab programs.

Respected by both his fellow astronauts and his NASA management colleagues, Slayton exerted a unique influence on America's manned space activities throughout the first era of the U.S. space program. Even as he chose the astronauts who would fly in space throughout the pioneering experimental flights, the lunar landing missions and aboard the first American space station, however, he still harbored the hope that he would some day receive medical clearance to take at least one spaceflight assignment himself.

He was finally accorded the chance to return to active duty in March 1972, after a thorough review of his health status and extensive medical examinations, and was declared medically able to fly in space just in time to become a crew member on the final mission of the U.S. space program's first era, the Apollo-Soyuz Test Project flight of July, 1975.

The final flight of an Apollo vehicle, the ASTP mission was a fitting conclusion to the intense "space race" competition between the United States and the Soviet Union that began in the late 1950s with the Soviet Sputnik 1 launch, and reached its climax with the American Apollo 11 first lunar landing in 1969. With ASTP, the U.S. and Soviet space programs would work together to plan and carry out a joint spaceflight in which an American Apollo and Soviet Soyuz would rendezvous and dock in orbit.

A seemingly simple concept, the mission required at least as great a leap of faith on the part of each nation's space program as it would require of the particular skills of the astronauts and cosmonauts who would meet in space. The administrators and engineers of each program had to overcome substantial barriers of language, culture and approach to develop a mutually acceptable mission profile, flight control procedures, a crew training regimen, and a new docking system that could accommodate the two different vehicles.

The crew members played an important part in blending the vastly different cultures of the U.S. and Soviet space programs. Acting as goodwill ambassadors and focusing on the traits they shared—including expert pilot skills, long military experience, and the occasional frustrations of strong-willed individuals working in a sometime stifling corporate environment—the U.S. and Soviet ASTP crews trained in both the United States and in Russia, and demonstrated a sincere willingness to work together.

The Apollo portion of the flight was commanded by Thomas Stafford, who had previously flown in space aboard Gemini 6 and Gemini 9 and orbited the Moon in Apollo 10. Slayton and Vance Brand rounded out the Apollo crew.

Soviet space hero Alexei Leonov was given the honor of commanding Soyuz 19, the Soviet portion of the ASTP mission. In March 1965, Leonov had been the first person ever to perform a spacewalk. He was joined by Valeri Kubasov, who had previously flown in space during Soyuz 6.

All five crew members seemed to enjoy the relatively relaxed training preparations for the flight, with language lessons emerging as the most challenging preflight chore, as the Americans learned Russian and the Russians learned English. Along the way, the veteran pilots engaged in several good-natured public relations activities, and by all accounts developed a sincere respect and genuine friendship as they prepared for the historic flight.

The ASTP mission began with the launch of Soyuz 19 from the Baikonur Cosmodrome on July 15, 1975, which, in a major departure from most Soviet flights, was broadcast live.

Then, after his long wait to one day lift off on a trip into orbit, Slayton launched with Stafford and Brand aboard the final Apollo space mission, about seven-and-a-half hours after Soyuz 19.

On July 17, the two spacecraft met in orbit and docked. Stafford and Leonov greeted each other at the Soyuz end of the docking module, and the two crews subsequently visited each other's vehicle over the course of a total of 47 hours of docked flight. The astronauts and cosmonauts shared several meals together, and exchanged symbolic gifts representative of the hopeful nature of the joint flight, including seeds that each group would later plant in their own country. All five crew members also took part in a televised news conference that was aired live in both the United States and the Soviet Union, and U.S. President Gerald Ford and Soviet leader Leonid Brezhnev each made congratulatory calls to their nation's crew members.

The two vehicles briefly undocked for a series of rendezvous experiments and then docked again, with the final separation taking place on July 21.

Slayton and his crew mates remained in orbit for three more days before returning to Earth on July 24, 1975. A problem during re-entry caused their Apollo command module to fill with toxic fumes, but Slayton, Stafford and Brand were successfully recovered after splashing down in the Pacific Ocean at the end of the final Apollo mission. They each required medical care, and were hospitalized for five days before they were able to return to normal activities. Once they were fully recovered, they joined their Soviet counterparts on an extensive public tour of the

United States and the Soviet Union, displaying the cooperative spirit that would later find larger expression in the docking flights of the American space shuttle and the Russian space station Mir, and in the subsequent participation of the United States and Russia in the multi-national International Space Station (ISS) program.

Slayton continued to serve NASA after his historic ASTP flight, initially as manager of the space shuttle Approach and Landing Test (ALT) program, and beginning in 1977, as manager of the shuttle's Orbital Flight Test operations and the shuttle ferry program.

Deke Slayton spent 9 days, 1 hour, 28 minutes and 23 seconds in space during the ASTP mission. He remained with NASA through the first launches of the space shuttle program, and then retired from the agency in 1982. He subsequently founded Space Services Inc. in Houston, Texas, to develop rockets intended to launch small commercial payloads.

On June 13, 1993 Slayton died in League City, Texas as the result of complications from a brain tumor. He is survived by his wife, Bobbie, and son, Kent.

Among the many honors he received during his long career, Slayton was a recipient of three NASA Distinguished Service Medals, the NASA Exceptional Service Medal, the NASA Outstanding Leadership Medal, the Robert J. Collier Trophy, the Ivan C. Kincheloe Award and the J.H. Doolittle Award of the Society of Experimental Test Pilots, the Haley Astronautics Award and a Special Presidential Citation of the American Institute of Aeronautics and Astronautics (AIAA), the Veterans of Foreign Wars National Space Award, the Wright Brothers International Manned Space Flight Award, and the American Heart Association's Heart of the Year Award.

SLOVAKIA (SLOVAK REPUBLIC): FIRST CITIZEN IN SPACE.

See Bella, Ivan

SMART-1

(September 27, 2003–September 3, 2006)

European Space Agency lunar orbiter

A project of the European Space Agency (ESA), the first flight of the Small Missions for Advanced Research in Technology program (SMART-1) was a lunar orbiter mission designed to test an innovative electric propulsion system that could influence the development of future planetary science missions. It was the first probe sent to the Moon by the ESA, and the first of the agency's planned tests of advanced spacecraft technologies.

The SMART-1 spacecraft was launched from Kourou, French Guiana on September 27, 2003 by a French Ariane 5 rocket, during a launch whose primary payload was two commercial satellites. Although its mission profile also called for extensive study of the Moon—including the gathering of data about lunar geology, topography and mineralogy—the main goal of the SMART-1 mission was to test a solar-powered ion drive as a means of propelling a vehicle in space.

The ion drive, a stationary plasma Hall-effect thruster (SPT), was designed to move the spacecraft by ionizing xenon gas and then accelerating and venting plasma at a high speed. The maximum discharge power of the device is 1.5 kilowatts.

Tested during Earth orbit maneuvers and during the 14 months prior to SMART-1's insertion into lunar orbit on November 13, 2004, the ion drive proved its suitability for future missions.

The spacecraft was equipped with a miniature visible/near-infrared spectrometer to study the lunar crust, and a miniature X-ray spectrometer to conduct astronomy observations and to study lunar chemistry. SMART-1 also carried a panchromatic camera and probes to measure the interplanetary plasma environment, and tested an experimental deep space telecommunications mechanism.

The active scientific experimentation phase of the mission began in February 2005, and the mission, originally planned for six months, was extended for an additional year to allow for continued mapping of the lunar surface.

Smart-1 reached the end of its mission on September 3, 2006, when the tiny craft was deliberately crashed on the lunar surface in the area known as the Lake of Excellence.

The total cost of the Smart-1 mission was an estimated $140 million.

SMITH, MICHAEL J.

(1945–1986)

U.S. Astronaut

Michael Smith was born in Beaufort, North Carolina on April 30, 1945. In 1963 he graduated from Beaufort High School, and he then attended the United States Naval Academy, where he received a bachelor of science degree in Naval science in 1967.

In 1968 he earned a master of science degree in aeronautical engineering at the U.S. Naval Postgraduate School.

After receiving training as a pilot of jet aircraft at Kingsville, Texas, he received his aviator wings in 1969. He then served as an instructor at the Advanced Jet Training Command (VT-21). In 1971 he was deployed aboard the USS *Kitty Hawk* (CV-63) for service as an A-6 Intruder pilot during the Vietnam War, while attached to Attack Squadron 52.

Following his return to the United States, he attended the U.S. Navy Test Pilot School at Patuxent River, Maryland, and then served as a test pilot at the Strike Aircraft Test Directorate while helping to develop the A-6E TRAM and CRUISE missile guidance systems. In 1976 he returned to the Navy Test Pilot School as an instructor, and subsequently served as maintenance and operations officer aboard the USS *Saratoga* during two tours of duty in the Mediterranean Sea, while assigned to Attack Squadron 75.

During his outstanding military career, Smith accumulated more than 4,867 hours of flying time in 28 different types of civilian and military aircraft. He was awarded the Navy Distinguished Flying Cross, three Air Medals, 13 Strike Flight Air Medals, the Navy Commendation Medal with "V," the Navy Unit Citation, and the Vietnamese Cross of Gallantry with Silver Star. He had attained the rank of captain in the United States Navy at the time of his death, in the explosion of the space shuttle Challenger in January 1986.

NASA selected him for astronaut training in 1980, and he became an astronaut the following year. His technical assignments at the agency included service as a commander in the Shuttle Avionics Integration Laboratory (SAIL), and he also served as deputy chief of the Aircraft Operations Division and as technical assistant to the director of the Flight Operations Directorate.

Smith was assigned as pilot of the space shuttle Challenger for STS-51L, and was also assigned to STS-61N, which was to have flown in the Fall of 1986 but was subsequently canceled after the Challenger accident.

The STS-51L flight was launched on January 28, 1986 at the Kennedy Space Center (KSC) in Florida.

Tragically, Smith and his fellow Challenger crew mates were killed when a fault in one of the shuttle's huge solid rocket boosters caused a fuel leak that ignited and caused a massive explosion just 73 seconds after liftoff. He was 40 at the time of his death.

Michael Smith is survived by his wife, Jane, and three children. He was posthumously awarded the Congressional Space Medal of Honor and the Defense Distinguished Service Medal.

SMITH, STEVEN L.

(1958–)

U.S. Astronaut

Steven Smith was born in Phoenix, Arizona on December 30, 1958. In 1977 he graduated from Leland High School in San Jose, California, and he then attended Stanford University, where he received a bachelor of science degree in electrical engineering in 1981 and a master of science degree in electrical engineering in 1982.

He began his professional career at IBM in 1982, as a technical group leader at the company's Large Scale Integration Technology Group in San Jose. He took a leave of absence from the company in 1985 to return to Stanford, where he graduated with a master's degree in business administration in 1987.

An outstanding athlete, Smith was named a high school and collegiate All-American in swimming and water polo seven times. He was a member of two Stanford water polo teams that won National Collegiate Athletic Association (NCAA) championships, serving as captain of the 1980 championship team. He also pursued his interests in sports and in community service as a board member of the Special Olympics in Texas.

After receiving his M.B.A., he returned to IBM to work as a product manager in the Hardware and Systems Management Group, a position that he held until 1989. He is a recipient of the IBM Outstanding Technical Achievement Award, and the IBM Outstanding Community Service Award.

In 1989 he began his association with NASA as a payload officer in the agency's Mission Operations Directorate. He became an astronaut in 1993.

Smith first flew in space as a mission specialist during STS-68, which launched on September 30, 1994. The sixty-fifth mission of the space shuttle program, STS-68 marked the second flight of the Space Radar Laboratory (SRL), part of an international effort to study global environmental change as part of NASA's Mission to Planet Earth program.

Designed by scientists from countries around the globe, the SRL flights made in-depth environmental studies of natural and man-made phenomena on the Earth's surface and in the planet's atmosphere.

The crew endured a pad abort during a first attempt at launch, on October 18. The countdown proceeded all the way down to less than two seconds before Endeavour was scheduled to lift off Pad A at the Kennedy Space Center (KSC), when the shuttle's on-board computers detected a problem in an oxidizer turbopump turbine in one of the Space Shuttle Main Engines (SSMEs) and shut down the three SSMEs, bringing the launch to an immediate halt. After repairs, the launch on September 30 proceeded without further delay.

Once in space, the STS-68 astronauts worked around-the-clock in two shifts to make studies across a range of academic disciplines, including ecology, hydrology, geology and oceanography. Using the SRL's Spaceborne Imaging Radar-C and the x-band Synthetic Aperture Radar, the crew used interferometry to collect extremely precise data from sites that had been imaged in less detail on the earlier SRL flight. Proving the efficacy of the technology, the interferometric images that were collected during passes over North America, the Amazon rain forests, and volcanoes on the Kamchatka Peninsula in Russia produced data that can be rendered into highly detailed topographic representations of the surveyed areas.

The crew also used the Measurement of Air Pollution from Satellites (MAPS) payload to study carbon monoxide discharged by forest fires in British Columbia. At the end of the flight, Endeavour returned to Earth at Edwards Air Force Base in California on October 11, 1994.

Smith began his second space mission, the second servicing of the Hubble Space Telescope (HST), with the night-time launch of STS-82, aboard the shuttle Discovery on February 11, 1997. Among the most complex missions undertaken during the entire shuttle program, the HST service and repair flights required enormous stamina on the part of alternating spacewalking teams, and minute attention to detail over long periods of working in space.

The crew made five spacewalks, including an unscheduled EVA at the end of the repair program, and used more than 150 tools during their work on the HST. Working in two alternating teams, mission specialists Smith and Mark Lee took turns with Gregory Harbaugh and Joseph Tanner to make the necessary spacewalks; the 4 crew members accumulated a total of 33 hours and 11 minutes in EVA during the flight.

Smith and Lee made the first EVA on February 14, logging 6 hours and 42 minutes while replacing several of the telescope's scientific instruments. Harbaugh and Tanner made their first spacewalk the following day, replacing the HST's Engineering and Science Tape Recorder, and installing the Optical Control Electronics Enhancement Kit and Fine Guidance Sensor. During their 7 hour, 27 minute EVA they also discovered cracks in the telescope's thermal insulation. Smith and Lee then added another 7 hours and 15 minutes to the EVA total on February 15, while replacing more instruments and components.

Harbaugh and Tanner completed the original mission profile with their final spacewalk on February 16, and Smith and Lee then made the unscheduled fifth EVA of 5 hours and 17 minutes to attach thermal insulation blankets to the telescope, to protect key data processing, electronics and telemetry instrumentation.

Smith and Lee spent 19 hours and 33 minutes in EVA during their three STS-82 spacewalks. Discovery returned to Earth in a night landing on February 21, 1997.

Smith made his third trip into space during the December, 1999 STS-103 HST servicing mission, as a mission specialist aboard the space shuttle Discovery. The third HST maintenance and repair flight, STS-103 launched on December 19 with commander Curtis Brown, pilot Scott Kelly and an international crew that included NASA astronauts Smith, Michael Foale, and John Grunsfeld, and European Space Agency (ESA) astronauts Jean-Francois Clervoy and Claude Nicollier.

Using the shuttle's robotic arm, the Remote Manipulator System (RMS), Clervoy captured the HST on December 21. Then, working in teams—Smith paired with Grunsfeld, and Foale with Nicollier—the spacewalking astronauts made three long EVAs that totaled 24 hours and 33 minutes over the course of three days while replacing key parts of the space telescope.

Smith and Grunsfeld spent 8 hours and 15 minutes in EVA during the first spacewalk, on December 22, 1999, and Foale and Nicollier made the second EVA on December 23. Smith and Grunsfeld then finished the HST servicing on December 24, adding another 8 hours and 8 minutes to their spacewalking total while restoring the telescope to optimal operating condition.

Following the successful modifications, the HST was redeployed on December 25. Discovery returned to Earth on December 27, 1999.

After his third flight in space, Smith became deputy chief of the Astronaut Office at NASA.

On April 8, 2002, he lifted off on a fourth spaceflight, during which he again put his expertise as a spacewalker to good use, as mission specialist aboard the shuttle Atlantis during STS-110—the thirteenth shuttle flight to the International Space Station (ISS). The STS-110 crew delivered and installed the space station's S-Zero (S0) Truss—an effort that required four spacewalks, which were all performed for the first time from the ISS Quest Airlock, rather than from the shuttle itself.

Smith and crew mate Rex Walheim made the first spacewalk of the flight on April 11, spending more than seven hours in EVA while attaching the truss to the ISS Destiny module. Jerry Ross and Lee Morin continued the work on April 13, and then Smith and Walheim made their second EVA on April 14, when they reconfigured the ISS Remote Manipulator System (RMS) Canadarm 2 so it would operate properly with the new S0 Truss segment. Ross and Morin finished the installation on April 16 when they added hardware to the truss to facilitate future spacewalks.

The STS-110 crew also worked with the ISS Expedition Four crew during the mission, transferring supplies and equipment from the shuttle to the station. Atlantis returned to Earth on April 19, 2002, after 171 orbits.

During his 4 space shuttle flights, Steven Smith has spent more than 40 days in space, including more than 49 hours in EVA during 7 spacewalks. At the end of the STS-110 mission he had accumulated more spacewalking time than any U.S. astronaut except Jerry Ross, and ranked third behind Ross and Russian cosmonaut Anatoly Solovyov for the world record for most career EVA time.

After his fourth spaceflight, Smith became manager of NASA's Automated Transfer Vehicle (ATV) Launch Package for the ISS Program.

SOLOVYOV, ANATOLY Y.

(1948–)

Russian Cosmonaut

A veteran of five missions in space, including four long-duration stays aboard the space station Mir and travel on the space shuttle, Anatoly Solovyov has performed 16 spacewalks—more than any other individual—and set a record for the most time spent in extravehicular activities (EVAs).

Solovyov was born in Riga, Latvia on January 16, 1948. In 1972 he graduated from the Lenin Komsomol Chernigov Higher Military Aviation School, and served as a senior pilot and group commander in the Soviet Air Force. During his military career, he rose to the rank of colonel.

In August 1976 he was selected for cosmonaut training at the Gagarin Cosmonaut Training Center (GCTC), where he completed his basic space training in 1979. He then received additional preparation for space missions aboard the Soyuz T-series of spacecraft, and for operations aboard the Salyut and Mir space stations.

He served as a member of the backup crew for the Soyuz T-6 flight to Salyut 7, which launched the first French citizen, Jean-Loup Chrétien, into space. He was also a backup during the aborted Soyuz T-10-1 mission, in which the prime crew, Vladimir Titov and Gennady Strekalov, narrowly escaped disaster during a launch pad fire, and during the Soyuz TM-3 flight in 1987, which sent the first citizen of Syria, Muhammed Faris, into space.

Solovyov's first space mission began on June 7, 1988 when he traveled to the Mir space station with fellow cosmonaut Viktor Savinykh and Aleksandr

Aleksandrov of Bulgaria on Soyuz TM-5. After a few days' stay at the station, Solovyov, Savinykh and Aleksandrov returned to Earth on June 17, 1988 in Soyuz TM-4, having logged more than nine days in space during the flight.

His next visit to the space station was a long-duration stay of six months, as a member of the Mir 6 crew. Lifting off on February 11, 1990 with Aleksandr Belandin, Solovyov's second trip to Mir was aboard Soyuz TM-9.

On July 17 Solovyov and Belandin made an EVA of seven hours to repair insulation on the Soyuz TM-9 vehicle. They made a second EVA, on July 26, to inspect the hatch on the space station's Kvant module. The second spacewalk added three-and-a-half hours to their total EVA time during the Mir 6 mission.

At the end of their long flight, Solovyov and Belandin were replaced by the Mir 7 crew, and returned to Earth in their repaired Soyuz TM-9 on August 9, 1990. From launch to landing, they had spent more than 179 days in space.

On July 27, 1992 Solovyov began his third space mission, as a member of the Mir 12 crew. He traveled to Mir aboard Soyuz TM-15 with Mir 12 crew mate Sergei Avdeyev and with visiting cosmonaut Michel Tognini of France. Tognini returned to Earth aboard Soyuz TM-14 on August 10, while Solovyov and Avdeyev settled in at the station to begin their long-duration stay.

In September, Solovyov and Avdeyev made 4 spacewalks, spending a total of more than 18 hours outside Mir while installing equipment on the station.

They returned to Earth aboard Soyuz TM-15 on February 1, 1993, after a total flight of more than 188 days.

Solovyov's remarkable long-duration spaceflight career continued with his assignment to the Mir 19 crew. He and Mir 19 crew mate Nikolai Budarin traveled to Mir with the STS-71 crew of the space shuttle Atlantis. At the space station, the STS-71 crew carried out the historic first docking of the U.S. space shuttle and the Russian space station.

Solovyov and Budarin remained aboard Mir when Atlantis undocked and returned to Earth at the end of the 10-day shuttle flight.

As part of their initial duties as the Mir 19 long-duration crew, Solovyov and Budarin transferred into their Soyuz craft to get a good view of the undocking of Atlantis and Mir on July 4, before they returned to Mir.

Solovyov and Budarin made three spacewalks during their stay aboard the station. On July 14 they spent more than five-and-a-half hours outside of Mir while deploying a solar array panel for the station's Spektr module. Their next task, the installation of the MIRAS spectrometer, required a total of more than eight-and-a-half hours during EVAs on July 19 and July 21.

At the end of their Mir 19 mission, they returned to Earth aboard Soyuz TM-21, landing on September 11, 1995 after spending more than 75 days in space.

Solovyov flew a remarkable fourth long-duration mission—his fifth flight overall—as commander of Mir 24. He and Pavel Vinogradov lifted off on Soyuz TM-26 on August 5, 1997.

The Mir 24 mission came at a time of great concern about the state of the Mir space station. A collision with an unmanned Progress cargo spacecraft the previous June had damaged the station's Spektr module and caused the entire complex to become depressurized for a brief period, endangering Mir 23 crew members Vasili Tsibliyev and Aleksandr Lazutkin and NASA astronaut Michael Foale. Although they were able to seal off the Spektr module and re-establish a safe environment, the damage made it necessary for mission planners on Earth to devise some means of fixing the station, beginning with the re-connection of the module's three undamaged solar array panels. The novel solution arrived at by ground controllers and the crew was to make a spacewalk *inside* the space station. Tsibliyev and Lazutkin were scheduled to make the repair, but Tsibliyev was subsequently diagnosed as being in poor health, and then while Foale was training to take his place, a mishap occurred when a cable was accidentally unplugged, leaving the station temporarily without power.

In light of the difficulties and the general feeling of disarray, mission controllers decided to wait until the Mir 24 crew arrived at the station, and assigned the internal EVA to Solovyov and Vinogradov. At the same time, acting out of a similar concern, NASA decided to replace Wendy Lawrence, the astronaut originally scheduled to replace Foale at the end of his stay, with her backup David Wolf. Lawrence was smaller than the minimum size specified to wear the Russian Orlan spacesuit, which would be required for the internal spacewalk, so Wolf, who had been scheduled for a later flight, was moved up so he could serve as a backup for the spacewalks planned to fix the station over the following months.

For Solovyov, the new urgency attending the start of his Mir 24 mission meant a substantial increase in EVAs. His spacewalking duties during the flight would be vital to the continued health of the station and its subsequent inhabitants.

He and Vinogradov made the internal spacewalk on August 22, thoroughly inspecting the damage to the Spektr module and connecting power cables from the craft's solar arrays to a new hatch specially designed to accommodate them. Foale watched the cosmonauts during their tense three-plus hours inside Spektr, acting as a communications link between them and mission control.

Foale joined Solovyov during the second EVA to inspect the module, the two spacewalkers this time venturing outside of the station to try to find the point where the damaged module's hull had been breached. While they were examining the craft, they also repositioned its solar arrays to increase the amount of energy the solar panels could absorb from the Sun. The second EVA lasted about six hours.

A happier task faced the crew on September 27, when they welcomed the STS-86 flight of the space shuttle Atlantis during the seventh shuttle-Mir docking mission. Wendy Lawrence and David Wolf were both aboard the shuttle for the visit, along with fellow STS-86 crew members commander James Wetherbee, pilot Michael Bloomfield and mission specialists Scott Parazynski, Jean-Loup Chrétien of France, and Vladimir Titov of the Russian Space Agency (RSA).

The shuttle and the space station were linked for 6 days while the crews transferred 10,400 pounds of equipment and supplies. Titov and Parazynski performed the first joint EVA by a Russian cosmonaut and an American astronaut, when they deployed a Solar Array Cap for the damaged Spektr module.

At the end of the docked flight, the shuttle returned to Earth with Foale aboard, while Wolf remained on Mir.

Solovyov made his next four EVAs with Vinogradov. During the first three spacewalks (on October 20, November 3 and November 6)—each more than six hours—the cosmonauts continued the exhausting repairs on the Spektr module's solar arrays, dismantling one panel and installing a new one. They spent an additional three hours spacewalking on January 8 and 9, 1998, to repair a leaking EVA hatch, and on January 14 Solovyov and Wolf performed a nearly four hour EVA inspection of the outside of the space station.

Solovyov's spacewalking total for Mir 24 was 34 hours and 54 minutes, during 7 EVAs.

Following the successful completion of their eventful flight, Solovyov and Vinogradov returned to Earth aboard Soyuz TM-26 on February 19, 1998, after a mission of more than 197 days.

Anatoly Solovyov has logged more than 651 days in space during his illustrious career as a cosmonaut. In his record-setting 16 spacewalks, he has accumulated a record 78 hours and 3 minutes in EVA.

SOLOVYOV, VLADIMIR A.

(1946–)

Soviet Cosmonaut

A veteran of two long-duration space missions and eight spacewalks, Vladimir Solovyov was born in Moscow, Russia on November 11, 1946. He attended the Bauman Higher Technical School in Moscow, where he graduated with an engineering degree in 1970.

Solovyov was chosen for training as a cosmonaut as a member of the Civilian Specialist Group Six selection of cosmonaut candidates in 1978.

He first flew in space as flight engineer for the Soyuz T-10/Salyut 7 long-duration mission in 1984. Launched from the Baikonur Cosmodrome on February 8, 1984, Solovyov traveled to Salyut 7 with commander Leonid Kizim and research engineer Oleg Atkov. They lived and worked aboard the station for nearly eight months, as the third Salyut 7 long-duration crew.

During their long stay, Solovyov and Kizim accomplished a spectacular repair job when they brought the space station's propulsion system back into good working order following damage it had sustained the previous summer receiving fuel from the unmanned Progress 17 cargo spacecraft. The cosmonauts devoted five of a total six EVAs to the repair work, beginning on April 23, 1984 with a spacewalk of more than four hours. They made a second EVA devoted to the orbital repair work on April 26, and added a third on April 29 and a fourth on May 4. In the course of just 12 days, Solovyov and Kizim accumulated more than 14 hours in EVA while working on the damaged propulsion system—and at that point, still had work to do before they could declare the system fixed.

They made a fifth spacewalk on May 18 to work on the station's solar array, and then completed their repairs on August 8, with a final EVA of five hours. Their exceptional, methodical repair work made the Salyut 7 propulsion system fully operational again.

Solovyov, Kizim and Atkov also welcomed two visiting crews to Salyut 7 during their stay at the station, and in each case the visits featured historic "firsts" in the history of spaceflight.

The Soyuz T-11 crew, which included Rakesh Sharma—the first citizen of India to fly in space—arrived in early April for a week-long visit. The flight was part of the Soviet Intercosmos program, which provided spaceflight opportunities for citizens of nations friendly to or aligned with the Soviet Union, and was also a Soyuz switching flight, in that the Soyuz T-11 craft was left at the station to replace the Soyuz T-10 vehicle that Solovyov and his crew mates had used to travel to the station in February. The necessity of replacing a docked Soyuz with a "fresh" vehicle arose from the fact that the Soyuz spacecraft could only remain in space for a limited period before its on-board systems began to deteriorate. Sharma and his Soviet crew mates returned to Earth in Soyuz T-10 on April 11.

The second visiting crew arrived on July 17 aboard Soyuz T-12. The all-Soviet crew was commanded by Vladimir Dzhanibekov, and also included research engineer Igor Volk and flight engineer Svetlana Savitskaya—who would during the short visit become the first woman in history to participate in a spacewalk.

As Solovyov and his crew mates looked on and Volk monitored the EVA from within Salyut 7, Savitskaya and Dzhanibekov spent 3 hours and 35 minutes outside the station on July 25, 1984, while testing an electron beam hand tool used for tasks such as cutting and soldering. The Soyuz T-12 crew left the station a few days after the historic EVA.

Then, with their own mission complete, Solovyov, Kizim and Atkov left Salyut 7 and returned to Earth on October 2, 1984 in Soyuz T-11. During his first spaceflight, Solovyov and his crew mates spent a then-record 236 days, 22 hours and 50 minutes in space, and he and Kizim accumulated more than 22 hours in EVA during their 6 spacewalks.

Solovyov again flew in space with Kizim during the Soyuz T-15/Mir 1 long-duration mission, which launched on March 13, 1986. A unique flight in the history of space exploration, the second space odyssey of Solovyov and Kizim included visits to both the newly-launched Mir orbital complex and the Salyut 7 space station where they had served their first long flight in orbit. It was the only instance in the long history of Soviet space stations in which a single crew served aboard separate stations during a single mission.

Solovyov and Kizim first headed to Mir, which had been launched unmanned on February 20, 1986. After docking with and entering the station, they began the initial checking out of its systems and equipment. They settled into a routine, receiving supplies from unmanned Progress cargo ships, carrying out scientific work, and performing the engineering and maintenance tasks necessary to the first test of Mir's ability to sustain a crew in orbit. During the first part of their mission, they remained aboard the station for a little over six weeks, until May 5.

Then, boarding Soyuz T-15 and undocking from the new space station, Solovyov and Kizim traveled to Salyut 7, where they were the first cosmonauts to enter the station since the hurried departure of the previous crew in November 1985. An illness that later turned out to be unrelated to the spaceflight itself had forced Vladimir Vasyutin and his long-duration crew mates Viktor Savinykh and Alexander Volkov to evacuate the station in November, and their quick exit had left their program of experimental work unfinished.

Six months later, Solovyov and Kizim were given the task of traveling to Salyut 7 to pick up where the previous crew had left off. They would ultimately spend 51 days aboard Salyut 7 while completing the work and closing out the station's operations. As part of their work, they made two experimental EVAs to test procedures for constructing large structures in space. On May 28, 1986 they made the first of the two spacewalks, deploying a large truss structure and then folding and returning it to the interior of the station, while spending nearly four hours outside Salyut 7. They tested the truss assembly procedure again during a 4 hour, 40 minute EVA on May 31.

Before the end of their long, unique flight, Solovyov and Kizim returned to Mir on June 25 for the final phase of their mission. Salyut 7 was placed in a "storage" orbit in the hope that it might again be used at some later time, but Solovyov and Kizim were ultimately the last crew to visit the station. Mir would be the focus of the Russian manned space program for the next decade and a half; Salyut 7 plummeted back to Earth on February 7, 1991, with most of the station burning up in the atmosphere during re-entry.

With the most challenging portion of their mission behind them, Solovyov and Kizim settled into a productive routine during their second stay of the flight at Mir, and returned to Earth on July 16, 1986 after a total flight—to Mir, to Salyut 7, back to Mir, and then back to Earth—of 125 days. Their Soyuz T-15 flight also marked the final use of the Soyuz T series of spacecraft, which was then replaced by the Soyuz TM series.

Vladimir Solovyov spent more than 361 days in space, including more than 31 hours in EVA during eight spacewalks, during his career as a cosmonaut.

He continued to serve the Soviet and Russian space program after leaving the cosmonaut corps, serving as Technical Flight Director at the Manned Space Flight Control Center. Continuing his education concurrently with his space program career, he earned a doctorate of technical sciences degree in 1995, and in 1997 became a professor.

SOVIET UNION: SPACE PROGRAM.

See Russian Federal Space Agency

SOYUZ PROGRAM

(1967–)

Soviet / Russian Manned Spaceflight Program

First envisioned as the ship that would take Russian cosmonauts into orbit around the Moon, the basic Soyuz spacecraft featured three parts: a Descent Module, which carried the crew into space and returned them to Earth at the end of a mission;

The Soyuz 19 spacecraft atop its launch vehicle at the Baikonur Cosmodrome, July 15, 1975. [NASA/courtesy of nasaimages.org]

the Orbital Module, in which the crew lived and worked while in space; and an Instrument Module, which housed the equipment necessary to sustain the flight, including the spacecraft's maneuvering engine.

In Russian "Soyuz" means "union"—which, at the time of the program's start in the 1960s, was a familiar way of referring to the Soviet Union.

The basic Soyuz design has gone through a large number of revisions over the course of its first four decades, and despite its initial failings, the spacecraft has proven a flexible and reliable mode of space transportation.

The initial Soyuz design was adapted for a variety of purposes during its first two decades, beginning with the Soviet Union's plans to mount a manned lunar landing mission, which were later abandoned, and also including flights to the first Soviet space stations. A specially adapted version of the Soyuz was used in the joint Soviet-American Apollo-Soyuz Test Project (ASTP) in July 1975.

The variety of design changes and varied uses of the original Soyuz have led to a fairly large number of subcategories describing various iterations (for example, the L1 Zond Moon craft; the Soyuz version A, B & C; the Salyut 1 "ferry" vehicle, etc.) but the broad classifications of Soyuz spacecraft generally group the original Soyuz and its adaptations as the spacecraft's first generation, which was flown from the late 1960s until the early 1980s. The last flight of the original Soyuz spacecraft was Soyuz 40, in May 1981 (which was also the last mission to the Salyut 6 space station).

In 1979 the first of the Soyuz T series spacecraft was launched, unmanned, on December 16. The Soyuz T represented a major overhaul of the basic Soyuz design, taking into account the many changes in the direction and purpose of the Soviet program since the spacecraft's inception in the late 1960s, when the U.S.S.R. was still focused on reaching the Moon. In the intervening years, the transportation of cosmonauts to orbiting space stations had emerged as the primary task for the Soyuz craft, and the Soyuz T redesign reflected the change.

The Soyuz T design allowed for the transport of three-person crews (the last previous crew of three prior to the introduction of the Soyuz T was the ill-fated Soyuz 11 in June 1971, which had resulted in a spacecraft malfunction that had cost the lives of the crew), and featured solar arrays rather than batteries, which allowed for longer missions.

Use of the Soyuz T was gradually phased in while the original Soyuz was still in use; the first manned Soyuz T was Soyuz T-2, in June 1980; the last in the series, Soyuz T-15, provided cosmonauts Leonid Kizim and Vladimir Solovyov the unique opportunity to be the first crew to visit two space stations during a single flight, as they served as the first resident crew of Mir and also made the last visit to the Salyut 7 space station, on May 5, 1986. (Salyut 7 fell from orbit in February 1991, after several years of unoccupied "storage" in space).

At about the same time that the basic Soyuz design was being reworked into the Soyuz T, another variant was put into use as the unmanned cargo delivery vehicle known as Progress. Designed to ferry equipment and supplies to orbiting space stations and then (in most cases) to burn up during re-entry into the Earth's atmosphere, the first Progress flight launched on January 20, 1978, and successfully delivered supplies to the Salyut 6 space station. Like their manned counterparts, the Progress vehicles have also been upgraded several times; subsequent versions of the original Progress design include Progress M, which made its first flight—to Mir—in 1989; and the Progress M-1 series, which first flew in February 2000.

Similar to the way in which the Soyuz T adapted the basic Soyuz spacecraft to fit the needs of a new era of Soviet spaceflight, the Soyuz TM series was a major upgrade to the systems and equipment of the Soyuz T, reflecting the myriad of advances in electronic systems that had begun to emerge by the mid-1980s. The first TM flight, Soyuz TM-1, was flown unmanned in May 1986 and docked with Mir (simultaneously with the Soyuz T-15 mission—Kizim and Solovyov were aboard Salyut 7 when the unmanned Soyuz TM-1 docked with Mir)—proving its worth as both a crew transport vehicle and as a delivery system for resupplying cosmonauts in their orbital outpost.

Among the many upgrades in the TM series, the new generation of Soyuz included state-of-the-art computer systems and improvements in the spacecraft's rendezvous, guidance, electrical and hydraulic systems.

The basic Soyuz TM craft was used to ferry cosmonauts to and from the Mir space station from the mid-1980s, starting with Soyuz TM-2 in 1987, until the last visit to Mir in 2000, and was then used to transport crews to and from the International Space Station (ISS). The last of the original TM series Soyuz craft was Soyuz TM-34, which visited the ISS in 2002.

In response to NASA concerns about adapting the TM-type Soyuz for use as a "lifeboat" that could return a crew to Earth in the event of an emergency aboard the ISS, Russian space officials re-engineered the TM design, resulting in a new iteration known as the Soyuz TMA series. Among the changes implemented in the new design, the TMA series includes improved displays in the spacecraft's cockpit and an improved parachute system, and reflecting the expanded role of the Soyuz spacecraft in the post-Cold War era of international cooperation in space, the TMA redesign also included changes designed to better accommodate astronauts of other nations that will fly aboard the resilient Russian vehicle.

Soyuz TMA-1 lifted off on October 30, 2002, inaugurating the new era of the venerable Soyuz.

Soyuz Flights:

Design	*First Manned Flight*	*Last Flight*
Soyuz	Soyuz 1 (04/23/67—04/24/67)	Soyuz 40 (05/14/81—05/22/81)
Soyuz T	T-2 (06/05/80—06/09/80)	T-15 (03/13/86—07/16/86)
Soyuz TM	TM-2 (02/05/87—07/30/87)	TM-34 (04/25/02—11/10/02)
Soyuz TMA	TMA-1 (10/30/02—05/04/03)	

Unmanned Progress Supply Spacecraft:

Design	*First Flight (Launch)*	*Last Flight (Launch)*
Progress	Progress 1 (01/20/78)	Progress 42 (05/06/90)
Progress M	Progress M-1 (08/23/89)	
Progress M1	Progress M1–1 (02/01/00)	

Although the design of its basic vehicles and equipment would eventually prove to be the basis of the single most consistent spacecraft in the first half century of manned spaceflight, the Soviet Soyuz program began in tragedy and disaster, at the height of the Cold War Moon landing competition between the Soviet Union and the United States.

Having fallen behind their American counterparts during the mid-1960s, when the Soviet Voskhod program was cut short after just two manned flights while the U.S. Gemini program rapidly moved forward, Soviet space officials were anxious to develop the means necessary for rendezvous and docking in orbit, and for transferring crews from one spacecraft to another.

Plans for the spacecraft that would become Soyuz had originally been approved in 1963, but the political upheaval involved in the removal of Soviet leader Nikita Khrushchev in October 1964 and the sudden death of Sergei Korolev, the "Chief Designer" of the Soviet space program, on January 14, 1966 had thrown the nation's space effort into disarray.

Having replaced Khrushchev, Leonid Brezhnev and Aleksei Kosygin appointed Vasily Mishin as the new head of the Soviet space program, and charged him with developing the vehicles, systems and equipment necessary for a lunar landing as quickly as possible. The development centered on the Soyuz spacecraft and the L-3 lunar landing vehicle.

The first test of the new Soyuz craft took place on November 28, 1966. Launched unmanned and known officially as Cosmos 133, the first Soyuz spent two days in orbit before Soviet mission controllers on Earth directed its return. During re-entry into the atmosphere, a manufacturing error in the spacecraft's heat shield caused the protective shield to fail. The result was heavy damage to the craft's Descent Module—which would have caused the loss of the crew if the spacecraft had been carrying cosmonauts during the flight.

For Soviet space officials, the Cosmos 133 flight was a discouraging start to the flight test phase of the Soyuz program. Then, as though the charred and gutted Descent Module was not evidence enough of the risks involved in preparing the Soyuz for manned flight, a tragic accident in the American Apollo program reinforced the gravity of the Soviets' problems.

On January 27, 1967, prior to the second Soyuz test flight, American astronauts Virgil Grissom, Edward White and Roger Chaffee were killed in a fire during a test of the systems and equipment of their Apollo command module. They had been scheduled to be the first crew to fly a manned Apollo, and their tragic loss imperiled the future of the entire American space program. As things eventually worked out, the tragedy also played a key role in advancing the American cause in space, as NASA redoubled its emphasis on safety; but the immediate effect of the Apollo 1 disaster was a period of investigation, introspection and inactivity for the American program.

Shaken by the American tragedy and the potential disaster hinted at by the failed Cosmos 133 flight, Soviet space officials favored a cautious pace for the development of the Soyuz spacecraft as a manned vehicle. Soviet political leaders, however, disagreed. Hoping to catch up with the Americans, who had advanced rapidly in their spaceflight expertise and systems during the Gemini program, the U.S.S.R.'s new political leadership under Leonid Brezhnev had envisioned the initial Soyuz flights as a means of quickly duplicating and then surpassing the U.S. Gemini achievements. Then, with the American program at a standstill after the Apollo 1 disaster, the Soviet leaders saw an even greater opportunity to "catch up" with their superpower enemy, and stressed the need for rapid development of the Soyuz.

The second unmanned Soyuz, labeled Cosmos 140, lifted off on February 7, 1967 and was brought back to Earth after two days in orbit. Free of the manufacturing defect that had led to the damage of the Descent Module during the first test, the Cosmos 140 Descent Module returned relatively intact. The flight was not without problems, but achieved an acceptable enough degree of success to allow for two subsequent test launches, which were designed to move the Soviet program closer to sending cosmonauts into orbit around the Moon.

The first two unmanned "L-1" flights (L-1 being the designation given to the spacecraft the U.S.S.R. hoped to fly to the Moon) began with Cosmos 146, which was launched on March 10, 1967. It spent a little over a week in orbit before being automatically brought back to Earth. The second unmanned L-1 test, Cosmos 154, lifted off on April 8 and spent 11 days in orbit.

As with the first two Soyuz test flights, the L-1 tests experienced a number of problems that indicated fundamental difficulties with the spacecraft's systems and equipment, including faults in critical areas such as its maneuvering and landing systems.

In the years since the fall of the Soviet Union, details that have emerged about the early days of the Soyuz program indicate that the engineers responsible for improving the spacecraft were hesitant to see it used for manned flights until the problems could be more thoroughly addressed. Despite their concerns, however, the Soviet Union's political leadership apparently felt a manned mission was a necessity in the early part of 1967.

Thus Soyuz 1, commanded by Voskhod veteran Vladimir Komarov, was launched on April 23, 1967. Komarov had flown in space with Konstantin Feoktistov and Boris Yegorov aboard Voskhod 1 in October 1964, and was given the honor—and the risk—of inaugurating the Soyuz program.

After safely reaching orbit aboard Soyuz 1, Komarov was to have been followed into space by Valeri Bykovsky, Aleksei Yeliseyev and Yevgeni Khrunov, who were scheduled to launch aboard Soyuz 2 on April 24. The mission profile for the Soyuz 1/Soyuz 2 flights called for the two spacecraft to rendezvous and dock in orbit, and then exchange crew members via a spacewalk. Yeliseyev and Khrunov would have returned to Earth with Komarov aboard Soyuz 1, and Bykovsky would return alone in Soyuz 2.

But Komarov experienced problems with Soyuz 1 shortly after launch, giving rise to concerns among Soviet space officials and engineers that their worst fears about the spacecraft's shortcomings might be unfolding right before their eyes. The trouble began with a fault in one of the craft's solar panels, which prevented the panel from deploying properly. As a result, Soyuz 1 had less electrical power than expected, and its computer system faltered.

After 15 increasingly tense orbits, mission controllers decided the spacecraft's failings were too severe for the flight to continue. Komarov was ordered to abort the mission, and to return to Earth. Two successive attempts at re-entry ended in failure, when Komarov and the ground crews were unable to orient the spacecraft properly for re-entry. With considerable courage and skill, the veteran cosmonaut was able to remain focused on the task at hand, despite the mounting tension of the situation, and he succeeded in orienting the spacecraft properly and firing its retrorockets during his eighteenth orbit.

Having overcome the first obstacle to a safe return, Komarov quickly faced another bone-rattling problem: the failed electrical systems and malfunctioning computer aboard Soyuz 1 led to an uncontrolled re-entry, with the Descent Module flailing wildly through the atmosphere. In an effort to by-pass the faulty computer and steady the spacecraft, mission control apparently had Komarov switch the spacecraft to ballistic re-entry mode, which, while giving the spacecraft a better chance of surviving the return to Earth intact by placing it in a constant spin, also increased the force with which it would re-enter the atmosphere. The increased force likely caused Komarov to black out prior to the landing; in any case, he was unable to halt the spacecraft's spin as it sped toward the ground, far off course, for an eventual impact near Orsk, in the Ural Mountains. The craft's parachute lines became entangled just prior to landing, and the Descent Module crashed violently, its retrorockets exploding on impact and engulfing the capsule in flames. Komarov was killed.

The untimely death of one of the nation's space heroes was a deeply felt tragedy for the Soviet Union, and a serious threat to the future of the U.S.S.R.'s space program. Komarov had been widely hailed for his achievements during the Voskhod program, and in the years following his death the story of his involvement in the Soyuz 1 flight only added to his legendary status among his fellow Russian citizens. As an experienced cosmonaut and pilot (he was a colonel in the Soviet Air Force), Komarov understood the risks involved with testing the new spacecraft, and apparently shared many of the apprehensions of others in the Soviet space program about the first flight. Although the married father of two apparently became virtually certain that Soyuz 1 was not ready to be flown into space with a cosmonaut on board, he reportedly declined to give up the assignment because he feared for the life of his backup—and close friend—Yuri Gagarin, who would be assigned to the dangerous mission if Komarov were to be removed from the flight.

Similar to the way in which the Apollo 1 tragedy resulted in a hiatus in the American drive to land astronauts on the Moon, the Soyuz 1 accident resulted in a halt in the Soviet lunar program.

In October 1967 the test flights whose success had been deemed by the engineering staff of the Soviet space program to be necessary to the start of manned Soyuz launches were finally achieved. The unmanned Cosmos 186 lifted off on October 27, followed on October 30 by the also unoccupied Cosmos 188. Mission controllers maneuvered the first craft into the vicinity of the second, and then Cosmos 186 successfully docked with Cosmos 188—just as Soyuz 1 and 2 would have in April, if the myriad problems with the Soyuz spacecraft had been ironed out before the Soyuz 1 flight was allowed to proceed.

The next attempt at a manned docking was made in October 1968. The unmanned Soyuz 2 was launched first, and on October 26, 1968 Georgi Beregovi lifted off aboard Soyuz 3, with a mission profile that was similar to the one envisioned for Komarov's flight. Despite several attempts at docking, Beregovoi returned to Earth on October 30 without having linked up with the unoccupied Soyuz 2; but his safe return augured well for the future of Soyuz, and for the Soviet program as a whole.

The long-sought manned rendezvous and docking was finally achieved with the joint flights of Soyuz 4 and Soyuz 5, in January 1969. Launched aboard Soyuz 4 on January 14, Vladimir Shatalov was able to rendezvous and dock with Soyuz 5, which had lifted off on January 15 with Boris Volynov, Aleksei Yeliseyev and Yevgeni Khrunov. As they had been scheduled to do during the aborted Soyuz 1/2 flight in 1967, Yeliseyev and Khrunov spacewalked from Soyuz 5 to Soyuz 4, and returned to Earth with Shatalov, while Volynov endured a difficult landing in Soyuz 5 on January 18. Far off course and badly shaken from the impact of the landing, Volynov was, fortunately, not seriously injured.

Even though development of the other pieces of the Soviet lunar landing puzzle was gradually abandoned in the years following the successful American Apollo Moon landings, the Soyuz survived its tragic beginning to become a flexible, reliable means of transporting cosmonauts into space. The basic Soyuz design served the Soviet program well throughout the 1970s, regularly transporting cosmonauts to the Salyut and Almaz (military Salyut) space stations. Subsequent versions served as the primary means of travel to the Mir space station, and with further modifications, Soyuz craft have also been instrumental in ferrying astronauts from many nations to the International Space Station (ISS).

NOTE ABOUT SOYUZ LAUNCH AND LANDING DATES: Soyuz flights in the following chronology are calculated by the length of time in space of the crew that originally launched aboard the spacecraft—NOT the amount of time the vehicle itself was in space.

Thus in the case of a spacecraft that was used to travel to a space station and then switched for another Soyuz for the return trip to Earth, the launch and landing dates of the crew in the launch vehicle are used, regardless of when the launch vehicle itself returned to Earth, or which vehicle the crew used for their return.

For example: Soyuz 27 commander Vladimir Dzhanibekov and flight engineer Oleg Makarov launched aboard Soyuz 27 on January 10, 1978, and traveled to the Salyut 6 space station. After a brief stay at Salyut 6, they returned to Earth aboard Soyuz 26 on January 16, 1978. Thus the entry for Soyuz 27 carries the dates January 10–16, 1978—representing the flight of the crew that launched aboard Soyuz 27, rather than the time that the spacecraft itself spent in orbit.

Similarly, Soyuz 27 was later used by Yuri Romanenko and Georgi Grechko to return to Earth on March 16, 1978, at the end of their long-duration stay aboard Salyut 6. They had traveled to the station in Soyuz 26 on December 10, 1977.

Therefore, although Soyuz 26 was in space from December 10, 1977 to January 16, 1978, and Soyuz 27 was in space from January 10, 1978 to March 16, 1978, the Soyuz 26 entry is dated December 10, 1977 (Romanenko and Grechko launch, in Soyuz 26) to March 16, 1978 (Romanenko and Grechko landing, in Soyuz 27).

SOYUZ MISSION 1: SOYUZ 1

(April 23–24, 1967)

Soviet Manned Spaceflight

Launched on April 23, 1967 with veteran cosmonaut Vladimir Komarov aboard, Soyuz 1 suffered serious difficulties with its electrical and control systems, and ended in tragedy.

Komarov had previously flown in space as commander of the historic Voskhod 1 flight, which was the first spaceflight to feature a three-person crew. His Soyuz 1 mission called for him to rendezvous and dock with a second Soyuz, which was scheduled to launch on April 24 with Valeri Bykovsky, Aleksei Yeliseyev and Yevgeni Khrunov. Yeliseyev and Khrunov would then make a spacewalk from their Soyuz to Soyuz 1, and return to Earth with Komarov, while Bykovsky returned aboard the second Soyuz.

Komarov's backup for the Soyuz 1 flight was Yuri Gagarin.

Despite problems with the four unmanned test flights of Soyuz spacecraft prior to Komarov's mission, the political leadership of the Soviet Union argued in favor of a manned flight as a means of "catching up" in the race with the United States to land a man on the Moon. Spurred by the success of its Gemini program, the American space effort had stalled after the tragic January 1967 Apollo 1 fire that had claimed the lives of astronauts Virgil Grissom, Edward White and Roger

Chaffee. Soviet leaders felt the halt in the American program provided an opportunity for the Russian lunar landing effort to duplicate and surpass the American Gemini achievements.

But Komarov experienced problems with the Soyuz shortly after launch. One of the spacecraft's solar panels failed to deploy properly, which cut the amount of electrical power available for crucial tasks. A malfunction in the craft's computer system followed, resulting in serious difficulties in controlling the flight.

Worried mission controllers decided to abort the complex docking mission, and canceled the scheduled launch of the second Soyuz. With Komarov's safe return as their only remaining priority, they approved his re-entry after fifteen orbits.

On his sixteenth trip around the Earth, Komarov attempted to orient the Soyuz for the proper re-entry trajectory, but the spacecraft's malfunctioning systems prevented him from doing so. He tried again on the next orbit, but again could not orient the spacecraft properly. The inability to set the Soyuz Descent Module on the proper path was a grave problem; the wrong trajectory could, on one hand, cause the spacecraft to skip off into space or, on the other, cause it to enter at too steep an angle and be destroyed by the force of re-entry into the Earth's atmosphere.

Despite the severity of the circumstances, Komarov was able to remain focused on the problem at hand, and working with mission controllers on the ground, he was able to orient the Soyuz craft properly on the eighteenth orbit.

Having overcome the first obstacle to his safe return, however, Komarov quickly faced another bone-rattling problem: the failed electrical systems and malfunctioning computer aboard Soyuz 1 caused his Descent Module to gyrate wildly, raising the prospect of the craft failing to survive re-entry even after all the effort to place it on the proper trajectory.

Aware of the fact that a ballistic re-entry—in which the capsule would be placed into a rapid spin—did not require the spacecraft's malfunctioning computer system, mission controllers apparently had Komarov place the craft into a spinning motion. While the maneuver did give him a better chance of returning to Earth with his spacecraft intact, it also increased the force with which he would re-enter the atmosphere. The increased force likely caused Komarov to black out prior to the landing; in any case, he was unable to halt the spacecraft's spin as it sped toward the ground, far off course, for an eventual impact near Orsk, in the Ural Mountains.

The craft's parachute lines became entangled just prior to landing, and the Descent Module crashed violently, its retrorockets exploding on impact and engulfing the capsule in flames.

Already considered a national hero for his previous flight aboard Voskhod 1, Vladimir Komarov was killed in the Soyuz 1 crash. A married father of two children, he was 40 years old at the time of his death.

As an experienced cosmonaut and pilot (he was a colonel in the Soviet Air Force), Komarov understood the risks involved with testing the new spacecraft, and apparently shared many of the apprehensions of others in the Soviet space program about the first flight. Although he apparently felt certain that Soyuz 1 was not ready to be flown into space with a cosmonaut on board, he reportedly declined to give up the assignment because he feared for the life of his backup—and close friend—Yuri Gagarin, who would automatically be assigned to the dangerous mission in his place if Komarov were to be removed from the flight.

The Soyuz 1 tragedy posed a serious threat to future development of the Soyuz spacecraft—and to the Soviet space program as a whole. After a series of design changes and further testing, the first successful linking of two unmanned Soyuz craft was achieved in October 1967, with the launch of Cosmos 186 on October 27 and Cosmos 188 on October 30. The two unoccupied spacecraft were successfully maneuvered and docked (with Cosmos 186 as the active craft, and Cosmos 188 as the passive target), fulfilling the basic mission profile that had been pursued in April. The successful docking of the two unoccupied vehicles ensured that there would be future Soyuz launches, but was of course bittersweet for those engineers and space agency officials who had envisioned the same course for Soyuz 1, if only the problems with the Soyuz spacecraft had been properly addressed before that tragic flight had been allowed to proceed.

Soyuz 1 returned to Earth after a flight of 1 day, 1 hour and 48 minutes.

SOYUZ MISSION 2: SOYUZ 2

(October 25–28, 1968)

Soviet Unmanned Spaceflight

In an attempt to achieve some measure of the rendezvous and docking envisioned in the mission profile of the first Soyuz flight—which had ended in disaster in April, 1967 with the loss of veteran cosmonaut Vladimir Komarov—the Soviets launched Soyuz 2 and Soyuz 3 in October 1968. The profile for Soyuz 1 had called for the linking of two manned spacecraft; in the case of Soyuz 2 and 3, the goal was modified to have one manned Soyuz (Soyuz 3) dock with an unmanned Soyuz target vehicle (Soyuz 2).

The Soyuz 2/Soyuz 3 flights followed a successful docking by two unmanned Soyuz craft known as Cosmos 186 and Cosmos 188 in October 1967.

Soyuz 2 was launched unmanned on October 25, 1968, followed by Soyuz 3, which launched on October 26 with Georgi Beregovoi aboard. Beregovoi made several attempts to dock with the Soyuz 2 craft, but could not. Soyuz 2 was de-orbited on October 28; Beregovoi landed in Soyuz 3 on October 30.

SOYUZ MISSION 3: SOYUZ 3

(October 26–30, 1968)

Soviet Manned Spaceflight

The primary goal of most of the early Soyuz flights was the same as the mission profile for the ill-fated Soyuz 1: to have two manned Soyuz spacecraft rendezvous and dock, and ultimately, to transfer crew members from one spacecraft to the other.

During Soyuz 1, which ended in a fiery crash that claimed the life of veteran cosmonaut Vladimir Komarov, the projected rendezvous and docking exercises had to be canceled because of failures aboard the first manned Soyuz.

A successful rendezvous and docking was achieved with the unmanned Cosmos 186 and Cosmos 188 in October 1967.

To some degree, the unmanned Soyuz 2/manned Soyuz 3 flights in October 1968 were a sort of bridge between the successful unoccupied Cosmos 186/Cosmos 188 link-up and the envisioned docking of two manned Soyuz craft.

Following the launch of the unmanned Soyuz 2 on October 25, 1968 Georgi Beregovoi lifted off aboard Soyuz 3 on October 26, with the objective of catching up to Soyuz 2 in orbit and docking with the unoccupied spacecraft.

Vladimir Shatalov served as Beregovoi's backup for Soyuz 3—the first manned Soviet flight since the disastrous Soyuz 1 accident in April 1967.

Although the procedures and goals of the Soyuz 2/Soyuz 3 flights were basically the same as the successful Cosmos 186/Cosmos 188 a year earlier, the docking process proved maddeningly difficult. Beregovoi made three separate attempts to dock his Soyuz 3 with Soyuz 2, and was frustrated each time to find that the two spacecraft simply would not join together as anticipated.

Wary of pushing the attempt any farther, mission controllers gave up the docking objective after the third try, and Soyuz 2 was de-orbited on October 28. Beregovoi remained in orbit for another 2 days, and returned to Earth on October 30, after a flight of 3 days, 22 hours and 51 minutes.

Despite the failed attempts at docking, Beregovoi's flight remained a substantial step forward for the Soyuz program, as he was able to safely spend nearly four days in orbit and then safely return to Earth in the spacecraft's Descent Module while facing none of the deadly system failures that had doomed Komarov in Soyuz 1. Even with the rendezvous and docking procedures still to be mastered on some later flight, Soviet space officials could at least breathe a sigh of relief at the intact return of their cosmonaut and space capsule, and all that the safe flight and landing implied for future missions.

SOYUZ MISSION 4: SOYUZ 4

(January 14–17, 1969)

Soviet Manned Spaceflight

With the Soyuz 4 and Soyuz 5 flights of January 1969, the Soviets achieved the first-ever docking of two manned spacecraft, and the transfer of crew members between the two craft. The flights were a major milestone for the Soviet space program; in fulfilling the mission profile of the tragic Soyuz 1 flight of April 1967, they honored the memory of Vladimir Komarov, who had been killed during Soyuz 1, and by successfully linking two manned Soyuz vehicles, they brought the Soviet program back to a competitive level with the American space program, which had achieved the docking of manned Gemini capsules with unmanned target vehicles.

Since the end of the Gemini project, the Americans had suffered their own tragic loss with the January 1967 Apollo 1 fire, which claimed the lives of astronauts Virgil Grissom, Edward White and Roger Chaffee; but in the intervening years, the U.S. program had rebounded with successful Apollo flights, including the December 1968 lunar orbital flight of Apollo 8. To keep alive any faint hope of beating their superpower rival to a manned landing on the Moon, the Soviets desperately needed the Soyuz 4/Soyuz 5 mission to work.

Vladimir Shatalov launched aboard Soyuz 4 on January 14, 1969. His backup for the mission was Anatoli Filipchenko.

Soyuz 5 was launched the following day, January 15, 1969, crewed by Boris Volynov, Aleksei Yeliseyev and Yevgeni Khrunov. On January 16, Shatalov rendezvoused with Soyuz 5, which served as the passive target vehicle, and docked with the spacecraft. Although the two Soyuzes were docked, they were not equipped for a direct crew transfer, so Yeliseyev and Khrunov made a 37 minute EVA while exiting Soyuz 5 and entering Soyuz 4. The two spacewalking cosmonauts had originally been scheduled to make the transfer during the ill-fated Soyuz 1 flight in April 1967, but their flight back then had been canceled after severe problems imperiled Komarov's Soyuz 1.

The successful transfer from Soyuz 5 to Soyuz 4 fulfilled the main goal of the program's early years. Shatalov, Yeliseyev and Khrunov returned to Earth aboard Soyuz 4 on January 17. Shatalov's flight took 2 days, 23 hours and 21 minutes; Yeliseyev and Khrunov (who had launched aboard Soyuz 5 a day later than Shatalov) spent a total of 1 day, 23 hours and 39 minutes in space during the historic flight.

Although the linking of Soyuz 4 and Soyuz 5 seemed primitive a few years later, in light of the American lunar landings and the routine travel of Soviet cosmonauts to the Salyut orbital space stations, it was a significant achievement for its time. Soviet propaganda of the day grandiosely hailed the mission as the world's first space station, while observers in the West dismissed the flights as technically comparable to the American Gemini flights of several years earlier. But the real achievement of the flights could be found in the increased expertise and confidence they provided the administrators and engineers responsible for the Soviet lunar race, and for development of the Soyuz and its related systems and equipment.

SOYUZ MISSION 5: SOYUZ 5

(January 15–18, 1969)

Soviet Manned Spaceflight

The Soviets achieved the first-ever docking of two manned spacecraft, and the transfer of crew members between the two craft, with the Soyuz 4 and Soyuz 5 flights of January 1969. The flights were a major milestone for the Soviet space program, and they fulfilled the mission profile of the tragic Soyuz 1 flight of April 1967, in which veteran cosmonaut Vladimir Komarov had been killed.

Boris Volynov was commander of Soyuz 5; Aleksei Yeliseyev served as flight engineer, and Yevgeni Khrunov was research engineer. The backup crew for the Soyuz 5 flight included Georgi Shonin, Victor Gorbatko, and Valeri Kubasov.

The Soyuz 4/Soyuz 5 flights also kept alive the Soviets' slim hopes of beating the United States to be first to land a human being on the Moon. During their mid-1960s Gemini program, the Americans had achieved the docking of manned Gemini spacecraft and unmanned target vehicles, but the American drive to the Moon then suffered a major setback with the Apollo 1 fire in January 1967, which claimed the lives of astronauts Virgil Grissom, Edward White and Roger Chaffee.

Thus by January 1969 each nation had suffered tragedy, and both were anxious to safely test new spacecraft (the Soviet Soyuz and the American Apollo). The United States had rebounded from the Apollo 1 fire with several successful flights, and in December 1968 had achieved the first manned lunar orbital mission, with Apollo 8. With Soyuz 4/Soyuz 5, the Soviets hoped to complete their recovery from the ill-fated Soyuz 1 flight.

Following the launch of Soyuz 4 a day earlier with Vladimir Shatalov aboard, Soyuz 5 launched on January 15, 1969 with Boris Volynov, Aleksei Yeliseyev and Yevgeni Khrunov. The Soyuz 5 spacecraft served as a target vehicle for Soyuz 4, which, on January 16, rendezvoused and docked with Soyuz 5.

Although the two spacecraft were docked, they were not equipped for a direct transfer of crew members from one craft to the other. Yeliseyev and Khrunov made the switch from Soyuz 5 to Soyuz 4 via an EVA that took about 37 minutes, joining Shatalov in Soyuz 4 and leaving Volynov alone aboard Soyuz 5.

Yeliseyev and Khrunov had originally been scheduled to perform their EVA during the Soyuz 1 flight in April 1967; their flight at that time had been canceled after severe problems arose with Komarov's Soyuz 1.

Shatalov, Yeliseyev and Khrunov returned to Earth aboard Soyuz 4 on January 17. Shatalov's flight took 2 days, 23 hours and 21 minutes; Yeliseyev and Khrunov (who had launched aboard Soyuz 5 a day later than Shatalov) spent a total of 1 day, 23 hours and 39 minutes in space during their historic flight.

Having successfully fulfilled his part of the mission, Volynov prepared to return to Earth aboard Soyuz 5 on January 18. His re-entry, however, quickly dampened the elation of Soviet mission controllers and engineers when it became apparent that Soyuz 5 was experiencing an entirely new malfunction: the spacecraft's Orbital Module, which should have been jettisoned prior to re-entry, had not separated from the Descent Module, which housed Volynov for the return trip to Earth.

Reminiscent of the horrific re-entry of Soyuz 1, the bone-rattling trip through the atmosphere seemed to spell disaster for Volynov until the connecting mechanism between the two modules melted and the force of the flight shook the Descent Module loose. The re-entry assumed a more normal attitude from that point on, but the incident caused Soyuz 5 to land far from its targeted landing site. Volynov was injured and badly shaken by the harsh conclusion to his flight, but he was able to exit the spacecraft and travel several miles from the impact site. He stayed at the home of civilian residents of the remote area in which he'd landed, sheltered from the cold and recovering from his injuries for several hours before he was recovered. His injuries were not life-threatening, and he returned to space in 1976, aboard Soyuz 21.

During his flight aboard Soyuz 5, Volynov spent 3 days and 54 minutes in space.

The successful linking of Soyuz 4 and Soyuz 5 and the subsequent transfer of crew members was a significant achievement for the Soviets. Even though the flights would shortly be eclipsed by the American Moon landings and, within a few years, by the routine travel of Soviet cosmonauts to the Salyut space stations, the flights fulfilled the main goal of the Soyuz program's early years.

Western observers tended to view the link-up as bringing the Soviets to a par with the technical abilities of the American Gemini program, and ranked the Soviet achievement inferior to the accomplishments of the U.S. Apollo program. Soviet propaganda probably exacerbated the harshness of Western reaction, as the Russian press hailed the Soyuz 4/Soyuz 5 mission as the world's first space station.

Whatever the interpretation inside or outside the Soviet Union, the real value of the successful rendezvous, docking and crew transfer—especially in light of the Soyuz 1 tragedy and the subsequent difficulties in achieving docking with Soyuz 3 and the unmanned Soyuz 2—was in the increased expertise and confidence the Soyuz 4/Soyuz 5 flight gave the administrators, engineers and cosmonauts responsible for the future of the Russian space program.

SOYUZ MISSION 6: SOYUZ 6

(October 11–16, 1969)

Soviet Manned Spaceflight

Georgi Shonin was commander of Soyuz 6; he was accompanied by flight engineer Valeri Kubasov, whose duties during the flight included an innovative welding experiment using a Vulkan smelting furnace.

Vladimir Shatalov and Aleksei Yeliseyev served as the backup crew for Soyuz 6.

The launch of Soyuz 6 on October 11, 1969 was the first of three Soyuz launches on three consecutive days, and the start of a remarkable week in space for the Soviet Union. Soviet space officials announced the purpose of the three simultaneous flights as being scientific in nature, with the Soyuz 6 cosmonauts tasked with conducting research in space medicine, the Soyuz 7 crew engaged in a photography mission, and the Soyuz 8 crew investigating the reflection of sunlight in the Earth's atmosphere.

Beyond the stated scientific objectives, however, the prospect of having three separate craft and their crews all in space at the same time was an impressive achievement for the Soviet Union. The United States had won the race to the Moon three months earlier, in July, and was preparing Apollo 12 to make the second lunar landing in November; the Soviets announced in October that they had no further plans for manned lunar flights (although they actually continued development and test flights related to the program until 1973, and at various times denied entirely that they had ever even planned to send cosmonauts to the Moon, although they clearly had done so throughout the 1960s).

A remarkable achievement in itself, the trio of Soyuz flights in October 1969 also hinted at the direction the Soviet program would eventually take in the future. In much the same way that the United States struggled to define the future of its space efforts in the post-Apollo period, the U.S.S.R. also had to determine new space goals once the race to the Moon was over. For some in the Soviet space program, the idea of putting several spacecraft into orbit at the same time seemed a logical precursor to putting cosmonauts aboard orbiting space stations—such as the Salyut (and Almaz—the military version of Salyut) stations of the 1970s and 1980s.

Kubasov performed the welding experiments on October 15, fulfilling the primary objective of the Soyuz 6 mission profile. He and Shonin then maneuvered their Soyuz near their fellow cosmonauts in Soyuz 7 and 8, and watched while those two crews performed rendezvous maneuvers.

Soyuz 6 returned to Earth after a flight of 4 days, 22 hours, and 43 minutes.

SOYUZ MISSION 7: SOYUZ 7

(October 12–17, 1969)

Soviet Manned Spaceflight

Anatoli Filipchenko made his first flight in space as commander of Soyuz 7. His crew for the flight included flight engineer Vladislav Volkov and research engineer Victor Gorbatko.

The backup crew for Soyuz 7 included Vladimir Shatalov, Pyotr Kolodin and Aleksei Yeliseyev.

Soyuz 7 lifted off on October 12, 1969, as part of a remarkable string of three Soyuz launches on three consecutive days. Soviet space officials announced the purpose of the three simultaneous flights as being scientific in nature, with the Soyuz 6 cosmonauts tasked with conducting research in space medicine, the Soyuz 7 crew on a photography mission, and the Soyuz 8 crew investigating the reflection of sunlight in the Earth's atmosphere.

Subsequent details about the joint operations of Soyuz 6, 7 and 8 have surfaced in the decades since the historic flights, indicating that Soyuz 7 and 8 were actually designed to test out a new docking system, which apparently malfunctioned and prevented the two spacecraft from linking up in orbit. Given the frequent lack of information (and the distribution of false information) characteristic of the leadership of the Soviet Union, it is difficult to say with certainty that Soyuz 7 and Soyuz 8 were intended to dock with each other, although evidence largely argues in favor of a scenario that envisions these flights as a follow-on to the first docking of two Soyuz spacecraft and the transfer of crew members, which had been achieved with the flights of Soyuz 4 and 5 in January 1969.

In any case, placing three separate spacecraft and their crews in space at the same time was an impressive achievement for the Soviet space program. A total of seven cosmonauts were involved in the launches—two each in Soyuz 6 and 8, plus Filipchenko, Volkov, and Gorbatko aboard Soyuz 7. The thrill of the feat was overlooked somewhat in the aftermath of the first American lunar landing, which had occurred three months earlier, in July; but the simultaneous launch and control of multiple craft in space was still an impressive demonstration of the Soviets' technical abilities, and an indicator of the direction their program would take in the post-Moon-race period, when they would pursue the Salyut (and the military Almaz) space station program throughout the 1970s and 1980s.

On the third day of their flight, on October 15, the Soyuz 7 crew participated in rendezvous exercises with Soyuz 8, while the Soyuz 6 crew looked on nearby. The Soyuz 7 crew had presumably completed their space photography duties during the earlier portion of the flight; and Filipchenko, Volkov, and Gorbatko landed in Soyuz 7 on October 17, after 4 days, 22 hours, and 40 minutes in space.

The call sign for Soyuz 7, "Buran" (which means "snowstorm" in Russian) was also the name given to the first Russian space shuttle in the 1980s. In much the same fashion, the first American space shuttle to fly in space (in 1981) was Columbia—the same name given to the command module of the July, 1969 Apollo 11 lunar landing mission.

SOYUZ MISSION 8: SOYUZ 8

(October 13–18, 1969)

Soviet Manned Spaceflight

Vladimir Shatalov was commander of the Soyuz 8 flight, accompanied by flight engineer Aleksei Yeliseyev.

Andrian Nikolayev and Vitali Sevastyanov served as the backup crew for Soyuz 8.

Shatalov and Yeliseyev launched on October 13, 1969, in the third Soyuz launch in three days, as part of a remarkable week in space for the Soviet space program. Soviet space officials announced the purpose of the three simultaneous flights as being scientific in nature, with the Soyuz 6 cosmonauts tasked with conducting research in space medicine, the Soyuz 7 crew on a photography mission, and the Soyuz 8 crew investigating the reflection of sunlight in the Earth's atmosphere.

Although the scientific goals likely were part of the Soyuz 6, 7 and 8 mission profiles, details about the flights have emerged in the ensuing years that indicate that Soyuz 7 and 8 were also scheduled to dock with each other and to transfer crew members from one spacecraft to the other. Initially achieved with the Soyuz 4 and 5 crew in January 1969, it is now thought that Soyuz 7 and 8 were designed to test a new version of the earlier docking system, and that the failure of the new equipment prevented Soyuz 7 and 8 from linking up in orbit.

Whatever the truth of the announced plans or the subsequently available information, putting three separate spacecraft and their crews in space at the same time was an impressive achievement for the Soviet Union. A total of seven cosmonauts were involved in the launches: Georgi Shonin and Valeri Kubasov aboard Soyuz 6; Anatoli Filipchenko, Vladislav Volkov, and Victor Gorbatko on Soyuz 7, and Shatalov and Yeliseyev aboard Soyuz 8.

The trio of launches demonstrated the capabilities of the Soviet program at a particularly difficult time, as the U.S.S.R. announced the same month that it was abandoning plans for manned lunar landings in light of the Americans' first landing on the Moon in July 1969. (The Soviets actually continued development and test flights related to their manned lunar effort until 1973). Impressive in their own right, the Soyuz 6, 7 and 8 flights also utilized many of the skills necessary to the next step in the Soviet program—the Salyut (and military Almaz) space station program of the 1970s and 1980s.

On October 15—the second day of their flight—Shatalov and Yeliseyev took part in a rendezvous exercise with Soyuz 7, while the Soyuz 6 crew observed the maneuvers. The last of the three spacecraft to launch, Soyuz 8 was also the last to land; Shatalov and Yeliseyev spent 4 days, 22 hours, and 51 minutes in space before returning to Earth on October 18, 1969.

SOYUZ MISSION 9: SOYUZ 9

(June 1–19, 1970)

Soviet Manned Spaceflight

Designed to gain medical data about the effects that longer duration stays in space might have on human beings—a particularly important consideration given the Soviet Union's plan to launch the first Salyut space station the following year—Soyuz 9 began its 18-day flight on June 1, 1970.

Vostok 3 veteran Andrian Nikolayev served as commander of Soyuz 9; he was accompanied by flight engineer Vitali Sevastyanov. The backup crew for the flight included Anatoli Filipchenko and Georgi Grechko.

Once in orbit, Nikolayev and Sevastyanov settled into a demanding routine of rigorous medical tests and strenuous exercise (to combat the loss of bone mass and decreased blood plasma volume that occurs during prolonged exposure to the microgravity environment). They also conducted a variety of experiments that included psychological tests, navigation exercises and photography, and every move they made during their busy, highly regimented 16 hour days was carefully monitored, as part of the medical data collection process.

In order to keep the spacecraft operating at peak electrical efficiency, the Soyuz had to be oriented so that its solar array panels remained constantly exposed to the Sun. As a result, the linked Descent Module and Orbital Module were "spin-stabilized," or set in a slow rotating motion, which caused a variety of unpleasant sensations for Nikolayev and Sevastyanov, and may well have contributed to their relatively severe post-flight discomfort.

Halfway through the flight, the cosmonauts were given a reprieve from their intense routine; they spent a one-day break relaxing and reading and playing chess in the orbiting Soyuz before returning to their previous testing and exercise schedule for the rest of the flight.

Although the cosmonauts had passed the 18-day flight in good spirits and with no obvious ill effects to their health, they were experiencing pronounced fatigue and weakness when the Soyuz 9 Descent Module was recovered after landing on June 19. Neither Nikolayev nor Sevastyanov suffered any permanent damage, but they both required 10 full days of bedrest in medical isolation before they could return to normal activities.

During their flight—the longest spaceflight up to that time—Nikolayev and Sevastyanov spent 17 days, 16 hours and 59 minutes in space.

Following the difficult early flights of the Soyuz program and given the large number of faults that had cropped up in the spacecraft and its systems and equipment during the initial manned flights, Soyuz 9 was an audacious proposition from the very beginning. In stark contrast to the impressive but generally one-dimensional "firsts" that had characterized the early years of the Soviet space program, flights like Soyuz 9 were crucial to expanding the knowledge and capabilities of the Soviet engineers and scientists responsible for the nation's space program, and were directly supportive of subsequent missions. In the case of Soyuz 9, Soviet space officials gathered key medical data and information about their spacecraft and its systems that was necessary for the safe long-duration occupation of the Salyut space stations, the first of which was launched on April 19, 1971.

SOYUZ MISSION 10: SOYUZ 10

(April 22–24, 1971)

Soviet Manned Spaceflight

Launched on April 22, 1971 to transport the cosmonauts expected to be the first resident crew of the Salyut 1 space station to the station, Soyuz 10 was commanded by veteran cosmonaut Vladimir Shatalov, who was accompanied by flight engineer Aleksei Yeliseyev and research engineer Nikolai Rukavishnikov.

Aleksei Leonov, Valeri Kubasov and Pyotr Kolodin served as the backup crew for the flight.

Shatalov and Yeliseyev had flown together in space before, during the historic Soyuz4/Soyuz 5 docking and crew transfer mission in January 1969, and again during Soyuz 8 in October 1969.

The Salyut 1 space station was composed of three linked compartments—the Transfer/Docking compartment, the Work Compartment, and the Instrument Module, which housed the station's propulsion system. The station's components were largely derived from the design of the Soyuz spacecraft. The Salyut 1 station was launched on April 19, 1971, two days before the launch of Soyuz 10.

The trip to the space station went smoothly for Shatalov, Yeliseyev, and Rukavishnikov, and when they arrived at Salyut 1, they were pleased to be able to rendezvous and dock with the station as planned. Unfortunately, a technical difficulty arose, and they were unable to transfer from their Soyuz craft into the

Salyut station. (It has since been suggested that the difficulty may have been a fault in one or the other of the spacecrafts' environmental control systems, or perhaps a mechanical defect).

Puzzled and disappointed by the frustrating setback after the flawless docking, mission controllers had the cosmonauts undock Soyuz 10 from the space station and try again to dock and transfer. Once again, the docking proceeded without any trouble, to the relief of both the crew and their support team on the ground; but the problem preventing them from entering Salyut 1 persisted.

After poring over all reasonable potential solutions to the dilemma, Soviet space officials came to the conclusion that the crew transfer would have to be abandoned. The flight was cut short, and Shatalov, Yeliseyev, and Rukavishnikov returned to Earth on April 24, 1971, after a flight of 1 day, 23 hours and 46 minutes.

Although their hopes of being the first occupants of the first manned orbiting space station had not been fulfilled, the Soyuz 10 crew did achieve an historic "first" when they became the first Soviet crew to make a night-time landing. Even that achievement was marred, however, when a malfunction in the Descent Module caused a vapor leak that nearly suffocated the crew. Fortunately, they survived the harrowing landing without permanent injury.

SOYUZ MISSION 11: SOYUZ 11

(June 6–29, 1971)

Soviet Manned Spaceflight

As the crew of Soyuz 11, commander Georgy Dobrovolsky, flight engineer Vladislav Volkov, and research engineer Viktor Patsayev were the first human beings to occupy the first manned orbiting space station, Salyut 1. The joy of their remarkable achievement turned to sorrow at the end of their mission, however, when they were killed because of a malfunction in their Soyuz capsule.

The tragic end of the Soyuz 11 flight was particularly difficult to accept in light of the remarkable achievement of the crew in occupying Salyut 1. The Soviet Union had launched the Salyut station on April 19, 1971, and had launched Soyuz 10 two days later with the intention of having the Soyuz 10 crew be the first to live and work aboard the station. A malfunction forced the early end of the Soyuz 10 flight, however, so the Soyuz 11 crew was given the honor of being the first to enter and occupy the new orbital outpost.

There were in the details of the Soyuz 11 tragedy a series of eerie parallels with the Americans' harrowing Apollo 13 flight, which had taken place a little more than a year earlier, in April 1970.

Just as the Apollo 13 crew had been forced to deal with an emergency crew substitution shortly before launch because of medical concerns, the original Soyuz 11 crew—Aleksei Leonov, Valeri Kubasov and Pyotr Kolodin—was forced to give up the flight when Kubasov was diagnosed with a lung condition. The option of replacing only Kubasov, who was the prime flight engineer, with his backup,

Volkov, was apparently considered by Soviet mission controllers, but in an ironic twist of fate whose consequences could not have been anticipated at the time, Soviet space officials felt it would be safer to replace the entire crew.

Each mission represented a harrowing ordeal for each nation's space program, and each proved another great example of the courage and idealism that characterized all those pioneers who flew in space, regardless of the vast differences in their nations' political ideologies. The Soyuz 11 tragedy and the Apollo 13 near-tragedy also served as reminders of just how close the two superpower enemies were in the risks they shared during their separate but similar space efforts, in much the same way that the January 1967 Apollo 1 fire and the April 1967 Soyuz 1 crash had forced both programs to re-evaluate their definitions of what constituted acceptable risk, and ultimately to re-orient themselves to better ensuring the safety of their space travelers.

Dobrovolsky, Volkov, and Patsayev lifted off aboard Soyuz 11 on June 6, 1971, and docked with Salyut 1 the following day. To the relief of ground controllers who had been frustrated in their attempts to "fix" the problems that had prevented the crew of Soyuz 10 from entering the station, the Soyuz 11 cosmonauts had no apparent difficulties in entering Salyut 1.

Largely derived from the Soyuz spacecraft, the Salyut space station was composed of three linked compartments—the Transfer/Docking compartment, the Work Compartment, and the Instrument Module, which housed the station's propulsion system.

During their stay aboard the station, Dobrovolsky, Volkov, and Patsayev tested out the Salyut's systems and equipment, performed navigational exercises, and conducted a variety of scientific experiments, including medical tests, astronomical and geological observations, and studies of cosmic rays, cloud formations and the Earth's atmosphere.

They also made daily television broadcasts, frequently highlighting the humorous human interest aspects of life in their orbital "home" to a large and fascinated Russian audience. They gave viewers a virtual tour of the Salyut, demonstrated their daily exercise regimen, and, on June 19, celebrated Patsayev's thirty-eighth birthday. The joy and apparent ease with which they performed their remarkable duties endeared the crew to millions of Russian citizens, and Dobrovolsky, Volkov and Patsayev were embraced as national heroes even before they had completed their stay aboard Salyut 1.

Subsequent details about the flight, closely held at the time, paint an even more heroic portrait of the three cosmonauts. Following the dissolution of the Soviet Union in 1991 and the resulting release of a more complete account of the Soyuz 11/Salyut 1 mission, it has become known that there was an electrical fire aboard the station on June 17, which the crew was able to extinguish successfully before it caused serious damage. The fire was not related to the Soyuz malfunction that later cost the cosmonauts their lives, but it did probably contribute to the timing of their exit from the station, as it has long been thought that they left earlier than originally planned.

On June 29, 1971, Dobrovolsky, Volkov, and Patsayev left Salyut 1 and boarded their Soyuz 11 for the return trip to Earth.

The multi-compartment Soyuz was designed to split into separate pieces just prior to re-entry, so that the Descent Module carrying the cosmonauts could land, while the rest of the spacecraft would be burnt up in the Earth's atmosphere. The dangers of the Orbital Module and Descent Module failing to split had been demonstrated during the difficult return of Soyuz 5, in which the two pieces had remained linked far longer than scheduled, nearly resulting in disaster for Soyuz 5 commander Boris Volynov.

To ensure a clean separation between the Orbital Module and Descent Module, the two compartments had been linked with a frame outfitted with explosive bolts. When the bolts were fired in the proper sequence, the Descent Module and its occupants would be cleanly separated from the expendable Orbital Module.

In the case of Soyuz 11, although the bolts fired as planned, they did so virtually simultaneously instead of in the carefully prescribed sequence. As a result, the two compartments were shoved abruptly apart with a sharp jolt, which in turn forced open a vent (officially known as a low altitude pressure equalization valve) that had been designed to let air into the Descent Module to stabilize it after the craft's parachute was deployed, just prior to landing.

While the vent would normally have allowed breathable air into the capsule after the cosmonauts had re-entered the Earth's atmosphere, its premature opening while they were still in space allowed their life-sustaining atmosphere within the Descent Module to escape. The crew was not wearing spacesuits—mission planners had been working under the assumption that they would not be necessary during re-entry when the capsule itself would provide the necessary protection—and therefore had no protection against the sudden depressurization.

Later investigation indicated that Patsayev was able to close the vent about halfway before the Descent Module became completely depressurized. The crew apparently lost consciousness and died less than a minute later.

Eerily, the rest of the re-entry process went as planned, and the Descent Module made an automatic return and landing. Even the half-open vent worked as planned once the capsule descended into the Earth's atmosphere, as it allowed air into the Descent Module prior to the landing. Mission controllers only realized that the crew had been lost when recovery teams reached the capsule.

The loss of Dobrovolsky, Volkov, and Patsayev was a shocking national tragedy for the Russian people. In a spectacular flight of 23 days, 18 hours and 22 minutes, the three cosmonauts had displayed the courage and joy of spaceflight, and given the ultimate measure of dedication in living out the ideals of their fellow citizens.

Soyuz 11 commander Georgy Dobrovolsky was 43 years old at the time of his death; Soyuz 11/Salyut 1 was his first space mission. Flight engineer Vladislav Volkov, 35, had flown in space before, aboard Soyuz 7 in October 1969. Research engineer Viktor Patsayev, who had celebrated his thirty-eighth birthday aboard Salyut 1, made his first flight in space during Soyuz 11.

SOYUZ MISSION 12: SOYUZ 12

(September 27–29, 1973)

Soviet Manned Spaceflight

Launched on September 27, 1973, Soyuz 12 was the first manned test of the Soyuz spacecraft following the tragic Soyuz 11 accident. Soyuz 12 was commanded by Vasili Lazarev, who was accompanied by flight engineer Oleg Makarov, who made his first trip into space during Soyuz 12.

Georgi Grechko and Alesksei Gubarev served as the backup crew for Soyuz 12.

After the Soyuz 11 disaster in June 1971, which had resulted in the deaths of cosmonauts Georgi Dobrovolsky, Vladislav Volkov, and Viktor Patsayev, the Soviet space program entered a long period of painful introspection and gradual recovery. The Soyuz spacecraft and its systems, equipment, and procedures were minutely studied and reworked to ensure the safety of future crews and the success of future missions.

As the Soviets gradually emerged from the post-Soyuz 11 period, their emphasis on redesigning the Soyuz was coupled ever more urgently with their development of the Salyut series of space stations. Publicly at stake for the Russian program was the national prestige involved in orbiting a successful Salyut before the Americans could launch their Skylab station; but the Soviets also had a more pressing interest in the Salyut development. In addition to civilian space stations, they were also engaged in the creation of a highly secretive military version (known within the Soviet space program as "Almaz," to differentiate it from the publicly revealed Salyut program).

Beginning in the summer of 1972, the Soviets conducted a number of unmanned test flights related to both the Soyuz spacecraft and the Salyut/Almaz space stations, with varying degrees of success.

The Soyuz 12 flight was basically a manned version of the unmanned tests. The short flight went well, and with their safe return to Earth, Lazarev and Makarov brought a renewal of confidence to the struggling Soviet program. The cosmonauts spent a total of 1 day, 23 hours and 16 minutes in space during their successful Soyuz 12 mission, and set the stage for future manned flights.

SOYUZ MISSION 13: SOYUZ 13

(December 18–26, 1973)

Soviet Manned Spaceflight

Launched December 18, 1973 with commander Pyotr Klimuk and flight engineer Valentin Lebedev, Soyuz 13 is thought to have involved tests of space reconnaissance technology that had originally been scheduled to be carried out aboard the Salyut 2 space station.

Lev Vorobyov and Valeri Yazdovsky served as the backup crew for the flight.

Soyuz 13 also played a role in an historic "first" in the history of world spaceflight, as it marked the first instance in which the Soviet Union and the United

States had crews in space at the same time (albeit on separate missions; the first joint Soviet-American mission was the Apollo-Soyuz Test Project flight, in July 1975). By the time that Klimuk and Lebedev lifted off aboard Soyuz 13 on December 18, the third crew of the American Skylab space station (Gerald Carr, William Pogue and Edward Gibson) had been in space for nearly a month, since November 16, 1973.

Apparently designed to serve as the first manned military space station, Salyut 2 had been launched unmanned on April 3, 1973. The station developed an unspecified malfunction and had to be de-orbited before it could be occupied. It is thought that the Salyut 2 station would have enabled the Soviet military to conduct high-resolution photo reconnaissance, and that the Soyuz 13 flight successfully tested the equipment for just such a mission.

After the tragic Soyuz 11 accident, the basic Soyuz spacecraft design had been carefully studied and reworked; one of the major design changes involved the replacement of the craft's solar panels with batteries for missions in which the Soyuz would be traveling to a space station. The switch to battery power for the relatively short "transport" flights was intended to lighten the spacecraft overall, and to increase its ability to maneuver when near a space station. The specified batteries generally produced enough electricity for two to three days—enough to cover the trip to and from the Salyuts.

In the case of Soyuz 13, however, there was no station to go to (Salyut 2 had failed earlier in the year and been de-orbited; and Salyut 3 would not be launched until June 24, 1974), so the Soyuz 13 craft was outfitted with solar panels that provided ample electric power for the flight, which lasted 7 days, 20 hours and 56 minutes.

SOYUZ MISSION 14: SOYUZ 14

(July 3–19, 1974)

Soviet Manned Spaceflight

Heartened by the success of the Soyuz 12 (September 1973) and Soyuz 13 (December 1973) missions and two successful unmanned tests of Soyuz/Salyut spacecraft, the Soviets launched the Salyut 3 space station on June 24, 1974, followed by the launch of Soyuz 14 on July 3, 1974.

Pavel Popovich served as commander of Soyuz 14, accompanied by flight engineer Yuri Artyukhin.

Gennadi Sarafanov and Lev Demin were the backup crew for the Soyuz 14/Salyut 3 flight.

In a little over two weeks aboard Salyut 3, Popovich and Artyukhin achieved the first entirely successful Soviet space station mission. They conducted a series of tests and participated in activities similar to those carried out by the Soyuz 11 crew in 1971, with one major difference, which was carefully concealed at the time. In addition to their publicly revealed medical tests, scientific studies, and spacecraft-related duties, Popovich and Artyukhin apparently also carried out the

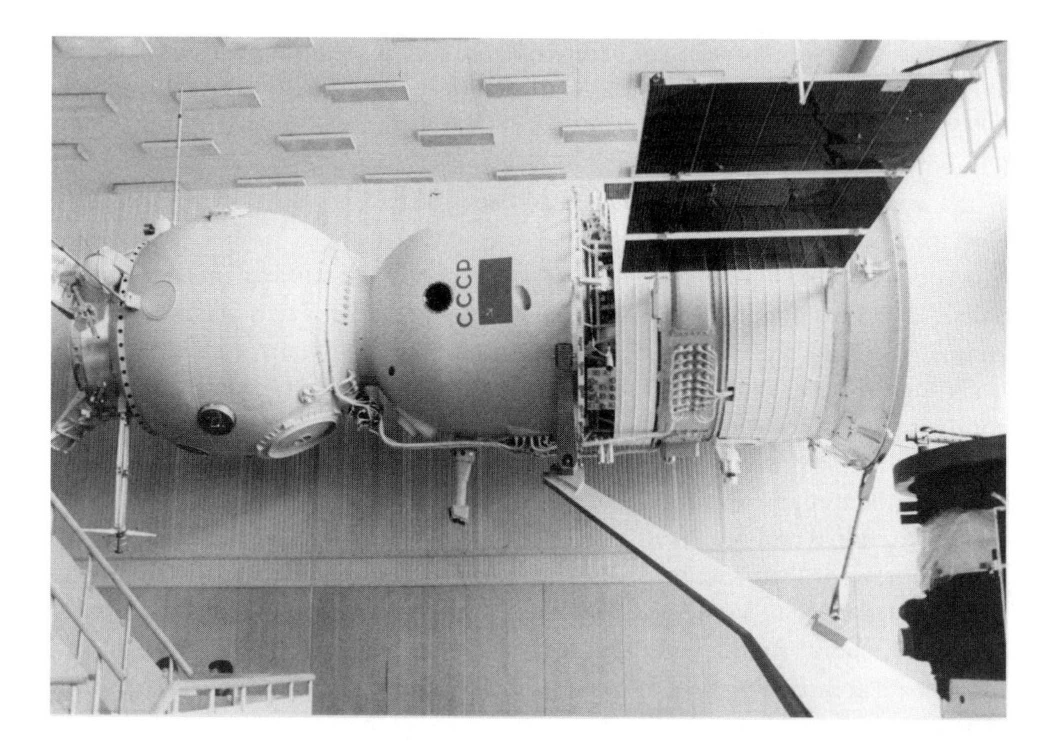

A replica of the Soyuz spacecraft at the Gagarin Cosmonaut Training Center in June, 1974 shows the Descent Module (center); the Orbital Module (left); and the Instrument Module (right, with solar panels). [NASA/courtesy of nasaimages.org]

military photo reconnaissance that had been intended for the failed Salyut 2 space station in April 1973 and which had apparently been successfully tested during the Soyuz 13 flight of December 1973. The imaging exercise was disguised as a standard photographic study.

The idea of placing military observers in space to spy on other nations from orbit had had a long history in the programs of both the U.S.S.R. and the United States In the mid-1960s the United States Air Force had developed plans for a Manned Orbiting Laboratory (MOL), and progressed to the point of selecting astronauts for the program before the project was abandoned near the end of the decade. For their part, the Soviets had developed virtually their entire space program as an outgrowth of their military aims, and, with the Salyut/Almaz stations, fulfilled their plans for manned spying from space even at a time when they were cooperating with their superpower enemy on the development of the joint Soviet-American Apollo-Soyuz Test Project (ASTP), which culminated in the July 1975 docking of an American Apollo and a Soviet Soyuz spacecraft.

During their Soyuz 14/Salyut 3 mission, Popovich and Artyukhin spent 15 days, 17 hours and 30 minutes in space. They landed safely on July 19, 1974.

SOYUZ MISSION 15: SOYUZ 15

(August 26–28, 1974)

Soviet Manned Spaceflight

Intended to dock with the Salyut 3 space station, Soyuz 15 launched on August 26, 1974, with commander Gennadi Sarafanov and flight engineer Lev

Demin. At 48, Demin was the oldest person to fly in space up to that time, and the first space traveler who was also a grandfather.

Boris Volynov and Vitali Zholobov served as the backup crew for the Soyuz 15 mission.

As things turned out, Sarafanov and Demin had to content themselves with the "firsts" involving Demin's age as being the only highlights of their short flight. A major malfunction occurred with the maneuvering system of their Soyuz craft just as they were about to dock with Salyut 3, and all the best efforts of the crew and mission handlers on the ground were unable to overcome the fault, despite several attempts.

A good deal of the available electrical power in Soyuz 15's batteries was expended during the docking tries, and, as a result, the spacecraft was starved for electricity during its return to Earth. Mission controllers were forced to limit communications with the crew, and the spacecraft's telemetry was turned off, in an effort to stretch the remaining electrical power as far as possible.

Fortunately for Sarafanov and Demin, the landing was achieved without major difficulties. Their disappointingly abbreviated flight to Salyut 3 lasted 2 days and 12 minutes.

SOYUZ MISSION 16: SOYUZ 16

(December 2–8, 1974)

Soviet Manned Spaceflight

Launched on December 2, 1974, Soyuz 16 was a manned test of the Soyuz systems, equipment and procedures that the Soviets would use during the Apollo-Soyuz Test Project (ASTP) flight in July 1975. Soyuz 16 was commanded by Anatoli Filipchenko, who was accompanied by flight engineer Nikolai Rukavishnikov.

Yuri Romanenko and Aleksandr Ivanchenkov served as the backup crew for the Soyuz 16 flight.

ASTP was developed over the course of three years, and utilized the talents of Soviet and American cosmonauts and astronauts, engineers and scientists and administrators to forge a joint space mission for the two space superpowers. The project was intended both as a symbol of goodwill and as a demonstration of each nation's willingness to view space exploration as an ideal best placed beyond the reach of Earthly conflicts.

Simple in concept but considerably challenging to implement, the ASTP idea required the Russian and American space agencies to develop a new docking module to facilitate the linking in orbit of a Soviet Soyuz spacecraft and an American Apollo. Despite their frequently similar goals and the parallel progress and tragedies that had marked the two space programs over the years, the Soviet and American approaches to spacecraft design and spaceflight were very different, and a great deal of cooperative effort was required to achieve the ASTP flight.

While the Americans tested the systems, equipment and procedures they planned to use during ASTP in Earth-based simulations, the Soviets flew several flights to test their portion of the ASTP mission.

The Soyuz 16 test flight followed two unoccupied tests (Cosmos 638, in April 1974, and Cosmos 672, in August 1974). Filipchenko and Rukavishnikov put Soyuz 16 through the actual tasks that the ASTP Soyuz crew would need to accomplish for their mission to be successful.

Working in the altered atmosphere of their Soyuz cabin—the mix of nitrogen and oxygen standard for Soviet flights had to be adjusted to more closely match the pure oxygen atmosphere of the Apollo craft to smooth the adjustment of the astronauts and cosmonauts who would visit each others' quarters during the actual ASTP flight—Filipchenko and Rukavishnikov maneuvered Soyuz 16 as the later ASTP would have to do to link with the Apollo craft, and they made a hands-on evaluation of the two-ton module that would facilitate the actual docking.

Having successfully tested the systems, equipment and procedures the Soviets would later use during the ASTP flight, Filipchenko and Rukavishnikov returned to Earth on December 8, 1974, after a flight of 5 days, 22 hours and 24 minutes.

The importance of the Soyuz 16 test flight is perhaps even more evident in consideration of the Soyuz flights that had preceded it, as opposed to the successful ASTP flight that followed. The first 15 Soyuz flights had been marred by two wrenching tragedies and the loss of four cosmonauts. Only 7 of the 15 flights could reasonably be said to have achieved most or all of their objectives, and the frustrating, persistent malfunctions of Soyuz hardware, systems and procedures were accompanied by similarly disappointing faults in the Salyut series of space stations, which were largely derived from the Soyuz design.

Although the Soyuz would eventually prove itself a reliable and safe mode of space transportation, its pre-ASTP record was not encouraging. In addition, the potentially devastating blow to Soviet prestige that would have occurred if the Soyuz end of the ASTP flight had not gone well makes clear the importance of the successful Soyuz 16 test flight, and the contribution that Filipchenko and Rukavishnikov made to the ASTP mission, and to the Soviet program as a whole.

SOYUZ MISSION 17: SOYUZ 17

(January 10–February 9, 1975)

Soviet Manned Spaceflight

The Soviet Union launched the Salyut 4 space station on December 26, 1974, and then launched Soyuz 17 on January 10, 1975 with commander Alexei Gubarev and flight engineer Georgi Grechko, who were given the task of becoming the station's first crew.

Vasili Lazarev and Oleg Makarov served as the backup crew for the Soyuz 17/Salyut 4 mission.

Gubarev and Grechko successfully docked with the station on the second day of their flight, and quickly settled into a rigorous routine of scientific and medical research. They made in-depth studies of the Sun, the Crab Nebula, and the Earth's atmosphere, and conducted astronomical research and photography. They also carried out extensive medical studies, which were augmented by biological research involving insects.

On February 9, 1975, Gubarev and Grechko concluded their successful stay aboard Salyut 4 and re-entered Soyuz 17 to return to Earth. The short flight and re-entry went as planned, but the Descent Module landed in the midst of a severe snowstorm. Fortunately, neither cosmonaut was hurt, and the exceptional mission ended happily after 29 days, 13 hours and 20 minutes.

In their superb flight and busy stay aboard Salyut 4, Gubarev and Grechko solidified the success of the Soviets' space station program, and reinforced both the particular benefits of the Salyut development and the broader value of operating space stations in Earth orbit.

SOYUZ MISSION 18: SOYUZ 18–1

(April 5, 1975)

Soviet Manned Spaceflight

On April 5, 1975, cosmonauts Vasili Lazarev and Oleg Makarov endured the first manned launch abort in the history of world spaceflight.

The two veteran cosmonauts (each had flown in space once before) were looking forward to becoming the second crew to visit the Salyut 4 space station, where they were expected to live and work in orbit for about two months—doubling the stay of the Soyuz 17/Salyut 4 crew. Lazarev had been assigned as the commander of the flight, with Makarov serving as flight engineer.

Pyotr Klimuk and Vitali Sevastyanov served as the backup crew for Lazarev and Makarov.

As the launch progressed and the moment came for separation of the launch vehicle rocket stages, an electrical fault prevented the stages from separating cleanly. Dragging its entire huge rocket stack along during its ascent, the Soyuz spacecraft shuddered violently, and ground controllers immediately engaged the launch escape system. The capsule containing the cosmonauts was shot away from the launch vehicle at a tremendous speed, and rose to an altitude of 180 kilometers—enough to qualify the harrowing brief trip as an official flight into space.

Lazarev and Makarov re-entered the atmosphere about 21 minutes after the aborted launch, and made a rough landing about nine hundred miles away from their Baikonur Cosmodrome launch site. Mission controllers tracking their progress worried that the cosmonauts would land in China, where they might have been retrieved by Chinese authorities, but the landing was said to have occurred on Soviet soil.

The landing site was nearly as precarious as the brief flight itself. The capsule ended up wedged in the branches of a tree on the side of a mountain. Recovery forces reached Lazarev and Makarov a short while after the landing, but removing the injured cosmonauts from the site required both time and patience. They both survived the harrowing ordeal, but Lazarev was injured severely, and later left the cosmonaut corps as a result of the accident. From launch to landing, the flight lasted 21 minutes.

With the next Soyuz launch, on May 24, 1975 also being referred to as Soyuz 18, the aborted April 5 flight became known as Soyuz 18-1.

SOYUZ MISSION 19: SOYUZ 18

(May 24–July 26, 1975)

Soviet Manned Spaceflight

Pyotr Klimuk served as commander of Soyuz 18, accompanied by flight engineer Vitali Sevastyanov. Both cosmonauts made their second flight in space during the Soyuz 18 mission, during which they successfully lived and worked aboard the Salyut 4 space station for more than two months.

Klimuk and Sevastyanov had previously served as the backup for Vasili Lazarev and Oleg Makarov when Lazarev and Makarov had endured the first-ever manned launch abort on April 5, 1975.

For the Soyuz 18/Salyut 4 flight, Vladimir Kovalyonok and Yuri Ponomaryov served as the backup crew.

Benefiting from the extensive modifications and checkout of the launch systems that had followed the April 5 launch abort, Klimuk and Sevastyanov lifted off on May 24, 1975 without incident, and docked with Salyut 4 on May 25.

Their stay aboard the station was marked by highly publicized scientific research and astronomical observations, and unpublicized concerns about the health of the Salyut's environmental controls. On the positive side, Klimuk and Sevastyanov made extensive spectrographic and photographic studies of the Sun, and conducted astronomical observations of a number of constellations. They also captured thousands of photographs of the sprawling land mass that constituted the Soviet Union at that time.

In addition to their intense research activities and regular regimen of exercise and spacecraft maintenance, the cosmonauts were also forced to endure a frightening health scare when an unidentified green mold began to grow on the walls of Salyut 4. The prospect of sharing their quarters with the unpleasant bacteria understandably distressed Klimuk and Sevastyanov, and, under normal circumstances, mission controllers might have considered the option of cutting short their stay aboard the station. The Soyuz 18/Salyut 4 flight happened to overlap with the joint Soviet-American Apollo-Soyuz Test Project (ASTP) flight, however, and the Soviets had assured the Americans that the two missions could proceed simultaneously without any conflict, as each would be handled by separate teams of controllers.

Fortunately for Klimuk and Sevastyanov, the green mold proved harmless to their health, and both ASTP and Soyuz 18/Salyut 4 concluded without further difficulties. After an uncomfortable but productive mission of 62 days, 23 hours and 20 minutes, Klimuk and Sevastyanov returned to Earth on July 26, 1975, five days after the landing of Soyuz 19, which had been the Soviet participant in the ASTP flight.

Even with the unpleasant circumstance aboard the station, the Soyuz 18/Salyut 4 flight was a major success for the Soviet space program. It demonstrated the Soviets' space station expertise, setting a mark for endurance that was second only to that achieved by the final crew of the U.S. Skylab, and it proved the abilities of Soviet mission controllers, who flawlessly handled the simultaneous ASTP and Soyuz 18/Salyut 4 flights.

SOYUZ MISSION 20: SOYUZ 19

(July 15–21, 1975)

Soviet Manned Spaceflight

Launched on July 15, 1975 with commander Alexei Leonov and flight engineer Valeri Kubasov, Soyuz 19 was followed into space less than eight hours later by the American Apollo spacecraft that it would meet and dock with on July 17, in the fulfillment of the joint Soviet-American Apollo-Soyuz Test Project (ASTP).

One of the nation's original space heroes, Leonov had been the first person in history to perform a spacewalk, during the March 1965 Voskhod 2 mission. Kubasov was also a veteran cosmonaut, having flown in space during Soyuz 6 in October 1969.

Anatoli Filipchenko and Nikolai Rukavishnikov served as the backup crew for the Soviet portion of the ASTP flight. A second capsule was kept in reserve in the event that anything should happen to the primary Soyuz, and Vladimir Dzhanibekov and Boris Andreyev were assigned as the primary crew of the second vehicle, with Yuri Romanenko and Aleksandr Ivanchenkov serving as their backup crew. The second Soyuz was not needed during the ASTP mission, and it was later used during Soyuz 22, in September 1976.

Developed after years of intense competition between the two space superpowers, the idea that would eventually become the ASTP flight had first been broached by NASA shortly after the United States achieved the first manned lunar landing with Apollo 11 in July 1969. In May 1972 the two nations formalized the idea in a written agreement signed by Soviet leader Leonid Brezhnev and American President Richard Nixon, during their first summit meeting.

Engineers and space officials in both countries had to overcome a variety of technical and cultural barriers to achieve the historic mission, which called for closely coordinated launches of a Soviet Soyuz and an American Apollo, the rendezvous and docking of the two craft in orbit, and visits by the crew of each spacecraft to the other's vehicle.

Although each program had previously overcome the challenges of rendezvous and docking in the context of their own space efforts, the joint mission required the development of an entirely new docking module, the design of which required careful cooperation by engineers on both sides. Despite some concern in each nation about the possible transfer of closely-held technological secrets, members of the two programs were able to develop the new docking module without undue interference.

Many philosophical differences were also addressed in the mission's planning and execution. The Soviets favored automated control of complex procedures like rendezvous and docking, while the Americans pursued an approach that emphasized the piloting skills of their astronauts. For ASTP, each crew would be given an opportunity to direct the docking of the two spacecraft, with the Americans making the initial attempt and later, after a brief separation, giving Leonov and Kubasov a chance to use the Soyuz as the active vehicle in a second attempt.

Not unexpectedly, given the vastly different political systems of the two nations involved, the Soviet and American approaches toward disseminating information about the flight also required some compromise. In the Soviet Union, information about space missions was tightly controlled, and generally released after a given event had taken place—and often, for propaganda purposes. In the United States, NASA relied on continuous live media coverage of its activities to keep public interest—and government funding—at a high level.

In the case of ASTP, the Soviets allowed unprecedented access to information about the flight, and although they refused to allow Western journalists to attend the launch, the lift-off and the in-flight activities of the cosmonauts and astronauts were broadcast live.

Among the technical hurdles to be overcome, the different atmosphere that was used in each spacecraft caused some concern. Soyuz vehicles operated with an atmosphere that was composed of a mix of oxygen and nitrogen; the American Apollo spacecraft used an atmosphere of oxygen only. Mission planners overcame the disparity by using the 10-foot docking module as a transitional chamber, where either crew could become accustomed to the environment of the other's spacecraft before entering.

Tracking and control issues also required frank and open exchanges between officials of the two programs. The American representatives became particularly concerned when they learned of the Soviets' plan to have the crew of Soyuz 18 occupy the Salyut 4 space station simultaneously with the ASTP flight. To avoid having to cut short the Soyuz 18/Salyut 4 mission, the Soviets set up two separate teams of controllers, and reassured their ASTP partners that the two flights could be handled efficiently and safely. The Americans agreed to the arrangement, and as things turned out, there was no conflict between the two missions.

The two programs' approach to training also provided a learning opportunity for the American and Russian crews. Where the Soviets generally tested new hardware or procedures for a relatively short period in an Earth-based simulator and then flew an actual flight in space to confirm the results, the Americans relied heavily on simulator trials for testing virtually every bit of equipment and every conceivable contingency before committing to any given spaceflight.

Happily, the crew members selected for the historic ASTP flight proved to be excellent goodwill ambassadors for their respective countries, and seemed to enjoy their joint training activities. Thomas Stafford served as the commander of the American ASTP crew, joined by Donald "Deke" Slayton and Vance Brand. Stafford was a veteran of the Gemini 6 and Gemini 9 missions, and had orbited the Moon aboard Apollo 10—the final rehearsal for the first manned lunar landing by Apollo 11. Slayton was one of the original Mercury 7 astronauts; he had been sidelined by a heart condition in the Mercury era, and waited years for the medical clearance necessary to fly in space, which he received shortly before his assignment to the ASTP flight. Brand had served as a member of the backup crew for the Apollo 15 lunar landing and for two Skylab missions, and would make his first flight in space during the ASTP mission.

Due to the unique nature of the ASTP flight, the training program differed from previous missions for both the cosmonauts and the astronauts. Language lessons were an integral part of the preparations, with Leonov and Kubasov learning English and Stafford, Slayton and Brand learning Russian. The two crews also visited the launch and control facilities in each country, and in the course of their training and travels, developed a sincere respect and friendship for each other.

Plans to televise the in-flight activities were imperiled shortly after Leonov and Kubasov lifted off on July 15, when a TV camera aboard their Soyuz malfunctioned. The two cosmonauts were able to fix the recalcitrant camera, however, and the unprecedented live Soviet broadcasts continued throughout the flight.

Stafford, Slayton and Brand launched later on July 15 aboard their Apollo, and deployed the two-ton docking module shortly after entering orbit.

The docking took place on July 17, with Stafford at the controls of the Apollo. Leonov congratulated his American counterpart on the smooth linkup of the two spacecrafts: "Well done, Tom," and a short while later, he met Stafford at the Soyuz end of the customized docking module. As the two veteran space commanders shook hands, Stafford marked the historic occasion with the Russian word for comrade, "Tovarish." Leonov replied, in English, "Very happy to see you."

Soyuz 19 and the ASTP Apollo were docked for a total of 47 hours. The two crews visited each other's quarters, and exchanged gifts symbolic of the hopeful nature of the cooperative flight, such as seed that each crew returned to their own country for later planting. The crew members also shared several meals, and participated in a news conference that was aired live in both the U.S.S.R. and the United States. Soviet President Leonid Brezhnev called to congratulate Leonov and Kubasov, and the Americans talked with U.S. President Gerald Ford.

As planned, the two spacecraft undocked on July 19 to conduct a series of rendezvous experiments, and Leonov then had a try at docking the Soyuz craft with the Apollo. The second docking was more difficult than the first, but Leonov was able to link the two vehicles.

Two days later, on July 21, 1975, Leonov and Kubasov completed their portion of the historic ASTP mission when they undocked from the Apollo and returned to Earth, after a flight of 5 days, 22 hours and 31 minutes in space.

Stafford, Slayton and Brand spent three more days in orbit before returning to Earth after a flight of 9 days, 1 hour, 28 minutes and 23 seconds. A problem during re-entry caused their capsule to fill with toxic rocket propellant fumes, and the astronauts required five days' recovery in a hospital before they could resume normal activities.

After their remarkable journey into the heavens, the two ASTP crews reunited on Earth for extensive tours of the Soviet Union and the United States, providing the citizens of both nations a glimpse of the cooperative spirit that would, decades later, find a fuller expression in the docking flights of the American space shuttle and the Russian space station Mir, and in the subsequent participation of the United States and Russia in the multinational development of the International Space Station (ISS).

SOYUZ MISSION 21: SOYUZ 20

(November 17, 1975–February 16, 1976)

Soviet Unmanned Spaceflight

Launched on November 17, 1975, Soyuz 20 was used in an unmanned test as part of the development of the Progress supply and refueling spacecraft.

The first generation of Salyut space stations featured a single docking port and could not be resupplied or refueled, which limited the length of time that crews could spend aboard the stations. Looking ahead to the Salyut 6 and 7 stations, which were designed with two docking ports, Soviet space officials recognized the benefit of developing a fully automated version of the basic Soyuz spacecraft that could be sent unmanned to future space stations to deliver fuel and supplies and equipment for longer manned missions. Controlled from the ground and generally designed to be used for "one way" trips (most were destroyed on re-entry, although a small number were recovered), the unmanned cargo vessels were given the name Progress.

The Soyuz 20 flight was an early test of the automated docking capabilities of the Soyuz craft, and also provided the spacecraft's designers with data about the effect of longer duration flights on the spacecraft's onboard systems. The automated docking went well, and Soyuz 20 remained linked to Salyut 4 for three months before it was undocked. It re-entered the Earth's atmosphere on February 16, 1976, after a total flight of 90 days, 12 hours and 47 minutes.

Post-flight evaluation indicated that the flight had stretched the craft's systems to the limit of their capabilities. As a result, when later space station crews were scheduled to stay in space for more than 90 days, additional Soyuz flights were scheduled so a fresh spacecraft could be switched with the one the original crew had used to travel to the station. The brief switching flights were used as an opportunity to bring cosmonauts from Soviet-bloc nations to the Salyut 6 and Salyut 7 stations, as part of the Intercosmos program.

With tests like the Soyuz 20/Salyut 4 flight, the Soviets maximized the use of their available hardware to glean valuable data about the Soyuz spacecraft, and "road-tested" the key components of their important new Progress design, which would play a key role in missions to Salyut 6 (which the first operational Progress spacecraft visited in January 1978) and Salyut 7.

SOYUZ MISSION 22: SOYUZ 21

(July 6–August 24, 1976)

Soviet Manned Spaceflight

As commander of the Soyuz 21/Salyut 5 flight—which was secretly designed as a military photo surveillance mission—Boris Volynov made his second trip into space. He was accompanied by flight engineer Vitali Zholobov.

Vyacheslav Zudov and Nikolai Rukavishnikov served as the backup crew for the flight.

The Salyut 5 space station had been launched on June 22, 1976. Although it was given the same designation, Salyut, as previous civilian space stations, Salyut 5 was actually a military station known within the Soviet hierarchy as part of the Almaz manned space surveillance program. Both the civilian and military versions of the space station were publicly referred to as Salyuts to confuse Western observers, and to hide the covert nature of the spying activities carried out during the Almaz flights.

Configured slightly differently than the standard (civilian) Salyut, the Salyut 5 Almaz is thought to have included high resolution camera equipment that the cosmonauts used to gather images of locations on Earth that were of particular interest to Soviet military officials. Salyut 5 was the third and last Almaz military station; the first, Salyut 2, had failed in orbit in 1972 before it could be occupied, and the second, Salyut 3, had been occupied once, in 1974.

Volynov and Zholobov launched aboard Soyuz 21 on July 6, 1976. During their 49 days aboard Salyut 5, they performed a number of scientific experiments, medical observations, and materials processing investigations, in addition to their carefully concealed photo surveillance activities.

Although few details of the flight were released at the time—in keeping with the usual tight control of information characteristic of the Soviet program, and probably, to divert attention from the classified photo-taking exercise—it has long been suspected that a malfunction aboard the Salyut station caused the mission to be cut short.

Volynov and Zholobov returned to Earth on August 24, 1976, after logging 49 days, 6 hours and 24 minutes in space.

SOYUZ MISSION 23: SOYUZ 22

(September 15–23, 1976)

Soviet Manned Spaceflight

As commander of Soyuz 22, Valeri Bykovsky made his second trip into space, lifting off on September 15, 1976 with flight engineer Vladimir Aksyonov.

Yuri Malyshev and Gennady Strekalov served as the backup crew for the Soyuz 22 flight.

The mission profile for Soyuz 22 called for the crew to make atmospheric and geological observations, and to test a sophisticated new camera system manufactured by the Carl Zeiss Company. The Soyuz 22 spacecraft had been the backup for Soyuz 19, which had been used in the Apollo-Soyuz Test Project in July 1975. It was the last of the original design to be used for a self-contained mission; subsequent flights all made use of the Soyuz as a transport vehicle to ferry crews to space stations.

Bykovsky and Aksyonov landed on September 23, 1976, after a flight of 7 days, 21 hours and 52 minutes.

SOYUZ MISSION 24: SOYUZ 23

(October 14–16, 1976)

Soviet Manned Spaceflight

Soyuz 23 commander Vyacheslav Zudov and flight engineer Valeri Rozdestvensky launched on October 14, 1976 with the intention of occupying the Salyut 5 (Almaz) military space station, on what was most likely designed as a military photo reconnaissance mission.

Victor Gorbatko and Yuri Glazkov served as the backup crew for the flight.

When Zudov and Rozdestvensky arrived at Salyut 5 and engaged the automated rendezvous and docking system that was supposed to link their Soyuz to the station automatically, they were dismayed to find the system not working. They had not been prepared for a manual docking; so, when the malfunctioning automated system persistently failed to respond, they were forced to abandon the mission. They returned to Earth on October 16, 1976.

Following an otherwise routine return trip and re-entry, the cosmonauts found themselves trapped in their Descent Module, which had descended in a night-time landing in a snowstorm and came to rest upside down in Lake Tengiz. With the capsule's escape hatch below the frozen surface, recovery crews were unable to fasten cables to the spacecraft, and Zudov and Rozdestvensky endured six long hours in the frigid, dark capsule before a rescue was improvised. The spacecraft was finally opened after a helicopter dragged the capsule nearly five miles across a frozen swamp, onto shore.

Despite their discomfort, Zudov and Rozdestvensky suffered no lasting ill effects from the harrowing ordeal. From launch to landing, their frustrating flight had lasted two days and seven minutes.

SOYUZ MISSION 25: SOYUZ 24

(February 7–25, 1977)

Soviet Manned Spaceflight

Launched on February 7, 1977, the Soyuz 24/Salyut 5 flight was apparently designed to investigate the condition of the Salyut 5 space station following the abbreviated 1976 Soyuz 21/Salyut 5 mission. As commander of Soyuz 24/Salyut 5, Victor Gorbatko made his second trip into space. He was accompanied by flight engineer Yuri Glazkov.

Anatoli Berezovoi and Mikhail Lisun served as the backup crew for the flight.

Although little is known of the details of their visit to the station—Salyut 5 was a military space station, known within the Soviet hierarchy as part of the Almaz military program of manned space surveillance—the length of their stay indicates that Gorbatko and Glazkov were able to overcome whatever difficulties they encountered.

It is thought that the station suffered a partial depressurization on February 21, just four days before Gorbatko and Glazkov left the station. They returned to Earth safely on February 25, 1977, after a flight of 17 days, 17 hours and 26 minutes.

Gorbatko and Glazkov were the last cosmonauts to occupy Salyut 5, and the station was the last of the Almaz military stations orbited by the Soviets. On February 26, 1977 mission controllers de-orbited the unmanned Salyut 5 Descent Module, which was likely filled with the film and other data amassed by the Salyut 5 cosmonauts who had carried out the Almaz photo surveillance missions.

SOYUZ MISSION 26: SOYUZ 25

(October 9–11, 1977)

Soviet Manned Spaceflight

Launched amid national celebrations honoring the twentieth anniversary of the world's first satellite, Soyuz 25 was designed to transport commander Vladimir Kovalyonok and flight engineer Valeri Ryumin to the Salyut 6 space station, where they would have become the station's first long-duration crew, if a balky docking mechanism hadn't gotten in the way.

Yuri Romanenko and Aleksandr Ivanchenko served as the backup crew for the Soyuz 25 flight.

While the Russian people recalled the glory of the Sputnik 1 launch, Soviet space officials readied Soyuz 25 for launch a few days after the anniversary celebrations. Kovalyonok and Ryumin lifted off on October 9, 1977 and headed for Salyut 6—the first of a new generation Russian space stations.

Salyut 6 had launched unmanned on September 29, 1977. Equipped with two docking ports (the first five Salyuts had each had one), the new station could receive supplies and equipment and be refueled via unmanned cargo spacecraft that could be controlled by mission managers on Earth and then be automatically docked with the station. Known as Progress spacecraft, the unmanned automated cargo ships were derived from the basic Soyuz design. The additional capabilities they provided meant that cosmonauts could make longer stays at the space station, since they could be resupplied when their consumables ran out, or when fuel, equipment or other supplies needed to be replaced.

As the first residents of Salyut 6, Kovalyonok and Ryumin would have tested the new concepts embodied in the plans for longer duration stays in space, as well as the new space station itself.

But it was not to be. Although their trip to the station was uneventful, and their initial attempts at docking appeared to be successful, the docking mechanism of the Soyuz failed to fully engage the docking module on Salyut 6. Working with flight controllers on the ground, Kovalyonok and Ryumin backed the Soyuz 25 craft away from the station, and another attempt at docking was made, using the automated docking system.

Despite the best efforts of the crew and their support teams, Soyuz 25 could not make a safe docking with Salyut 6, and the first flight to the new next-generation space station had to be cut short. Kovalyonok and Ryumin returned to Earth on October 9, 1977, after a flight of just 2 days and 45 minutes.

SOYUZ MISSION 27: SOYUZ 26

(December 10, 1977–March 16, 1978)

Soviet Manned Spaceflight

Carrying the first successful long-duration crew to the Salyut 6 space station, Soyuz 26 lifted off on December 10, 1977 with commander Yuri Romanenko and flight engineer Georgi Grechko.

Vladimir Kovalyonok and Aleksandr Ivanchenko served as the backup crew for the Soyuz 26 flight.

Although they had originally been scheduled to be the station's first resident crew, Soyuz 25 cosmonauts Vladimir Kovalyonok and Valeri Ryumin had been unable to dock properly with Salyut 6 during their flight to the station in October 1977.

During Soyuz 26, Romanenko and Grechko were given the task of examining—and if necessary, making repairs to—the Salyut 6 docking mechanism. To carry out their orbital check-up duties, Romanenko and Grechko would need to dock their Soyuz 26 spacecraft at the space station's second docking port (Salyut 6 was the first station to have two docking ports; the first five Salyuts had been equipped with one). Fortunately, their attempt to link to the station went well, and they were able to board Salyut 6 and begin their three month stay without any difficulty.

To make a hands-on examination of the primary docking port, Grechko made a spacewalk of 1 hour and 28 minutes in the then-new Orlan spacesuit on December 20, 1977, while Romanenko looked on from the station's airlock. Grechko could find no problems in the docking mechanism; the module was properly configured and connected, and revealed no obvious damage that could present an obstacle to docking.

The good news of the results of Grechko's inspection of the docking port was nearly obliterated by a frightening development involving Romanenko. When the Soyuz 26 commander floated out of the airlock for a better view of Grechko's activities at the docking port, he apparently failed to attach the tether that was designed to prevent him from drifting off into space. The two cosmonauts became aware of the situation before Romanenko floated too far from the station; and neither was injured in the scramble to grasp the tether and secure it, but the incident temporarily overshadowed the docking port's clean bill of health for the startled cosmonauts.

Fortunately, the harrowing EVA mishap was the only frightening moment the cosmonauts had to endure during their otherwise superb long-duration mission. They quickly settled into a productive routine, testing the new space station's systems and equipment, conducting medical and scientific experiments and regularly exercising to ward off the effects of their extended stay in orbit.

The two docking ports on Salyut 6 (and the subsequent Salyut 7) made long-duration missions aboard the station possible because the second port could be used to deliver additional supplies and fuel via unmanned cargo carriers known as Progress spacecraft, which were flown under the control of mission managers on Earth and automatically docked with the station for brief periods before being undocked and (in most cases) de-orbited over the Pacific Ocean.

But while the Salyut and the cosmonauts could be maintained in orbit for longer periods of time, via the Progress supply and refueling flights, the Soyuz spacecraft used to transport the crew to the station was limited to a period of 90 days in space before its onboard systems began to deteriorate. To overcome the spacecraft's endurance limitations, a new Soyuz would need to be flown before the end of each 90-day interval, and switched with the Soyuz docked at the station.

Soviet space officials translated the necessity of the brief switching flights into an opportunity to send cosmonauts from Soviet-bloc nations (and from other countries allied with or friendly to the Soviet Union) into space. The resulting flights were an extension of the previously unmanned cooperative venture known as the Intercosmos Program.

Before the start of the Intercosmos flights, however, a test of the Soyuz-switching concept was necessary. The first test of the Soyuz-switching flights was achieved by Soyuz 27 in January 1978, when Soyuz 27 cosmonauts Vladimir Dzhanibekov and Oleg Makarov visited with Romanenko and Grechko for several days, and then returned to Earth aboard Soyuz 26, leaving Soyuz 27 docked to the station for the later return trip by Romanenko and Grechko, who would remain aboard Salyut 6 until March.

Also in January, the first Progress cargo spacecraft docked with the station, bringing supplies and equipment and achieving the first-ever refueling of an orbiting space station.

On March 2, 1978, the first Intercosmos flight launched with Soviet commander Alexei Gubarev and Czechoslovakian cosmonaut Vladimir Remek, aboard Soyuz 28. Gubarev and Remek visited with Romanenko and Grechko for about a week. Because Romanenko and Grechko had already switched their Soyuz 26 for the Soyuz 27 spacecraft, Gubarev and Remek returned to Earth aboard the spacecraft they had used to travel to the station, Soyuz 28.

At the end of their remarkable long-duration mission aboard Salyut 6, Romanenko and Grechko returned to Earth aboard Soyuz 27 on March 16, 1978. Their stay of 96 days and 10 hours in space set a new record for the longest space mission up to that time, breaking the previous mark set by the final crew of the American Skylab space station.

SOYUZ MISSION 28: SOYUZ 27

(January 10–16, 1978)

Soviet Manned Spaceflight

Launched on January 10, 1978 with commander Vladimir Dzhanibekov and flight engineer Oleg Makarov, Soyuz 27 became the first Soyuz to be switched for another Soyuz at the Salyut 6 space station.

Vladimir Kovalyonok and Aleksandr Ivanchenko served as the backup crew for the Soyuz 27 flight.

The necessity of replacing one Soyuz docked at the space station with a second craft that had been in space for a shorter period of time became an issue with the launch of Salyut 6, on September 29, 1977. Representing a major step forward for the Soviet space station program, Salyut 6 was equipped with two docking ports, where the first five Salyuts had each had one port. The additional docking capability allowed cosmonauts on long-duration missions to be resupplied, and the space station itself refueled, via unmanned, automated Progress spacecraft.

But while the Progress craft could keep the Salyut station and the cosmonaut crews in space for longer periods of time, the Soyuz spacecraft the cosmonauts used to travel to the station had a maximum lifespan in orbit of about 90 days before its onboard systems would begin to deteriorate. To overcome the spacecraft's endurance limitations, a new Soyuz had to be flown to the station before the end of each 90-day interval, and switched with the Soyuz docked at the station.

Dzhanibekov and Makarov made good use of their brief trip to Salyut 6, where they visited the first Salyut 6 long-duration crew, Yuri Romanenko and Georgi Grechko. Dzhanibekov and Makarov performed a variety of scientific experiments and other tests, and then left their Soyuz 27 craft for Romanenko and Grechko, who used it to return to Earth at the end of their record-setting long-endurance stay, in March 1978.

Dzhanibekov and Makarov returned to Earth aboard Soyuz 26 on January 16, 1978, after a flight of 5 days, 22 hours and 59 minutes.

SOYUZ MISSION 29: SOYUZ 28

(March 2–10, 1978)

Soviet Manned Spaceflight

The first manned flight of the Soviet Intercosmos program, Soyuz 28 lifted off on March 2, 1978, with commander Alexei Gubarev and guest cosmonaut Vladimir

Remek—the first citizen of Czechoslovakia to fly in space (Czechoslovakia has since become the Czech Republic following the dissolution of the Soviet Union).

Nikolai Rukavishnikov and Oldrich Pelcak served as the backup crew for the Soyuz 28 mission.

The short flights that became the basis for the manned Intercosmos program became a necessity with the launch of the Salyut 6 space station on September 29, 1977. In contrast to the first five Salyuts, which were each equipped with a single docking port, Salyut 6 (and the subsequent Salyut 7) featured two docking ports. As a result, Salyut 6 and Salyut 7 could be resupplied and refueled by unmanned, automated Progress spacecraft, which in turn meant that cosmonauts could live and work aboard the station for longer periods of time.

But while the Progress vehicles could keep the Salyut station and the cosmonaut crews in space for longer periods of time, the Soyuz spacecraft the cosmonauts used to travel to the station had a maximum lifespan in orbit of about 90 days before its onboard systems would begin to deteriorate. To overcome the spacecraft's endurance limitations, a new Soyuz had to be flown to the station before the end of each 90-day interval, and switched with the old Soyuz at the station.

Soviet space officials seized upon the necessity of the brief switching flights as an opportunity to send cosmonauts from Soviet-bloc nations (and from other countries allied with or friendly to the Soviet Union) into space. The resulting flights were an extension of the previously unmanned multinational cooperative venture known as the Intercosmos program.

During their Soyuz 28 Intercosmos flight, Gubarev and Remek visited the first long-duration crew of Salyut 6, Yuri Romanenko and Georgi Grechko, for a little more than a week. During his time at the space station, Remek performed several scientific experiments that had been designed by Czech scientists, and all four cosmonauts participated in television broadcasts highlighting the cooperative nature of the mission.

Although the usual mission profile for an Intercosmos flight would normally call for the visiting crew members to leave their Soyuz at the space station and return to Earth in the Soyuz that had been docked at the Salyut for some time, Romanenko and Grechko had already hosted the Soyuz 27 crew in January, and had switched their original Soyuz 26 craft with the Soyuz 27 spacecraft. Because they already had a new Soyuz that had only been at the station for a relatively short period, they didn't need the Soyuz 28 craft, so Gubarev and Remek returned to Earth aboard Soyuz 28 on March 10, 1978, after a flight of 7 days, 22 hours and 16 minutes.

SOYUZ MISSION 30: SOYUZ 29

(June 15–November 2, 1978)

Soviet Manned Spaceflight

Launched aboard Soyuz 29 on June 15, 1978, commander Vladimir Kovalyonok and flight engineer Aleksandr Ivanchenkov traveled to the Salyut 6 space station for a remarkable four-and-a-half month mission as the second long-duration Salyut 6 crew, and set a new record for the longest stay in space up to that time.

Vladimir Lyakhov and Valeri Ryumin served as the backup crew for the Soyuz 29/Salyut 6 mission.

During their stay aboard the space station, Kovalyonok and Ivanchenkov conducted a variety of scientific experiments, including medical and biological investigations and materials processing exercises. They also made extensive observations with an advanced telescope that could operate in infrared and ultraviolet wavelengths, which they used to study the Earth's atmosphere, the Moon, Venus, Mars and Jupiter, as well as other astronomical bodies and phenomena. Other scientific work included photography of Earth resources, and mapping photography.

Salyut 6 was refueled and resupplied three times during their stay, via the unmanned, automated Progress spacecraft Progress 2, Progress 3 and Progress 4.

On June 27, 1978, Kovalyonok and Ivanchenkov were visited by the crew of Soyuz 30, Soviet commander Pyotr Klimuk and Miroslav Hermaszewski—the first citizen of Poland to fly in space. The Soyuz 30 flight was part of the Intercosmos program, which the Soviets had devised as a means of providing opportunities for citizens from Soviet-aligned countries to make their nation's first flight in space. The Soyuz 30 crew visited with the Soyuz 29/Salyut 6 cosmonauts for about a week.

On July 29, Kovalyonok and Ivanchenkov made an EVA of two hours and five minutes, with Ivanchenkov retrieving some scientific experiment packages from the exterior of the station and Kovalyonok following his progress from the station's airlock. The two cosmonauts had a scary moment when a meteor passed by, blinding them both for several seconds, but neither was injured, and Ivanchenkov successfully retrieved the experiments as planned.

A second Intercosmos crew—Soviet commander Valeri Bykovsky and Sigmund Jähn of East Germany (Germany was still divided into East and West at the time of Soyuz 31, only being reunited in the 1990s after the fall of the Soviet Union)—arrived at the station on August 26, aboard Soyuz 31. Following the established pattern for the Intercosmos flights, Bykovsky and Jähn visited with Kovalyonok and Ivanchenkov for about a week, during which they made broadcasts touting the cooperative nature of the flight. When they left the station on September 3, 1978, Bykovsky and Jähn left behind their Soyuz 31 vehicle and returned to Earth aboard Soyuz 29.

The switching of the two Soyuz spacecraft had been the genesis of the Intercosmos flights. The onboard systems of the Soyuz began to deteriorate after 90 days in space and therefore required that a Soyuz docked at the space station during a long-duration stay would have to be replaced with a new Soyuz before the end of the 90-day interval.

At the end of their remarkable long-duration mission aboard Salyut 6, Kovalyonok and Ivanchenkov returned to Earth aboard Soyuz 31, on November 2, 1978. During their record-setting Soyuz 29/Salyut 6 flight, they spent

139 days, 14 hours and 48 minutes in space—the longest space mission up to that time.

SOYUZ MISSION 31: SOYUZ 30

(June 27–July 5, 1978)

Soviet Manned Spaceflight

Pyotr Klimuk made his third flight in space as commander of Soyuz 30, which launched on June 27, 1978. He was joined by Miroslav Hermaszewski, who with his flight aboard Soyuz 30 became the first citizen of Poland to fly in space.

Valeri Kubasov and Zenon Jankowski served as the backup crew for the Soyuz 30 flight.

Soyuz 30 was part of the Soviet Intercosmos Program, which made use of brief flights to the Salyut 6, Salyut 7 and Mir space stations to provide opportunities for Soviet-bloc nations (and other countries allied with or friendly to the Soviet Union) to send one of their citizens into space.

During a typical Intercosmos flight, an experienced Russian cosmonaut and the visiting international cosmonaut would make a visit of about a week to the resident space station crew. During his stay, the visiting cosmonaut would carry out a program of experiments devised by scientists in his country, and all the cosmonauts aboard the station would participate in ceremonies and television broadcasts touting the cooperative nature of the mission.

Short Soyuz switching flights became necessary with the Salyut 6 space station, which launched on September 29, 1977. Salyut 6 (and the subsequent Salyut 7) featured two docking ports, where earlier Salyuts had been equipped with only a single port. The additional docking mechanism made long-duration missions aboard the stations possible, because the second port could be used to deliver additional supplies and fuel via unmanned cargo carriers known as Progress spacecraft, which had been derived from the basic Soyuz design and adapted for automatic flight controlled by mission handlers on Earth. The Progress ships were periodically flown to the station and docked automatically for brief periods, and were then de-orbited over the Pacific Ocean at the end of their flight.

But while the redesigned Salyut and the cosmonauts themselves could be maintained in orbit for longer periods of time because of the Progress supply and refueling flights, the Soyuz spacecraft used to transport the crew to the station was limited to a period of 90 days in space before its onboard systems began to deteriorate. To overcome the limitations of the Soyuz craft's limited endurance, a new spacecraft would need to be flown before the end of each 90-day interval, and switched with the Soyuz docked at the station.

Soviet space officials seized on the necessity of the brief switching flights as an opportunity to send cosmonauts from nations friendly to the Soviet Union into space. The resulting flights were an extension of the previously unmanned cooperative venture known as the Intercosmos Program. The manned portion of the program would result in guest cosmonauts making brief visits to the Salyut 6, Salyut 7 and Mir space stations.

During their Soyuz 30 Intercosmos flight, Klimuk and Hermaszewski visited Soyuz 29/Salyut 6 long-duration crew members Vladimir Kovalyonok and Alexander Ivanchenkov, who were in the midst of a then-record four-and-a-half month stay aboard Salyut 6.

Although the usual protocol for the brief Intercosmos flights called for the visiting crew to leave their Soyuz at the station and return to Earth in the vehicle that had previously been docked there, another Intercosmos mission was scheduled to visit Kovalyonok and Ivanchenkov a short while later, so Klimuk and Hermaszewski returned to Earth aboard the same spacecraft they had used to travel to Salyut 6. They landed in Soyuz 30 on July 5, 1978, after a flight of 7 days, 22 hours and 3 minutes.

SOYUZ MISSION 32: SOYUZ 31

(August 26–September 3, 1978)
Soviet Manned Spaceflight

Valeri Bykovsky made his third flight in space as commander of Soyuz 31, which launched on August 26, 1978. He was accompanied by Sigmund Jähn, who with the Soyuz 31 flight became the first citizen of East Germany to fly in space (Germany was still divided into East and West at the time of Soyuz 31, only being reunited in the 1990s after the fall of the Soviet Union).

Victor Gorbatko and Eberhard Kollner served as the backup crew for the Soyuz 31 flight.

Soyuz 31 was part of the Intercosmos Program, which made use of brief flights to the Salyut 6, Salyut 7 and Mir space stations to provide opportunities for Soviet-bloc nations (and other countries allied with or friendly to the Soviet Union) to send one of their citizens into space.

During a typical Intercosmos flight, an experienced Russian cosmonaut and the visiting international cosmonaut would make a visit of about a week to the resident crew of one of the space stations, and during the stay the visiting cosmonaut would carry out a program of scientific research designed by scientists in his native country, and would participate with his Soviet hosts in ceremonies and television broadcasts touting the cooperative nature of the mission.

Short flights to the space stations in the midst of long-duration missions became necessary with Salyut 6, which launched on September 29, 1977. Salyut 6 (and the subsequent Salyut 7) featured two docking ports, while earlier Salyuts had been equipped with only a single port. The additional docking mechanism made long-duration missions aboard the stations possible, because the second port could be used to deliver additional supplies and fuel via unmanned cargo carriers known as Progress spacecraft, which had been derived from the basic Soyuz design and adapted for automatic flight controlled by mission handlers on Earth. The Progress ships were periodically flown to the station and docked automatically for brief periods, and were then de-orbited over the Pacific Ocean at the end of their flight.

But while the redesigned Salyut and the cosmonauts themselves could be maintained in orbit for longer periods of time thanks to the Progress supply and refueling flights, the Soyuz spacecraft used to transport the crew to the station was limited to a period of 90 days in space before its onboard systems began to deteriorate. To overcome the constraints imposed by the Soyuz craft's limited endurance, a new spacecraft would need to be flown before the end of each 90-day interval, and switched with the Soyuz docked at the station.

Soviet space officials seized on the necessity of the brief switching flights as an opportunity to send cosmonauts from nations friendly to the Soviet Union into space. The resulting flights were an extension of the previously unmanned Intercosmos Program, in which Soviet scientists and engineers worked cooperatively on unmanned satellite projects with representatives from other nations.

During their Soyuz 31 Intercosmos flight, Bykovsky and Jähn visited Soyuz 29/Salyut 6 long-duration crew members Vladimir Kovalyonok and Alexander Ivanchenkov, who were in the midst of a then-record four-and-a-half month stay aboard Salyut 6.

At the end of their week-long visit, Bykovsky and Jähn returned to Earth aboard the Soyuz 29 spacecraft that Kovalyonok and Ivanchenkov had used to travel to Salyut 6 at the start of their journey on June 15, 1978. Soyuz 31 remained docked to the station, and was used by Kovalyonok and Ivanchenkov during their return to Earth on November 2, 1978.

With the advent of the switching flights, mission controllers faced an interesting dilemma regarding the call signs assigned to the various flights of Soyuz spacecraft. Because different crews would use a given spacecraft during the trip to and from the space station, mission planners had to decide how to coordinate the call signs for the "upload" and "download" flights. The dilemma was resolved in favor of the originally assigned call sign remaining with the spacecraft; thus Soyuz 29 was given the call sign "Foton" for both the June trip of Kovalyonok and Ivanchenkov to Salyut 6 and for the return to Earth of Bykovsky and Jähn in September; and Soyuz 31 was assigned the call sign "Yastreb" during both its August flight to the station and for its November return.

Bykovsky and Jähn landed in Soyuz 29 on September 3, after a flight of 7 days, 20 hours and 49 minutes.

SOYUZ MISSION 33: SOYUZ 32

(February 25–August 19, 1979)

Soviet Manned Spaceflight

Launched aboard Soyuz 32 on February 25, 1979, commander Vladimir Lyakhov and flight engineer Valeri Ryumin traveled to the Salyut 6 space station, where they lived and worked for a then-record six months, as the station's third long-duration crew.

The Soyuz 32 flight was Ryumin's second space mission; he had previously flown in space aboard Soyuz 25, in October 1977.

Leonid Popov and Valentin Lebedev served as the backup crew for the Soyuz 32 flight.

Lyakhov and Ryumin had been aboard Salyut 6 for about a month when the station suffered a malfunction in one of its three fuel tanks. The failure did not endanger the crew—they could leave on their Soyuz 32 craft at any time—but it did render the station's main engine useless, which meant that any correction to its orbit would have to be made by the Soyuz or by an unmanned Progress spacecraft.

Further difficulties arose in April, when Nikolai Rukavishnikov and Bulgarian cosmonaut Georgi Ivanov traveled to the station aboard Soyuz 33 to visit with Lyakhov and Ryumin and to switch their Soyuz 33 with the Soyuz 32 that Lyakhov and Ryumin had used to travel to Salyut 6 in February.

The necessity of switching the two spacecraft arose from the fact that a Soyuz could only safely remain in space for a period of about 90 days before its onboard systems would begin to deteriorate. With the advent of the Salyut 6 and later space stations which had more than one docking port and could therefore be resupplied and refueled, missions in which cosmonauts would be on the station for more than 90 days became possible. To ensure that the crew aboard the station had a reliable means of returning to Earth in the event of an emergency, a replacement Soyuz had to be flown before the end of each 90-day interval, and switched with the Soyuz docked at the station.

Soviet space officials seized on the necessity of the brief switching flights as an opportunity to send cosmonauts from nations friendly to the Soviet Union into space. The resulting flights were an extension of the previously unmanned cooperative venture known as the Intercosmos Program, and Soyuz 33 was intended as both an Intercosmos flight that would be the first space trip for a citizen of Bulgaria and also as a switching mission that would replace Soyuz 32.

When Rukavishnikov and Ivanov reached Salyut 6, however, a major malfunction occurred in Soyuz 33's main engine. Their planned visit had to be abandoned, and their flight cut short; relying on their spacecraft's backup engine, they were fortunate to survive their ordeal and return to Earth safely on April 12.

As had been the case with the failed Salyut fuel tank, the Soyuz 33 crisis did not directly endanger Lyakhov and Ryumin, but resulted in further consternation for the two cosmonauts as they continued their medical, scientific and maintenance activities aboard the space station. Concerns about the state of their Soyuz 32 craft—which by early June had reached its 90-day safe operations limit—were alleviated when the unmanned Soyuz 34 was launched on June 6. With the new Soyuz safely docked, Lyakhov and Ryumin prepared Soyuz 32 for an unmanned return to Earth; the Soyuz 32 Descent Module re-entered Earth's atmosphere and was recovered on June 13, 1979.

Because of the Soyuz 33 ordeal, no additional manned flights were made to Salyut 6 during the six months that Lyakhov and Ryumin served aboard the station. They did receive visits from three unmanned Progress cargo spacecraft

(Progress 5, 6 and 7), however, which brought the fuel and supplies necessary to sustain the two cosmonauts throughout their long-duration stay. The last of the three Progress flights delivered a radio telescope which was to provide yet another bit of drama for the crew before their long mission finally came to an end.

Although the KRT-10 radio telescope worked well and enabled the cosmonauts to carry out a series of observations that had been part of their planned mission profile, they were unable to remove the instrument from Salyut 6's rear docking port when they were finished using it. Wires trailing from the telescope's large dish antenna became entangled in the docking mechanism, and after some consultation between mission controllers and the crew, it was decided that Lyakhov and Ryumin should make a spacewalk to inspect the situation close-up, and hopefully, to remove the jammed equipment.

On August 15, the cosmonauts ventured outside the station—Ryumin equipped with wire cutters—and in an EVA of 1 hour and 23 minutes, they managed to free the entangled antenna, ensuring the continued usefulness of the crucial docking port.

Their long, often precarious mission complete, Lyakhov and Ryumin returned to Earth four days after their superb spacewalk. They landed on August 19 aboard Soyuz 34, after a then-record flight of 175 days and 36 minutes.

SOYUZ MISSION 34: SOYUZ 33

(April 10–12, 1979)

Soviet Manned Spaceflight

Nikolai Rukavishnikov made his third flight in space as commander of Soyuz 33, which launched on April 10, 1979. He was accompanied by Georgi Ivanov, who with the Soyuz 33 flight became the first citizen of Bulgaria to fly in space. Because of a major malfunction in their Soyuz 33 craft, the two cosmonauts were unable to visit the Salyut 6 space station as originally planned, and were fortunate to survive and land safely.

Yuri Romanenko and Aleksander Aleksandrov served as the backup crew for the Soyuz 33 flight.

Soyuz 33 was part of the Intercosmos Program, which made use of brief flights to the Salyut 6, Salyut 7 and Mir space stations to provide opportunities for Soviet-bloc nations (and other countries allied with or friendly to the Soviet Union) to send one of their citizens into space.

During a typical Intercosmos flight, an experienced Russian cosmonaut and the visiting international cosmonaut would make a visit of about a week to a resident space station crew, and during the stay the visiting cosmonaut would normally carry out a program of scientific research designed by scientists from his country, and would participate with his Soviet hosts in ceremonies and television broadcasts touting the cooperative nature of the mission.

Short flights to the space stations in the midst of long-duration missions became necessary with Salyut 6, which launched on September 29, 1977. Salyut 6 (and the subsequent Salyut 7) featured two docking ports, where earlier Salyuts had been equipped with only a single port. The additional docking mechanism made long-duration missions aboard the stations possible, because the second port could be used to deliver additional supplies and fuel via unmanned cargo carriers known as Progress spacecraft, which had been derived from the basic Soyuz design and adapted for automatic flight controlled by mission handlers on Earth. The Progress ships were periodically flown to the station and docked automatically for brief periods, and were then de-orbited over the Pacific Ocean at the end of their flight.

But while the redesigned Salyut and the cosmonauts themselves could be maintained in orbit for longer periods of time thanks to the Progress supply and refueling flights, the Soyuz spacecraft used to transport the crew to the station was limited to a period of 90 days in space before its onboard systems began to deteriorate. To overcome the limitations of the Soyuz craft's limited endurance, a new spacecraft would need to be flown before the end of each 90-day interval, and switched with the Soyuz docked at the station.

Soviet space officials seized on the necessity of the brief switching flights as an opportunity to send cosmonauts from nations friendly to the Soviet Union into space. The resulting flights were an extension of the previously unmanned cooperative venture known as the Intercosmos Program, in which Soviet scientists and engineers would cooperate with their counterparts from other nations to develop unmanned scientific satellites.

During their Soyuz 33 Intercosmos flight, Rukavishnikov and Ivanov were scheduled to visit with Vladimir Lyakhov and Valeri Ryumin, who were in the midst of a then-record stay of six months aboard the Salyut 6 space station. The Soyuz 33 flight was also intended to provide Lyakhov and Ryumin with a new Soyuz, as Rukavishnikov and Ivanov were to have left Soyuz 33 docked at the station and returned to Earth aboard Soyuz 32, which Lyakhov and Ryumin had used to travel to Salyut 6 in February.

When Rukavishnikov and Ivanov reached the station on April 10, however, a major malfunction occurred in Soyuz 33's main engine. The loss of the spacecraft's main engine in the midst of the flight was an unprecedented crisis; it meant that the cosmonauts would have to rely on the vehicle's backup engine for an immediate return to Earth.

Rukavishnikov and Ivanov endured a very long day in space while mission controllers and emergency teams on the ground plotted their best course for a safe return. Finally, on April 12, they were cleared to fire the backup engine, which fortunately worked without incident. They began their re-entry preparations a short while later.

The change in plans—and the novel use of the backup engine—necessitated a bone-rattling ballistic re-entry, with the cosmonauts enduring forces as great as 10 Gs. Their Descent Module touched down about 180 miles from their originally intended landing site, and recovery forces found them shaken but unharmed. Rukavishnikov and Ivanov had spent a total of 1 day, 23 hours and 1 minute in space during their harrowing ordeal.

Even as mission handlers breathed a sigh of relief at the safe return of the Soyuz 33 crew, they still faced the problem of how to replace Soyuz 32, which remained docked at the space station and had to be switched with a new Soyuz before its 90-day safe operations limit expired. The solution settled upon was the unmanned launch of Soyuz 34, on June 6; Lyakhov and Ryumin used the Soyuz 34 vehicle in their return, on August 19, 1979.

Although the Soyuz 33 flight had been cut short and could not achieve the objectives of its original mission profile, the fact that Rukavishnikov and Ivanov returned safely was a major achievement in itself, and an indication of the capabilities of the Soviet space program at the end of the 1970s.

SOYUZ MISSION 35: SOYUZ 34

(June 6–August 19, 1979)

Soviet Unmanned Spaceflight

Launched unmanned on June 6, 1979, Soyuz 34 was sent to the Salyut 6 space station as a replacement for the Soyuz 32 vehicle that cosmonauts Vladimir Lyakhov and Valeri Ryumin had used to travel to the station in February 1979.

When Soyuz 34 arrived in June, Lyahkov and Ryumin were in the midst of a then-record six month stay aboard Salyut 6. Their Soyuz 32 vehicle had to be replaced because it had reached the end of its 90-day safe operations limit; the onboard systems of the Soyuz craft tended to deteriorate after 90 days in space.

Soyuz 33 had been intended as the switching flight for Soyuz 32 in April, but the Soyuz 33 craft had experienced a failure in its main engine, and its crew, Nikolai Rukavishnikov and Georgi Ivanov, were fortunate to survive their ordeal and safely return to Earth. Their early return left Lyakhov and Ryumin without a replacement for Soyuz 32.

With the arrival of Soyuz 34, Lyakhov and Ryumin prepared Soyuz 32 for an unmanned return to Earth. The Soyuz 32 Descent Module re-entered Earth's atmosphere and was recovered on June 13, 1979, and Lyakhov and Ryumin returned aboard Soyuz 34 on August 19, 1979. From launch to landing, the Soyuz 34 spacecraft spent a total of 73 days, 18 hours and 14 minutes in space.

SOYUZ MISSION 36: SOYUZ T-1

(December 16, 1979–March 26, 1980)

Soviet Unmanned Spaceflight

Launched unmanned on December 16, 1979 and docked with the Salyut 6 space station for three months, Soyuz T-1 was an operational test of the redesigned Soyuz spacecraft.

The Soyuz T represented a major overhaul of the basic Soyuz design, taking into account the many changes in the direction and purpose of the Soviet program since the spacecraft's inception in the late 1960s, when the U.S.S.R. was still focused on reaching the Moon. In the intervening years the transportation of cosmonauts to orbiting space stations had emerged as the primary task for the Soyuz craft, and the Soyuz T redesign reflected the change.

The Soyuz T design allowed for the transport of three person crews (the last previous crew of three prior to the introduction of the Soyuz T was the ill-fated Soyuz 11 in June, 1971, which had resulted in a spacecraft malfunction that had cost the lives of the crew), and featured solar arrays rather than batteries, which allowed for longer missions.

Use of the Soyuz T was gradually phased in while the original Soyuz was still in use; the first manned Soyuz T was Soyuz T-2, in June, 1980; the last in the series, Soyuz T-15, provided cosmonauts Leonid Kizim and Vladimir Solovyov the unique opportunity to be the first crew to visit two space stations during a single flight, as they served as the first resident crew of Mir and also made the last visit to the Salyut 7 space station, on May 5, 1986.

SOYUZ MISSION 37: SOYUZ 35

(April 9–October 11, 1980)

Soviet Manned Spaceflight

Launched aboard Soyuz 35 on April 9, 1980, commander Leonid Popov and flight engineer Valeri Ryumin traveled to the Salyut 6 space station, where they lived and worked for a then-record 184 days, as the station's fourth long-duration crew.

The mission was the third spaceflight for Ryumin, who had already served 175 days aboard Salyut 6 as a member of the station's third long-duration crew, with Vladimir Lyakhov. His remarkable involvement in Soyuz 35/Salyut 6 was the result of an injury suffered by Valentin Lebedev, who had originally been assigned as flight engineer to accompany Leonid Popov.

Over the years since the inception of the Soyuz program, Soviet space officials had evolved a set of rules governing the assignment of crew members for a given flight. Experience was a key factor in determining the makeup of any crew, and was deemed particularly important in the case of long-duration missions aboard a space station. As a result, the flight rules called for at least one experienced cosmonaut to be assigned to the Soyuz 35/Salyut 6 flight.

In the original crew, Lebedev was the experienced crew member; he had previously flown in space aboard Soyuz 13, in December, 1973. Popov would make his first flight in space as commander of Soyuz 35.

The original plan for the flight was thrown into disarray, however, when Lebedev was injured while preparing for the mission. His backup was Boris Andreyev—who, like Popov, had never flown in space before; Popov's backup, Vyacheslav Zudov, had commanded the Soyuz 23 crew in October 1976. Mission controllers were faced with the prospect of replacing both Popov and Lebedev, or violating their own carefully designed flight rules by sending Popov and Andreyev on the long-duration flight with neither cosmonaut having previously flown in space.

While flight planners mulled their options, Ryumin presented himself as a potential solution. Having already achieved a long-duration stay on Salyut 6, he volunteered to return to the station with Popov, thereby retaining as much of the original plan for Soyuz 35/Salyut 6 as possible while still adhering to the flight rules calling for an experienced cosmonaut on each mission. Soviet space officials approved of the idea, and Popov and Ryumin became the primary crew.

Ironically, Popov and Lebedev had served as the backup crew for Lyakhov and Ryumin during the long-duration Soyuz 32/Salyut 6 flight in 1979.

Once they'd progressed past the substantial pre-flight confusion, Popov and Ryumin quickly settled into a productive, busy routine aboard Salyut 6. In April and May they unloaded supplies from the unmanned Progress 8 and Progress 9 cargo supply vehicles, and on May 26 received their first visitors at the station, when Soyuz 36 cosmonauts Valeri Kubasov and Bertalan Farkas—the first citizen of Hungary to fly in space—arrived at Salyut 6 for a brief stay, as part of the Soviet Intercosmos program of providing short flights in space to representatives of nations friendly to the Soviet Union.

The manned Intercosmos flights were the result of a need that arose from the fact that a Soyuz could only safely remain in space for a period of about 90 days before its onboard systems would begin to deteriorate. With the advent of the Salyut 6 and later space stations which had more than one docking port and could therefore be resupplied and refueled, missions in which cosmonauts would be on the station for more than 90 days became possible. To ensure that the crew aboard the station had a reliable means of returning to Earth in the event of an emergency—and a spacecraft capable of bringing them safely home at the end of their mission—a replacement Soyuz had to be flown before the end of each 90-day interval, and switched with the Soyuz docked at the station. Soviet space officials had translated the necessity of the brief switching flights into an opportunity to send cosmonauts from nations friendly to the Soviet Union into space.

During a typical Intercosmos flight, an experienced Russian cosmonaut and the visiting international cosmonaut would make a visit of about a week to the resident space station crew, and the visiting cosmonaut would conduct a scientific program designed by scientists from his home country. All the cosmonauts would also participate in ceremonies and television broadcasts touting the cooperative nature of the mission.

In keeping with the routine established for the flights, Kubasov and Farkas returned to Earth aboard Soyuz 35 on June 3, 1980, leaving their Soyuz 36 spacecraft docked at the space station.

Popov and Ryumin continued their routine of scientific and medical research, regular exercise (necessary to combat the bone density and blood plasma losses that result from long-duration stays in space), and space station maintenance duties.

On June 5, 1980 they received a visit from fellow Russian cosmonauts Yuri Malyshev and Vladimir Aksyonov, who had successfully flown the Soyuz T-2 to the station in the first manned test of the new Soyuz T-series of spacecraft. The second generation of Soyuz vehicles, the Soyuz T-craft were gradually phased into the Soviet program, and replaced the original Soyuz after the flight of Soyuz 40 in May 1981. Malyshev and Aksyonov spent four days at the station, and then returned to Earth aboard Soyuz T-2 on June 9, 1980.

The unmanned Progress 10 was launched June 29 on a resupply flight to Salyut 6, and in July Popov and Ryumin received another Intercosmos visit that turned out to be the most controversial and politically charged mission of the entire program.

Launched on July 23, 1980 aboard Soyuz 37, the sixth flight of the Intercosmos program paired veteran Soviet commander Victor Gorbatko (who was making his third trip into space) and Pham Tuan—the first citizen of Vietnam to fly in space. Following the usual routine for the international flights, the visiting crew spent about a week with Popov and Ryumin aboard Salyut 6, mixing ceremonies and speeches with broadcasts touting the cooperative nature of the flight.

Such formalities were normally dismissed as simple propaganda in the West, even as they were trumpeted by the Soviet Union as the progenitor of long-term international cooperation in space by friendly nations of the Communist bloc. The Soyuz 37 flight, however, came at a time when the United States was in the midst of an angry protest of the Soviet Union's invasion of Afghanistan—a protest embodied in the American boycott of the 1980 summer Olympics, which were hosted by the Soviets in Moscow.

Launched concurrently with the Olympics dispute, the Soyuz 37 flight featured harsher than usual anti-U.S. propaganda, probably spurred as much by the involvement of Pham Tuan—a pilot in the Vietnamese Air Force who had fought against the United States during the Vietnam War—as it was by Soviet anger over the Olympic boycott.

Unprecedented in the history of space exploration, the deep bitterness elicited by the brief flight was, thankfully, short-lived and largely forgotten in the weeks after Gorbatko and Tuan returned to Earth, aboard Soyuz 36, on July 31, 1980. The practical aspect of the flight—the switching of Soyuz 37, which was left at Salyut 6 for Popov and Ryumin to use during their return to Earth at the end of their long-duration mission, and the Soyuz 36 craft the visiting crew used for their return—was successfully achieved.

On September 18, 1980 another Intercosmos crew—Soviet cosmonaut Yuri Romanenko and Arnaldo Tamayo-Mendez, the first citizen of Cuba to fly in space—docked their Soyuz 38 craft at the station. Although the Cuban political regime of Fidel Castro was also considered virulently hostile to the United States, the Soyuz 38 Intercosmos mission—and Western reaction to it—proceeded in calmer fashion than the previous Intercosmos flight, which involved Vietnamese cosmonaut Pham Tuan.

Because Popov and Ryumin already had a new Soyuz available (Soyuz 37, which had been switched for Soyuz 36 in July), Romanenko and Tamayo-Mendez returned to Earth on the same spacecraft they'd used to travel to the station, landing in Soyuz 38 on September 26, 1980.

The unmanned Progress 11 was launched on September 28, 1980, and was docked to Salyut 6 for several months before being de-orbited in December.

At the end of their long and occasionally tumultuous long-duration stay aboard Salyut 6, Popov and Ryumin left the station and returned to Earth on October 11, 1980. During their remarkable Soyuz 35/Salyut 6 mission, they spent a then-record 184 days, 20 hours and 12 minutes in space.

With the completion of his third space mission, Valeri Ryumin had accumulated an amazing total of 361 hours, 21 hours and 33 minutes in space.

SOYUZ MISSION 38: SOYUZ 36

(May 26–June 3, 1980)
Soviet Manned Spaceflight

Valeri Kubasov made his third flight in space as commander of Soyuz 36, which launched on May 26, 1980. He was accompanied by Bertalan Farkas, who with the Soyuz 36 flight became the first citizen of Hungary to fly in space.

Vladimir Dzhanibekov and Bela Magyari served as the backup crew for the Soyuz 36 flight.

Soyuz 36 was part of the Intercosmos Program, which made use of brief flights to the Salyut 6, Salyut 7 and Mir space stations to provide opportunities for Soviet-bloc nations (and other countries allied with or friendly to the Soviet Union) to send one of their citizens into space.

During a typical Intercosmos flight, an experienced Russian cosmonaut and the visiting international cosmonaut would make a visit of about a week to the resident space station crew, and the visiting cosmonaut would conduct scientific activities devised by scientists from his home country. The visiting and resident cosmonauts would also participate in ceremonies and television broadcasts touting the cooperative nature of the mission.

Short flights to the space stations in the midst of long-duration missions became necessary with Salyut 6, which launched on September 29, 1977. Salyut 6 (and the subsequent Salyut 7) featured two docking ports, where earlier Salyuts had been equipped with only a single port. The additional docking mechanism made long-duration missions aboard the stations possible, because the second port could be used to deliver additional supplies and fuel via unmanned cargo carriers known as Progress spacecraft, which had been derived from the basic Soyuz design and adapted for automatic flight controlled by mission handlers on Earth. The Progress ships were periodically flown to the station and docked automatically for brief periods, and were then de-orbited over the Pacific Ocean at the end of their flight.

But while the redesigned Salyut and the cosmonauts themselves could be maintained in orbit for longer periods of time thanks to the Progress supply and refueling flights, the Soyuz spacecraft used to transport the crew to the station was limited to a period of 90 days in space before its onboard systems began to deteriorate. To overcome the limitations of the Soyuz craft's limited endurance, a new spacecraft would need to be flown before the end of each 90-day interval, and switched with the Soyuz docked at the station.

Soviet space officials seized on the necessity of the brief switching flights as a means of providing spaceflight opportunities for cosmonauts from nations friendly to the Soviet Union. The resulting flights were an extension of the previously unmanned Intercosmos Program, in which Soviet scientists and engineers worked on the development and launch of unmanned scientific satellites with their counterparts from other nations.

During their Soyuz 36 Intercosmos flight, Kubasov and Farkas visited with Salyut 6 long-duration crew members Leonid Popov and Valeri Ryumin, who were in the midst of a record-setting 184-day stay aboard the station.

In keeping with the routine established for the flights, Kubasov and Farkas left their Soyuz 36 spacecraft docked at Salyut 6 when they left the station and returned to Earth, aboard Soyuz 35, on June 3, 1980. From launch to landing, they had spent 7 days, 20 hours and 46 minutes in space.

SOYUZ MISSION 39: SOYUZ T-2

(June 5–9, 1980)

Soviet Manned Spaceflight

Launched on June 5, 1980, commander Yuri Malyshev and flight engineer Vladimir Aksyonov served as the crew of Soyuz T-2, the first manned test flight of the new Soyuz T series of spacecraft.

The Soyuz T represented a major overhaul of the basic Soyuz design, taking into account the many changes in the direction and purpose of the Soviet program since the spacecraft's inception in the late 1960s, when the U.S.S.R. was still focused on reaching the Moon. In the intervening years the transportation of cosmonauts to orbiting space stations had emerged as the primary task for the Soyuz craft, and the Soyuz T redesign reflected the change.

The first Soyuz T test, the unmanned Soyuz T-1, had launched on December 16, 1979 and was docked with Salyut 6 for three months.

During their Soyuz T-2 flight, Malyshev and Aksyonov also docked with Salyut 6, where they visited with the space station's fourth long-duration crew, Leonid Popov and Valeri Ryumin, who were in the midst of a then-record 184-day stay aboard the station.

Leonid Kizim and Oleg Makarov served as the backup crew for the Soyuz T-2 flight.

Following the successful test flight by Malyshev and Aksyonov, use of the Soyuz T was gradually phased in while the original Soyuz was still in use. The Soyuz T replaced the original Soyuz after the flight of Soyuz 40 in May, 1981; the final flight of the Soyuz T series, Soyuz T-15 in 1986, provided cosmonauts Leonid Kizim and Vladimir Solovyov the unique opportunity to be the first crew to visit two space stations during a single flight, as they served as the first resident crew of Mir and also made the last visit to the Salyut 7 space station, on May 5, 1986.

At the end of their Soyuz T-2 test flight, Malyshev and Aksyonov returned to Earth on June 9, 1980, after a flight of 3 days, 22 hours and 19 minutes.

SOYUZ MISSION 40: SOYUZ 37

(July 23–31, 1980)

Soviet Manned Spaceflight

Victor Gorbatko made his third flight in space as commander of Soyuz 37, which launched on July 23, 1980 and docked with the Salyut 6 space station for about a week. He was accompanied by Pham Tuan, who with the Soyuz 37 flight became the first citizen of Vietnam to fly in space.

Valeri Bykovsky and Bui Thanh Liem served as the backup crew for Soyuz 37.

The sixth flight of the Soviet Intercosmos program, which provided opportunities for representatives from Soviet-bloc nations (and other countries allied with or friendly to the Soviet Union) to fly in space, the Soyuz 37 mission was the most controversial and politically-charged flight in the project's history.

Short flights to the space stations in the midst of long-duration missions had become a necessity with Salyut 6, which launched on September 29, 1977. Salyut 6 (and the subsequent Salyut 7) featured two docking ports, where earlier Salyuts had been equipped with only a single port. The additional docking mechanism made long-duration missions aboard the stations possible, because the second port could be used to deliver additional supplies and fuel via unmanned cargo carriers known as Progress spacecraft, which had been derived from the basic Soyuz design and adapted for automatic flight controlled by mission handlers on Earth. The Progress ships were periodically flown to the station and docked automatically for brief periods, and were then de-orbited over the Pacific Ocean at the end of their flight.

But while the redesigned Salyut and the cosmonauts themselves could be maintained in orbit for longer periods of time thanks to the Progress supply and refueling flights, the Soyuz spacecraft used to transport the crew to the station was limited to a period of 90 days in space before its onboard systems began to deteriorate. To overcome the limitations of the Soyuz craft's limited endurance, a new spacecraft would need to be flown before the end of each 90-day interval, and switched with the Soyuz docked at the station.

Soviet space officials seized on the necessity of the brief switching flights as a means of providing guest cosmonauts from nations friendly to the Soviet Union a chance to fly in space. The resulting flights were an extension of the previously unmanned Intercosmos Program, in which Soviet scientists and engineers developed unmanned scientific satellites in cooperation with their counterparts from other countries.

During their Soyuz 37 Intercosmos flight, Gorbatko and Tuan followed the established routine for the international flights, visiting with Salyut 6 long-duration crew members Leonid Popov and Valeri Ryumin, who were in the midst of a then-record 184-day stay aboard the space station and participating in a series of ceremonies and speeches touting the cooperative nature of the flight, which were broadcast on Soviet television.

Such formalities were normally dismissed as simple propaganda in the West, even as they were trumpeted by the Soviet Union as the progenitor of long-term international cooperation in space by friendly nations of the Communist bloc. The Soyuz 37 flight, however, came at a time when the United States was in the midst of an angry protest of the Soviet Union's invasion of Afghanistan. To protest the military action, the United States refused to participate in the 1980 summer Olympics, which were hosted by the Soviets in Moscow.

Launched concurrently with the Olympics dispute, the Soyuz 37 flight featured harsher than usual anti-U.S. propaganda, probably spurred as much by the involvement of Pham Tuan—a pilot in the Vietnamese Air Force who had fought against the United States during the Vietnam War—as it was by Soviet anger over the Olympic boycott.

Unprecedented in the history of space exploration, the virulent bitterness that resulted from the brief flight was, fortunately, short-lived and largely forgotten in the weeks after Gorbatko and Tuan returned to Earth. They landed in Soyuz 36 on July 31, 1980, after a flight of 7 days, 20 hours and 42 minutes.

Despite the ugly rhetoric surrounding the mission, the practical purpose of the flight—the switching of the Soyuz 36 that had been docked at the station and Soyuz 37, which was left at Salyut 6 for Popov and Ryumin to use during their return to Earth at the end of their long-duration mission—was successfully achieved.

SOYUZ MISSION 41: SOYUZ 38

(September 18–26, 1980)

Soviet Manned Spaceflight

Yuri Romanenko made his second flight in space as commander of Soyuz 38, which launched on September 18, 1980. He was accompanied by Arnaldo Tamayo-Mendez, who with the Soyuz 38 flight became the first citizen of Cuba to fly in space.

Yevgeni Khrunov and Jose Lopez-Falcon served as the backup crew for the Soyuz 38 flight.

Soyuz 38 was part of the Intercosmos Program, which made use of brief flights to the Salyut 6, Salyut 7 and Mir space stations to provide opportunities for Soviet-bloc nations (and other countries allied with or friendly to the Soviet Union) to send one of their citizens into space.

During a typical Intercosmos flight, an experienced Russian cosmonaut and the visiting international cosmonaut would make a visit of about a week to the resident space station crew, and the visiting cosmonaut would participate in scientific research (often designed by scientists from his home country) and participate with his Soviet hosts in ceremonies and television broadcasts touting the cooperative nature of the mission.

Short flights to the space stations in the midst of long-duration missions had become necessary with Salyut 6, which launched on September 29, 1977. Salyut 6 (and the subsequent Salyut 7) featured two docking ports, where earlier Salyuts had been equipped with only a single port. The additional docking mechanism made long-duration missions aboard the stations possible, because the second port could be used to deliver additional supplies and fuel via unmanned cargo carriers known as Progress spacecraft, which had been derived from the basic Soyuz design and adapted for automatic flight controlled by mission handlers on Earth. The Progress ships were periodically flown to the station and docked automatically for brief periods, and were then de-orbited over the Pacific Ocean at the end of their flight.

But while the redesigned Salyut and the cosmonauts themselves could be maintained in orbit for longer periods of time thanks to the Progress supply and refueling flights, the Soyuz spacecraft used to transport the crew to the station was limited to a period of 90 days in space before its onboard systems began to deteriorate. To overcome the Soyuz spacecraft's endurance limitations, a new spacecraft would need to be flown to the space station before the end of each 90-day interval, and switched with the Soyuz docked at the station.

Soviet space officials seized on the necessity of the brief switching flights as an opportunity to offer cosmonauts from nations friendly to the Soviet Union a chance to fly in space. The resulting flights were an extension of the previously unmanned Intercosmos Program, in which Soviet scientists and engineers cooperated on the development of unmanned scientific satellites with their counterparts from other nations. The manned portion of the program resulted in spaceflight opportunities to the Salyut 6, Salyut 7 and Mir space stations over the course of a ten-year period.

During their Soyuz 38 Intercosmos flight, Romanenko and Tamayo-Mendez visited with Salyut 6 long-duration crew members Leonid Popov and Valeri Ryumin, who were in the midst of a record-setting 184-day stay aboard the station. In keeping with the usual routine for the flights, the four cosmonauts participated in ceremonies and speeches touting the cooperative nature of the mission, which were aired on Soviet television.

The rituals attracted more attention than usual from Western observers because they followed those of the tumultuous Soyuz 37 Intercosmos flight, which had been carried out amidst a U.S. boycott of the 1980 summer Olympics in Moscow and featured a Vietnamese cosmonaut and strident anti-American propaganda.

The Cuban political regime of Fidel Castro was also considered hostile to the United States, but the Soyuz 38 Intercosmos mission—and Western reaction to it—proceeded in calmer fashion than the previous Intercosmos flight.

Because Popov and Ryumin already had a new Soyuz available (Soyuz 37, which had been switched for Soyuz 36 in July), Romanenko and Tamayo-Mendez returned to Earth in the same spacecraft they'd used to travel to the station. They landed in Soyuz 38 on September 26, 1980, after a flight of 7 days, 20 hours and 43 minutes.

SOYUZ MISSION 42: SOYUZ T-3

(November 27–December 10, 1980)

Soviet Manned Spaceflight

Launched on November 27, 1980, the flight of Soyuz T-3 was the first Soviet space mission to feature a three-man crew since the tragic flight of Soyuz 11 in June, 1971, which had resulted in the death of cosmonauts Georgy Dobrovolsky, Vladislav Volkov, and Viktor Patsayev.

Leonid Kizim made his first flight in space as commander of Soyuz T-3. He was accompanied by flight engineer Oleg Makarov (who made his fourth trip into space during the Soyuz T-3 flight), and research engineer Gennadi Strekalov (who like Kizim was flying in space for the first time). The mission profile for the three cosmonauts called for them to spend slightly under two weeks aboard the Salyut 6 space station while they performed maintenance and repair tasks.

Vasili Lazarev, Viktor Savinykh and Valeri Polyakov served as the backup crew for the Soyuz T-3 flight.

Kizim, Makarov and Strekalov had a superb flight and a productive stay aboard Salyut 6. Their refurbishing of the station ensured that it could continue to be safely operated, and could support additional crews. At the end of their successful mission, the three cosmonauts returned to Earth aboard their Soyuz T-3 craft, following a flight of 12 days, 19 hours and 8 minutes.

SOYUZ MISSION 43: SOYUZ T-4

(March 12–May 26, 1981)

Soviet Manned Spaceflight

As commander of the Soyuz T-4/Salyut 6 long-duration mission, Vladimir Kovalyonok made his third flight in space. He was accompanied by flight engineer Viktor Savinykh. The two cosmonauts would live and work together aboard Salyut 6 for two-and-a-half months as the fifth (and last) Salyut 6 long-duration crew, and Kovalyonok would increase his career time in space to a remarkable total of more than 216 days.

Vyacheslav Zudov and Boris Andreyev served as the backup crew for the Soyuz T-4/Salyut 6 mission.

Kovalyonok and Savinykh quickly settled into a productive routine aboard Salyut 6. A variety of tasks had been worked out for resident crews to perform over the course of the station's remarkable life span, including scientific experiments, photography, medical and biological investigations, space station maintenance and technology experiments, and a regular program of exercise to lessen the bone density and blood plasma losses that accumulate over the course of a long-duration stay in space.

The cosmonauts were also busied with the unloading of supplies that were delivered to the station by the unmanned cargo spacecraft Progress 12, which had launched on January 24 and remained docked at the station until March 20.

On March 22, 1981, Kovalyonok and Savinykh received their first visitors at the station, when Soyuz 39 cosmonauts Vladimir Dzhanibekov and Jugderdemidiyn Gurragcha—the first citizen of Mongolia to fly in space—arrived at Salyut 6 for a brief stay, as part of the Soviet Intercosmos program of providing short flights in space to representatives of nations friendly to the Soviet Union.

The manned Intercosmos flights were the result of a need that arose from the fact that a Soyuz could only safely remain in space for a period of about 90 days before its onboard systems would begin to deteriorate. With the advent of the Salyut 6 and later space stations which had more than one docking port and could therefore be resupplied and refueled, missions in which cosmonauts would be on the station for more than 90 days became possible. To ensure that the crew aboard the station had a reliable means of returning to Earth in the event of an emergency—and a spacecraft capable of bringing them safely home at the end of their mission—a replacement Soyuz had to be flown before the end of each 90-day interval, and switched with the Soyuz docked at the station. Soviet space officials had translated the necessity of the brief switching flights into spaceflight opportunities for cosmonauts from nations friendly to the Soviet Union.

During a typical Intercosmos flight, an experienced Russian cosmonaut and the visiting international cosmonaut would make a visit of about a week to the resident crew of one of the space stations, and during the stay would participate in ceremonies and television broadcasts touting the cooperative nature of the mission. The visiting cosmonaut would also typically participate in a scientific program designed by scientists from his home country.

Although the normal routine for Intercosmos flights called for the visiting crew to leave their Soyuz craft at the station when they left, the usual switching of spacecraft was not necessary during the Soyuz T-4/Salyut 6 mission, because at a planned duration of about 75 days, Kovalyonok and Savinykh would return to Earth before the expiration of the 90-day Soyuz safe operations limit.

Dzhanibekov and Gurragcha landed in Soyuz 39 on March 30, 1981.

A second visiting crew arrived aboard Soyuz 40 on May 14, 1981. Commanded by Soviet cosmonaut Leonid Popov, who was accompanied by Dumitru Prunariu—the first citizen of Romania to fly in space—the Soyuz 40 Intercosmos flight was the last manned mission to the Salyut 6 space station, and the time that a Soyuz spacecraft of the original design was flown in space.

After visiting with Kovalyonok and Savinykh for about a week, Popov and Prunariu returned to Earth aboard Soyuz 40 on May 22, 1981.

Four days later, on May 26, 1981, Kovalyonok and Savinykh concluded their long-duration mission when they undocked from Salyut 6 and returned to Earth aboard Soyuz T-4 after a total flight of 74 days, 17 hours and 37 minutes.

Savinykh had begun his spaceflight career with the Soyuz T-4/Salyut 6 flight; and Kovalyonok, by the end of the flight, had accumulated 216 days, 9 hours and 12 minutes in space during three missions (Soyuz 25, Soyuz 29/Salyut 6, and Soyuz T-4/Salyut 6).

Salyut 6 was not occupied again. In June, the unmanned Cosmos 1267 module was flown to the station and automatically docked, in a test of equipment and procedures designed for use in later space stations. The two spacecraft remained linked for more than a year, until they were de-orbited over the Pacific Ocean on July 29, 1982.

SOYUZ MISSION 44: SOYUZ 39

(March 22–30, 1981)

Soviet Manned Spaceflight

Vladimir Dzhanibekov made his second flight in space as commander of Soyuz 39, which launched on March 22, 1981. He was accompanied by Jugderdemidiyn

Gurragcha, who with the Soyuz 39 flight became the first citizen of Mongolia to fly in space.

Vladimir Lyakhov and Maidarjabyn Ganzoring served as the backup crew for the Soyuz 39 fight.

Soyuz 39 was part of the Intercosmos Program, which made use of brief flights to the Salyut 6, Salyut 7 and Mir space stations to provide opportunities for Soviet-bloc nations (and other countries allied with or friendly to the Soviet Union) to send one of their citizens into space.

During a typical Intercosmos flight, an experienced Russian cosmonaut and the visiting international cosmonaut would make a visit of about a week to the space station's resident crew, and during the stay they would participate in ceremonies and television broadcasts touting the cooperative nature of the mission. The visiting cosmonaut also typically participated in a program of scientific research, often developed by scientists from his home country.

Short flights to the space stations in the midst of long-duration missions had become necessary with Salyut 6, which launched on September 29, 1977. Salyut 6 (and the subsequent Salyut 7) featured two docking ports, where earlier Salyuts had been equipped with only a single port. The additional docking mechanism made long-duration missions aboard the stations possible, because the second port could be used to deliver additional supplies and fuel via unmanned cargo carriers known as Progress spacecraft, which had been derived from the basic Soyuz design and adapted for automatic flight controlled by mission handlers on Earth. The Progress ships were periodically flown to the station and docked automatically for brief periods, and were then de-orbited over the Pacific Ocean at the end of their flight.

But while the redesigned Salyut and the cosmonauts themselves could be maintained in orbit for longer periods of time thanks to the Progress supply and refueling flights, the Soyuz spacecraft used to transport the crew to the station was limited to a period of 90 days in space before its onboard systems began to deteriorate. To overcome the Soyuz spacecraft's endurance limitations, a new spacecraft would need to be flown before the end of each 90-day interval, and switched with the Soyuz docked at the station.

Soviet space officials translated the necessity of the brief switching flights into spaceflight opportunities for cosmonauts from nations friendly to the Soviet Union. The resulting flights were an extension of the previously unmanned Intercosmos Program, in which Soviet scientists and engineers cooperated on the development of unmanned scientific satellites with their counterparts from other countries.

During their Soyuz 39 Intercosmos flight, Dzhanibekov and Gurragcha visited with Salyut 6 long-duration crew members Vladimir Kovalyonok and Viktor Savinykh, who were in the midst of a two-and-a-half month stay aboard the station. In keeping with the usual routine for the flights, the four cosmonauts participated in ceremonies and speeches touting the cooperative nature of the mission, which were aired on Soviet television.

Although the normal routine for Intercosmos flights called for the visiting crew to leave their Soyuz craft at the station when they left, the usual switching of spacecraft was not necessary during the Soyuz 39/Salyut 6 mission, because at a planned duration of about 75 days, Kovalyonok and Savinykh would return to Earth before the expiration of the 90-day Soyuz safe operations limit.

As a result, Dzhanibekov and Gurragcha returned to Earth in Soyuz 39, landing on March 30, 1981, after a flight of 7 days, 20 hours and 42 minutes.

SOYUZ MISSION 45: SOYUZ 40

(May 14–22, 1981)

Soviet Manned Spaceflight

Leonid Popov made his second flight in space as commander of Soyuz 40, the last flight to the Salyut 6 space station, which launched on May 14, 1981. He was accompanied by Dumitru Prunariu, who with the Soyuz 40 mission became the first citizen of Romania to fly in space.

Yuri Romanenko and Dumitru Dediu served as the backup crew for the Soyuz 40 flight.

Soyuz 40 was part of the Intercosmos Program, which made use of brief flights to the Salyut 6, Salyut 7 and Mir space stations to provide opportunities for Soviet-bloc nations (and other countries allied with or friendly to the Soviet Union) to send one of their citizens into space.

During a typical Intercosmos flight, an experienced Russian cosmonaut and the visiting international cosmonaut would make a visit of about a week to the space station resident crew. The visiting cosmonaut would typically carry out a limited program of scientific research (often devised by scientists from his home country), and would participate with his Soviet hosts in ceremonies and television broadcasts touting the cooperative nature of the mission.

Short flights to the space stations in the midst of long-duration missions had become necessary with Salyut 6, which launched on September 29, 1977. Salyut 6 (and the subsequent Salyut 7) featured two docking ports, where earlier Salyuts had been equipped with only a single port. The additional docking mechanism made long-duration missions aboard the stations possible, because the second port could be used to deliver additional supplies and fuel via unmanned cargo carriers known as Progress spacecraft, which had been derived from the basic Soyuz design and adapted for automatic flight controlled by mission handlers on Earth. The Progress ships were periodically flown to the station and docked automatically for brief periods, and were then de-orbited over the Pacific Ocean at the end of their flight.

But while the redesigned Salyut and the cosmonauts themselves could be maintained in orbit for longer periods of time thanks to the Progress supply and refueling flights, the Soyuz spacecraft used to transport the crew to the station was limited to a period of 90 days in space before its onboard systems began to deteriorate. To overcome the Soyuz craft's endurance limitations, it was necessary to replace the Soyuz docked at the station with a new spacecraft before the end of each 90-day interval.

Soviet space officials translated the necessity of the brief switching flights into spaceflight opportunities for cosmonauts from nations friendly to the Soviet Union. The resulting flights were an extension of the previously unmanned Intercosmos Program, in which Soviet scientists and engineers cooperated on the development of unmanned scientific satellites with their counterparts from other nations. The manned portion of the program provided opportunities for visiting cosmonauts to receive spaceflight training and to make brief visits to the Salyut 6, Salyut 7 or Mir space stations.

During their Soyuz 40 Intercosmos flight, Popov and Prunariu visited with Salyut 6 long-duration crew members Vladimir Kovalyonok and Viktor Savinykh, who were in the midst of a two-and-a-half month stay aboard the station. In keeping with the usual routine for the flights, the four cosmonauts participated in ceremonies and speeches touting the cooperative nature of the mission, which were aired on Soviet television.

Although the normal routine for Intercosmos flights called for the visiting crew to leave their Soyuz craft at the station when they left, the usual switching of spacecraft was not necessary during the Soyuz T-4/Salyut 6 mission, because at a planned duration of about 75 days, Kovalyonok and Savinykh would return to Earth before the expiration of the 90-day Soyuz safe operations limit.

As a result, Popov and Prunariu returned to Earth aboard Soyuz 40, landing on May 22, 1981, after a flight of 7 days, 20 hours and 41 minutes.

The Soyuz 40 flight was the last mission using a Soyuz spacecraft of the original design. After Soyuz 40 the original Soyuz version was replaced by the Soyuz T series spacecraft (the Soyuz T had been gradually phased in since the launch of the unmanned Soyuz T-1 in December 1979; the first manned test of the new vehicle had been Soyuz T-2, in June 1980).

Four days after Popov and Prunariu returned to Earth, they were followed by their Salyut 6 hosts Kovalyonok and Savinykh, who had concluded their long-duration mission aboard the station.

Salyut 6 was not occupied again. In June 1981, the unmanned Cosmos 1267 module was flown to the station and automatically docked, in a test of equipment and procedures designed for use in later space stations. The two spacecraft remained linked for more than a year, until they were de-orbited over the Pacific Ocean on July 29, 1982.

SOYUZ MISSION 46: SOYUZ T-5

(May 13–December 10, 1982)

Soviet Manned Spaceflight

Launched on May 13, 1982 aboard Soyuz T-5, commander Anatoli Berezovoi and flight engineer Valentin Lebedev served as the first long-duration crew of the Salyut 7 space station, and set a new record for the longest stay in space up to that time.

Vladimir Titov and Gennady Strekalov served as the backup crew for the Soyuz T-5/Salyut 7 mission.

During their remarkable seven months aboard Salyut 7—which had launched unmanned on April 19, 1982—Berezovoi and Lebedev settled into a productive routine similar to that established by the long-duration crews that had served aboard Salyut 6. The standard tasks included scientific experiments, photography, medical and biological investigations, space station maintenance and technology experiments, and a regular program of exercise to lessen the bone density and blood plasma losses that accumulate over the course of a long-duration stay in space.

The unmanned cargo spacecraft Progress 13 was launched on May 23, and brought supplies and fuel to the station.

On June 24, 1982, Berezovoi and Lebedev received their first visitors at the station, when the Soyuz T-6 crew arrived at Salyut 7 for a brief stay as part of the Soviet Intercosmos program. The manned portion of the Intercosmos program provided short flights in space to representatives of nations friendly to the Soviet Union. The Soyuz T-6 Intercosmos flight represented a departure from past missions in the program, as it featured a crew of three cosmonauts rather than the two crew members standard on earlier flights, and the visiting cosmonaut was the first from a Western European nation. The Soyuz T-6 crew consisted of Soviet commander Vladimir Dzhanibekov, Soviet flight engineer Aleksandr Ivanchenkov, and Jean-Loup J. M. Chrétien—the first citizen of France to fly in space.

The manned Intercosmos flights had arisen from the fact that a Soyuz spacecraft could only safely remain in space for a period of about 90 days before its onboard systems would begin to deteriorate. With the advent of the Salyut 6 and later space stations which had more than one docking port and could therefore be resupplied and refueled, missions in which cosmonauts would be on the station for more than 90 days became possible. To ensure that the crew aboard the station had a reliable means of returning to Earth in the event of an emergency—and a spacecraft capable of bringing them safely home at the end of their mission—a replacement Soyuz had to be flown before the end of each 90-day interval, and switched with the Soyuz docked at the station. Soviet space officials had translated the necessity of the brief switching flights into a chance for guest cosmonauts from nations friendly to the Soviet Union to fly in space.

During a typical Intercosmos flight, the visiting crew would stay aboard the space station for about a week, and participate in a series of ceremonies and television broadcasts with the resident crew, touting the cooperative nature of the mission. The visiting international cosmonaut would also typically conduct some scientific activities, often developed by scientists from his home country.

Although the normal routine for Intercosmos flights called for the visiting crew to leave their Soyuz craft at the station when they left, the usual switching of spacecraft was not necessary in the case of Soyuz T-6, since another flight was scheduled to arrive shortly afterward and well within the required safe operations limit.

Dzhanibekov, Ivanchenkov, and Chrétien returned to Earth aboard Soyuz T-6 on July 2, after a flight of 7 days, 21 hours and 51 minutes. Their departure from the station was followed a little over a week later by the launch of the unmanned

Progress 14 cargo ship, which launched on July 10 with additional supplies for Berezovoi and Lebedev and fuel for Salyut 7.

On July 30, 1982, Lebedev ventured outside Salyut 7 to retrieve scientific experiments attached to the exterior of the station, while Berezovoi monitored his progress from the station's airlock. The first EVA from Salyut 7, their spacewalk lasted 2 hours and 33 minutes.

A second visiting crew arrived at the station on August 19, 1982. Unlike the Intercosmos flights, Soyuz T-7 featured three Soviet cosmonauts, and included research engineer Svetlana Savitskaya—the second woman to fly in space (Valentina Tereshkova had been first, during Vostok 6 in June 1963). As commander of Soyuz T-7 Leonid Popov made his third spaceflight, and he and Savitskaya shared the trip with flight engineer Aleksandr Serebrov.

Popov, Serebrov and Savitskaya visited with Berezovoi and Lebedev aboard Salyut 7 for about a week, leaving their Soyuz T-7 craft docked at the station when they departed. They landed in Soyuz T-5 on August 27.

Berezovoi and Lebedev received and unloaded two more resupply spacecraft, Progress 15, which launched on September 18, and Progress 16, which launched on October 31.

Then, at the conclusion of their remarkable stay aboard the station, Berezovoi and Lebedev returned to Earth aboard Soyuz T-7 on December 10, 1982, after 211 days, 9 hours and 5 minutes in space.

SOYUZ MISSION 47: SOYUZ T-6

(June 24–July 2, 1982)

Soviet Manned Spaceflight

Vladimir Dzhanibekov made his third spaceflight as commander of Soyuz T-6, which launched on June 24, 1982 and flew to the Salyut 7 space station. He was accompanied by flight engineer Aleksandr Ivanchenkov and Jean-Loup J. M. Chrétien—the first citizen of France to fly in space.

Leonid Kizim, Vladimir Solovyev and French cosmonaut Patrick Baudry were the backup crew for the Soyuz T-6 flight.

Soyuz T-6 was part of the Intercosmos Program, which made use of brief flights to the Salyut 6, Salyut 7 and Mir space stations to provide opportunities for countries allied with or friendly to the Soviet Union to send one of their citizens into space.

The Soyuz T-6 Intercosmos flight represented a departure from past missions in the program, as it featured a crew of three cosmonauts rather than the two crew members standard on earlier flights, and the visiting cosmonaut was the first from a Western European nation.

The manned Intercosmos flights had arisen from the fact that a Soyuz spacecraft could only safely remain in space for a period of about 90 days before its onboard systems would begin to deteriorate. With the advent of the Salyut 6 and later space stations which had more than one docking port and could therefore be resupplied and refueled, missions in which cosmonauts would be on the station for more than 90 days became possible. To ensure that the crew aboard the station had a reliable means of returning to Earth in the event of an emergency—and a spacecraft capable of bringing them home safely at the end of their mission—a replacement Soyuz had to be flown before the end of each 90-day interval, and switched with the Soyuz docked at the station. Soviet space officials had translated the necessity of the brief switching flights into spaceflight opportunities for cosmonauts from nations friendly to the Soviet Union.

During a typical Intercosmos flight, the visiting crew would stay aboard the space station for about a week, and would participate in a series of ceremonies and television broadcasts with the resident crew, touting the cooperative nature of the mission. The visiting cosmonaut also typically carried out a program of scientific research, often devised by scientists from his home country.

Although the normal routine for Intercosmos flights called for the visiting crew to leave their Soyuz craft at the station when they left, the usual switching of spacecraft was not necessary in the case of Soyuz T-6, since another flight was scheduled to arrive shortly afterward and well within the required safe operations limit for the Soyuz spacecraft.

Dzhanibekov, Ivanchenkov, and Chrétien returned to Earth aboard Soyuz T-6 on July 2, after a flight of 7 days, 21 hours and 51 minutes.

SOYUZ MISSION 48: SOYUZ T-7

(August 19–27, 1982)

Soviet Manned Spaceflight

Leonid Popov made his third flight in space as commander of Soyuz T-7, which launched on August 19, 1982. He was accompanied by flight engineer Aleksandr Serebrov, and research engineer Svetlana Savitskaya—the second woman to fly in space (Valentina Tereshkova had been first, during Vostok 6 in June, 1963).

Vladimir Vasyutin, Viktor Savinykh and Irina Pronina served as the backup crew for the Soyuz T-7 flight.

Short flights to the space station in the midst of long-duration missions had become necessary with Salyut 6, which was the first Salyut to feature two docking ports (the first five stations in the series had been equipped with a single port). The additional docking mechanism made long-duration missions aboard possible, because the second port could be used to deliver additional supplies and fuel via unmanned cargo carriers known as Progress spacecraft, which had been derived from the basic Soyuz design and were adapted for automatic flight controlled by mission handlers on Earth. The Progress ships were periodically flown to the station and docked automatically for brief periods, and were then de-orbited over the Pacific Ocean at the end of their flight.

But while the redesigned Salyut and the cosmonauts themselves could be maintained in orbit for longer periods of time thanks to the Progress supply and refueling flights, the Soyuz spacecraft used to transport the crew to the station was limited to a period of 90 days in space before its onboard systems began to deteriorate. To overcome the Soyuz craft's limited endurance, a new spacecraft would need to be flown before the end of each 90-day interval, and switched with the Soyuz docked at the station.

During their Soyuz T-7 flight, Popov, Serebrov and Savitskaya visited with the resident crew aboard Salyut 7 for about a week. At the end of their stay, they accomplished the main purpose of their mission when they left their Soyuz T-7 craft docked at the station for the resident crew to use in its return to Earth in December.

Popov, Serebrov and Savitskaya landed in Soyuz T-5 on August 27, after a flight of 7 days, 21 hours and 52 minutes.

SOYUZ MISSION 49: SOYUZ T-8

(April 20–22, 1983)

Soviet Manned Spaceflight

Launched on April 20, 1983, the Soyuz T-8 flight was intended to deliver commander Vladimir Titov and flight engineers Gennady Strekalov and Aleksandr Serebrov to the Salyut 7 space station, where, if a mishap at launch hadn't prevented them from doing so, they would have lived and worked while testing out a new module that had recently been added to the station.

Vladimir Lyakhov, Viktor Savinykh and Aleksandr Aleksandrov served as the backup crew for the Soyuz T-8 flight.

Representing a major step forward for the Soviet space station program, the Cosmos 1443 module had been launched unmanned on March 2, 1983 and was automatically docked with Salyut 7 on March 10. The module effectively doubled the living space aboard the station.

The concept of expanding an existing space station in orbit had previously been tested in the June 1981 docking of the unmanned Cosmos 1267 module and Salyut 6, which had already hosted its last crew. Cosmos 1267 and Salyut 6 had remained linked for more than a year before they were de-orbited over the Pacific Ocean in July 1982. The successful test led the way to the Cosmos 1443/Salyut 7 mission.

Titov, Strekalov and Serebrov would have been the first crew to occupy the expanded station, but the radar antenna on their Soyuz T-8—which was necessary for automatic docking with Salyut 7—had been torn loose during launch. Neither the crew nor mission controllers were aware of the problem during the early part of the flight, but it became clear when the cosmonauts arrived at the station that an automatic docking would not be possible. Titov attempted to make a manual rendezvous and docking with the station, but the effort proved fruitless, and the cosmonauts were forced to abandon the attempt and return to Earth. They landed on April 22, 1983, after a flight of 2 days and 18 minutes.

SOYUZ MISSION 50: SOYUZ T-9

(June 27–November 23, 1983)

Soviet Manned Spaceflight

Launched on June 27, 1983 aboard Soyuz T-9, commander Vladimir Lyakhov and flight engineer Aleksandr Aleksandrov lived and worked aboard the Salyut 7 space station for nearly five months as the station's second long-duration crew.

Vladimir Titov and Gennady Strekalov served as the backup crew for the Soyuz T-9/Salyut 7 mission.

Lyakhov and Aleksandrov were the first crew to occupy Salyut 7 after it had been expanded by the addition of the Cosmos 1443 module. (The Soyuz T-8 crew had been unable to dock with the station in April 1983 because of an equipment mishap at launch).

Representing a major step forward for the Soviet space station program, the Cosmos 1443 module had been launched unmanned on March 2, 1983 and was automatically docked with Salyut 7 on March 10. The module effectively doubled the living space aboard the station.

The concept of expanding an existing space station in orbit had previously been tested in the June 1981 docking of the unmanned Cosmos 1267 module and Salyut 6, which by that time had already hosted its final crew. Cosmos 1267 and Salyut 6 had remained linked for more than a year before they were de-orbited over the Pacific Ocean in July 1982. The successful test had led the way to the Cosmos 1443/Salyut 7 mission.

During their remarkable—and occasionally precarious—five month stay aboard Salyut 7, Lyakhov and Aleksandrov performed many of the tasks that had become a routine part of long-duration missions since the pioneering days of the Soviet space station program. Lyakhov was particularly well-acquainted with the regimen of experiments, observations and exercise, as he had previously served as commander of Soyuz 32/Salyut 6, the third long-duration flight to the Salyut 6 space station, in 1979.

Standard tasks for resident crews aboard the Salyut stations included scientific experiments, photography, medical and biological investigations, space station maintenance and technology experiments, and the regular program of exercise to lessen the bone density and blood plasma losses that accumulate over the course of a long-duration stay in space.

Lyakhov and Aleksandrov experienced a scary moment on July 25, when a micro-meteorite struck one of the station's windows. They were not injured, but the incident served to reinforce the dangerous nature of living and working in space for extended periods.

The entire program of long-duration missions had first been made possible with Salyut 6, which featured two docking ports, while the first five Salyut stations had been equipped with just one port. The extra docking capacity had made it possible to resupply and refuel the Salyut 6 and Salyut 7 stations via unmanned cargo carriers known as Progress spacecraft, which had been derived from the basic Soyuz design and were adapted for automatic flight controlled by mission handlers on Earth. The Progress ships were periodically flown to the station and docked automatically for brief periods, and were then de-orbited over the Pacific Ocean at the end of their flight.

But while the redesigned Salyut stations and the cosmonauts themselves could be maintained in orbit for longer periods of time thanks to the Progress supply and refueling flights, the Soyuz spacecraft used to transport the crew to the station was limited to a period of 90 days in space before its onboard systems began to deteriorate. To overcome the Soyuz craft's limited endurance, it was necessary to switch the Soyuz docked at the station for a new spacecraft before the end of each 90-day interval.

Both the Progress resupply flights and the Soyuz switching flights required the use of one of the station's two docking ports—which normally would be the port not occupied by the resident crew's spacecraft (the Soyuz T-9 that Lyakhov and Aleksandrov had used to travel to the station). With the addition of the Cosmos 1443 module, however, the second docking port was also occupied. So, although the extra room it afforded was a welcome addition, Cosmos 1443 was undocked on August 14, to accommodate the arrival of Progress 17, which launched on August 17, 1983.

Where the occasion of a Progress cargo ship arriving at the station would normally be a welcome break in the cosmonauts' routine and provide the reassurance of the station's continued safe and healthy operation, the arrival of Progress 17 represented the start of a string of calamitous events for Lyakhov and Aleksandrov. The Progress refueling process was automated, and normally proceeded without incident; but in the case of Progress 17, the introduction of fuel into the Salyut 7 propulsion system caused a fuel line aboard the station to leak, which in turn caused the system to fail.

As things turned out, the failure did not put Lyakhov and Aleksandrov in imminent danger, but during the initial confusing period in which the accident took place, mission controllers had the cosmonauts move out of Salyut 7 and into their Soyuz T-9 vehicle until it was determined that they could safely re-occupy the station.

Ironically, a similar failure had occurred during Lyakov's Soyuz 32/Salyut 6 mission in 1979. On that occasion, he and flight engineer Valeri Ryumin had been aboard Salyut 6 for about a month when the station suffered a malfunction in one of its three fuel tanks, which had resulted in a similar failure.

As was the case then, the mishap involving Progress 17 and Salyut 7 meant that any correction to the space station's orbit would have to be made by a Soyuz or Progress spacecraft.

In a further eerie parallel with his Salyut 6 experience, Lyakhov saw the next Soyuz flight to the station—with the Soyuz spacecraft that was to be switched with Soyuz T-9 to ensure his safe return to Earth—end with a perilous abort.

In the case of Soyuz 32/Salyut 6, the Soyuz 33 flight that would have provided the new Soyuz had to aborted when the Soyuz 33 vehicle's main engine failed.

Following the Progress 17 accident, Lyakhov and Aleksandrov awaited the arrival of Vladimir Titov and Gennady Strekalov, who were due to launch on September 26, 1983 and were intended to replace them on Salyut 7.

But rather than traveling to Salyut 7, Titov and Strekalov endured a fire and a near-death escape during their launch attempt. The fire began about a minute-and-a-half before the scheduled lift-off, when a valve designed to shut off the propellant line failed to close, which in turn allowed the rocket's fuel to ignite. Grown to a sizable inferno within just 12 seconds, before the automatic abort sequence could take hold, the fire trapped Titov and Strekalov in their Soyuz.

Fortunately for the cosmonauts, two quick-thinking launch controllers realized the dire circumstances and sent the necessary radio commands to abort the flight manually. As a result, Titov and Strekalov were shot away from the flaming rocket by the launch escape system, and endured as much as 17 Gs of force during a five-and-a-half minute, two-and-a-half-mile flight (G is defined as a unit of gravitational force equal to that which is exerted by the Earth's gravitational field; thus 17 Gs was equal to 17 times the amount of force exerted by the Earth's gravitational field). Just seconds after the Soyuz separated from it, the launch vehicle exploded. Miraculously, neither cosmonaut was seriously injured.

While they were undoubtedly pleased to hear of their fellow cosmonauts' miraculous escape from injury, Lyakhov and Aleksandrov faced the unpleasant fact that Titov and Strekalov would not be arriving in a replacement Soyuz vehicle—which meant that their own Soyuz T-9 vehicle would be well beyond its 90-day safe operations limit when they used it to return to Earth.

Faced with a situation that they could do little to correct, the two cosmonauts continued their mission as best they could, and carried out their activities, including several spacewalks, with admirable courage. They received additional supplies via Progress 18, which launched on October 20, 1983 and thankfully experienced a far less tumultuous encounter with the station than had its predecessor.

On November 1, 1983 Lyakhov and Aleksandrov made the first of two spacewalks to add panels to the Salyut 7's solar arrays (a task originally planned for Titov and Strekalov), spending 2 hours and 50 minutes in space. They added another 2 hours and 55 minutes to their EVA total when they finished the installation on November 3.

Finally, on November 23, 1983, Lyakhov and Aleksandrov undocked their well-over-the-limit Soyuz T-9 from Salyut 7 and began their return to Earth. Although the flight represented a major gamble, given the fact that the Soyuz craft's onboard systems generally tended to deteriorate after 90 days in space—the return flight proceeded without incident, and Lyakhov and Aleksandrov landed safely after 149 days, 10 hours and 46 minutes in space.

SOYUZ MISSION 51: SOYUZ T-10–1

(September 26, 1983)

Soviet Manned Spaceflight

As commander Vladimir Titov and flight engineer Gennady Strekalov prepared to be launched on September 26, 1983 aboard what was originally designated Soyuz T-10 (and later re-named Soyuz T-10–1), they were nearly killed when the rocket that was to have propelled them into space caught fire.

They had been scheduled to travel to the Salyut 7 space station, where they would have become the station's third long-duration crew; as things turned out, they were fortunate to survive a perilous escape without serious injury.

Leonid Kizim and Vladimir Solovyev served as the backup crew for the Soyuz T-10–1 flight.

The fire began about a minute-and-a-half before Titov and Strekalov were scheduled to lift-off. Although the cause was not immediately obvious, later investigation revealed that a valve designed to shut off the launch vehicle's propellant line had failed to close, which allowed the rocket's fuel to ignite.

Grown to a sizable inferno within just 12 seconds, before the automatic abort sequence could take hold, the fire trapped Titov and Strekalov in their Soyuz. Fortunately for the cosmonauts, two quick-thinking launch controllers recognized the severity of the circumstances and issued the radio commands necessary to abort the flight.

But while the launch escape system solved the immediate problem of removing the capsule containing the cosmonauts from the top of the explosive pile of rocket stages, it also had the unfortunate effect of catapulting them away from the launch pad with enormous force. As a result, Titov and Strekalov endured as much as 17 Gs of force during a five-and-a-half minute flight.

Just seconds after they were separated from it, the launch vehicle exploded. Titov and Strekalov landed two-and-a-half miles from the launch pad.

Miraculously, neither cosmonaut was seriously injured.

The initial relief felt by all involved upon learning that Titov and Strekalov had survived the rocket failure and massive fire was tempered somewhat by the realization that the Soyuz T-9 craft docked at Salyut 7—which Titov and Strekalov would have replaced with their Soyuz—would have to remain at the station beyond its 90-day Soyuz safe operations limit. Because the spacecraft's onboard systems tended to deteriorate after 90 days in space, a Soyuz docked at a space station for use by a long-duration crew would normally be switched with a second craft before 90 days passed.

The failed launch meant that Salyut 7 long-duration crew members Vladimir Lyakhov and Aleksandr Aleksandrov would have to return to Earth aboard the same Soyuz T-9 that they had used to travel to the station—despite the fact that the spacecraft would be in space for more than 149 days by the time of their return. Fortunately, despite the risks involved, Lyakhov and Aleksandrov landed in Soyuz T-9 without incident on November 23, 1983—so all four cosmonauts touched by the nearly disastrous launch failure ultimately survived the incident without serious injury.

Unlike the Soyuz 18 pad abort in April of 1975, the Soyuz T-10 accident—and the subsequent brief flight by Titov and Strekalov—was not classified as a spaceflight because the spacecraft failed to reach the sufficient altitude. The failed mission was re-named Soyuz T-10-1.

SOYUZ MISSION 52: SOYUZ T-10

(February 8–October 2, 1984)

Soviet Manned Spaceflight

Launched aboard Soyuz T-10 on February 8, 1984, commander Leonid Kizim, flight engineer Vladimir Solovyov and research engineer Oleg Atkov lived and worked aboard the Salyut 7 space station for nearly eight months as the station's third long-duration crew.

Vladimir Vasyutin, Viktor Savinykh and Valeri Polyakov served as the backup crew for the Soyuz T-10/Salyut 7 mission.

Shortly after their arrival at Salyut 7, Kizim, Solovyov, and Atkov received supplies via the unmanned Progress 19 cargo spacecraft, which was launched on February 21. Regular Progress flights were a necessity for sustaining the crew on the station for long periods of time; the Progress 19 launch was followed by Progress 20 in April, Progress 21 on May 8, Progress 22 on May 28, and Progress 23 on August 14. In each case, the unmanned cargo supply ship was flown to the station under the control of mission managers on Earth, and automatically docked with Salyut 7. The crew then unloaded the supplies, after which the unmanned Progress spacecraft was automatically undocked and de-orbited over the Pacific Ocean.

On April 3, Kizim, Solovyov, and Atkov received their first visitors at the station, when the Soyuz T-11 crew—commander Yuri Malyshev, flight engineer Gennady Strekalov, and Rakesh Sharma, who with the Soyuz T-11 flight became the first citizen of India to fly in space—arrived at Salyut 7 for a brief stay, as part of the Soviet Intercosmos program of providing short flights in space to representatives of nations friendly to the Soviet Union.

The manned Intercosmos flights had arisen from the fact that a Soyuz spacecraft could only safely remain in space for a period of about 90 days before its onboard systems would begin to deteriorate. With the advent of the Salyut 6 and later space stations which had more than one docking port and could therefore be resupplied and refueled, missions in which cosmonauts would be on the station for more than 90 days became possible. To ensure that the crew aboard the station had a reliable means of returning to Earth in the event of an emergency—and a spacecraft capable of bringing them safely home at the end of their mission—a replacement Soyuz had to be flown before the end of each 90-day interval, and switched with the Soyuz docked at the station. Soviet space officials had translated the necessity of the brief switching flights into a chance to offer spaceflight opportunities to cosmonauts from nations friendly to the Soviet Union.

During a typical Intercosmos flight, the visiting crew would stay aboard the space station for about a week, and participate in a series of ceremonies and television broadcasts with the resident crew, touting the cooperative nature of the mission. The visiting cosmonaut would also typically participate in scientific research or other activities designed by representatives from his home country.

In keeping with the routine established for the switching flights, Malyshev, Strekalov and Sharma left their Soyuz T-11 craft docked at Salyut 7 at the end of their flight, for Kizim, Solovyov, and Atkov to use during their return to Earth on October 2. Malyshev, Strekalov and Sharma landed in Soyuz T-10 on April 11, 1984.

Like previous long-duration crews, Kizim, Solovyov, and Atkov conducted the regular activities of cosmonauts living in space for an extended period, including scientific experiments, photography, medical and biological investigations (which were helped along by the presence of cosmonaut Oleg Atkov, a medical doctor), space station maintenance and technology experiments, and a regular program of exercise to lessen the bone density and blood plasma losses that accumulate over the course of a long-duration stay in space.

But the Soyuz T-10/Salyut 7 crew had a major additional challenge not faced by previous crews: they were assigned the job of repairing the space station's propulsion system, which had been damaged the previous summer during a refueling operation involving the unmanned Progress 17 and subsequently failed.

Kizim and Solovyov began their orbital repair duties with a 4 hour and 20 minute spacewalk on April 23, 1984, and followed up with a second EVA on April 26 that lasted 4 hours and 56 minutes, a third on April 29 that kept them outside the station for 2 hours and 45 minutes, and a fourth on May 4 that also lasted 2 hours and 45 minutes. In the course of just 12 days, they spent 14 hours and 46 minutes in EVA while working on the damaged propulsion system, and still had more work to do.

As if their remarkable repair work wasn't enough, Kizim and Solovyov also made a three hour and five minute EVA on May 18 to add a solar array to the station. They then took a break from their extraordinary spacewalking activities while they and Atkov hosted a second visiting crew, and witnessed an historic "first" in the history of space exploration.

Soyuz T-12 arrived at Salyut 7 on July 17 with commander Vladimir Dzhanibekov (who was making his fourth flight in space), flight engineer Svetlana Savitskaya (on her second flight), and research engineer Igor Volk. On her previous visit to Salyut 7, as a member of the Soyuz T-7 crew in 1982, Savitskaya had become the second woman in history to fly in space (Valentina Tereshkova had been first, during Vostok 6 in June 1963).

On July 25, 1984 Savitskaya became the first woman to make a spacewalk, when she and Dzhanibekov spent 3 hours and 35 minutes outside Salyut 7. During their historic EVA, Savitskaya and Dzhanibekov tested a specialized electron-beam hand tool for brazing, cutting, soldering and welding.

Dzhanibekov, Savitskaya and Volk left the station and returned to Earth aboard Soyuz T-12 on July 29.

Kizim and Solovyov finished their repair work on August 8, making a final EVA of five hours. Their remarkable work made the Salyut 7 propulsion system fully operational again.

With their long mission finally complete, Kizim, Solovyov, and Atkov returned to Earth in Soyuz T-11 on October 2, 1984. They had been in space for a then-record 236 days, 22 hours and 50 minutes, and Kizim and Solovyov had spent a total of 22 hours and 51 minutes in EVA during 6 spacewalks.

A great personal achievement for the three cosmonauts who had served as its crew, the Soyuz T-10/Salyut 7 mission also marked a substantial advance for the Soviet space program, demonstrating the capacity of the program's mission managers, engineers and support teams to overcome difficult challenges to maintain a continued presence in orbit.

SOYUZ MISSION 53: SOYUZ T-11

(April 3–11, 1984)

Soviet Manned Spaceflight

Yuri Malyshev served as commander of the Soyuz T-11 flight, which launched on April 3, 1984. He was accompanied by flight engineer Gennady Strekalov and Rakesh Sharma, who with the Soyuz T-11 flight became the first citizen of India to fly in space.

Anatoli Berezovoi and Georgi Grechko served as the backup crew for Soyuz T-11 (Sharma did not have a backup for the mission).

Soyuz T-11 was part of the Soviet Intercosmos program, which made use of brief flights to the Salyut 6, Salyut 7 and Mir space stations to provide opportunities for countries allied with or friendly to the Soviet Union to send one of their citizens into space.

The manned Intercosmos flights had arisen from the fact that a Soyuz spacecraft could only safely remain in space for a period of about 90 days before its onboard systems would begin to deteriorate. With the advent of the Salyut 6 and later space stations which had more than one docking port and could therefore be resupplied and refueled, missions in which cosmonauts would be on the station for more than 90 days became possible. To ensure that the crew aboard the station had a reliable means of returning to Earth in the event of an emergency—and a spacecraft capable of bringing them safely home at the end of their mission—a replacement Soyuz had to be flown before the end of each 90-day interval, and switched with the Soyuz docked at the station. Soviet space officials had translated the necessity of the brief switching flights into spaceflight opportunities for cosmonauts from nations friendly to the Soviet Union.

During a typical Intercosmos flight, the visiting crew would stay aboard the space station for about a week, and participate in a series of ceremonies and television broadcasts with the resident crew, touting the cooperative nature of the mission. The visiting cosmonaut would also typically carry out a scientific program or other activities designed by representatives from his home country.

In a highlight unique to the Soyuz T-11 visit, Rakesh Sharma demonstrated the use of yoga as a means of adapting to the weightless environment aboard Salyut 7.

In keeping with the routine established for the Soyuz switching flights, Malyshev, Strekalov and Sharma left their Soyuz T-11 vehicle docked at Salyut 7 when they left the station after visiting with long-duration crew members Leonid Kizim, Vladimir Solovyov and Oleg Atkov for about a week, so the resident cosmonauts could use the newer Soyuz during their return to Earth on October 2.

Malyshev, Strekalov and Sharma landed in Soyuz T-10 on April 11, 1984, after a flight of 7 days, 21 hours and 41 minutes.

SOYUZ MISSION 54: SOYUZ T-12

(July 17–29, 1984)

Soviet Manned Spaceflight

On July 25, 1984, while she and her Soyuz T-12 crew mates visited the Salyut 7 space station's third long-duration crew, Svetlana Savitskaya became the first woman to make a spacewalk.

As commander of Soyuz T-12, Vladimir Dzhanibekov made his fourth flight in space. He was accompanied on the historic mission by flight engineer Savitskaya (who was making her second spaceflight) and research engineer Igor Volk.

Vladimir Vasyutin, Yekaterina Ivanova and Viktor Savinykh served as the backup crew for the Soyuz T-12 mission.

Dzhanibekov, Savitskaya and Volk arrived at Salyut 7 aboard Soyuz T-12 on July 17, at the start of a brief visit with the station's third long-duration crew, Leonid Kizim, Vladimir Solovyov and Oleg Atkov.

On her previous visit to Salyut 7, as a member of the Soyuz T-7 crew in 1982, Savitskaya had become the second woman in history to fly in space (Valentina Tereshkova had been first, during Vostok 6 in June 1963).

On July 25, Dzhanibekov and Savitskaya ventured outside Salyut 7 and conducted a spacewalk of 3 hours and 35 minutes. The historic first for Savitskaya was also Dzhanibekov's first EVA; he would make a second spacewalk the following year.

The cosmonauts tested an electron beam device known as the Universalny Rabochy Instrument (URI), or Universal Hand Tool, which could be used for brazing, cutting, soldering or welding.

Savitskaya cut six sample materials with the URI, and carried out soldering and spray coating experiments. Dzhanibekov also tried out the tool, and together the cosmonauts replaced experiments on the outside of Salyut 7.

After achieving the unprecedented EVA, Dzhanibekov and Savitskaya and their Soyuz T-12 crew mate Volk visited with Kizim, Solovyov, and Atkov for several more days, and then returned to Earth aboard Soyuz T-12 on July 29, 1984, after a flight of 11 days, 19 hours and 14 minutes.

SOYUZ MISSION 55: SOYUZ T-13

(June 6–September 26, 1985)

Soviet Manned Spaceflight

As commander of Soyuz T-13, Vladimir Dzhanibekov made his fifth flight in space. He was accompanied by Viktor Savinykh (who was making his second spaceflight); together the cosmonauts resuscitated and then made a remarkable stay aboard the Salyut 7 space station, which had failed in orbit several months earlier.

Leonid Popov and Aleksandr Aleksandrov served as the backup crew for the Soyuz T-13 flight.

After supporting three long-duration crews and four visiting crews since its launch in April 1982, the Salyut 7 space station suddenly failed in orbit on February 5, 1985. At the time, it appeared that the station's useful life had come to an end, even though it remained it orbit for several months while Soviet space officials pondered the possibility of sending a crew to examine—and if possible, to repair—the apparently moribund station.

With just such a mission in mind, Dzhanibekov and Savinykh launched aboard Soyuz T-13 and made a cautious rendezvous with Salyut 7. They took an extra day to assess the station's condition before attempting to dock; because the systems controlling its flight were not in working order, the station was slowly rolling. The docking required great skill, but Dzhanibekov had had previous experience in manually docking with Salyut 7—he had performed the task during the Soyuz T-6 flight in June 1982.

Once again the veteran commander was able to link his Soyuz craft to the space station; the successful docking took place on June 8, with the help of a laser range-finder.

Slowly examining the inside of the cold, dark shell that had been Salyut 7, Dzhanibekov and Savinykh searched for some obvious cause for the station's failure. Over the course of several forays into the cocooned station, they gradually narrowed the source of the problem to a fault in one of the batteries that stored the power derived from the station's solar arrays. A control mechanism that had been designed to prevent the batteries from draining had failed, and caused all eight of the station's batteries to lose their charge.

Dzhanibekov and Savinykh were able to re-circuit the batteries so they bypassed the failed control mechanism, and electricity soon began to course through the station's systems again. The cosmonauts then made a careful assessment of the damage wrought by the four frigid months of Salyut 7's unintended hibernation, and relayed their needs for tools and equipment to mission controllers on the ground.

Replacement parts and the equipment and supplies necessary for returning the station to fully operational status were sent via the unmanned cargo supply spacecraft Progress 24, which launched on June 21, 1985.

On July 19, mission controllers launched the Cosmos 1669 module, which when docked with Salyut 7 would effectively double the station's available living and work space. The module also doubled as a supply ship; it could hold even more than a Progress spacecraft, which had been specifically designed for resupply purposes.

Salyut 7 had previously been docked with a similar module, Cosmos 1443, in March 1983, and the concept of expanding an existing space station in orbit had been successfully tested in the June 1981 docking of Cosmos 1267 and Salyut 6.

In the case of Cosmos 1669, the new module was a welcome addition for Dzhanibekov and Savinykh. The restored and expanded Salyut 7 offered much greater comfort than the cosmonauts had found at the station in the first days of their rescue mission, when the brutal frigid conditions inside the dark station limited them to stays of little more than an hour or so at a time.

On August 2, 1985, Dzhanibekov and Savinykh continued their refurbishment of Salyut 7 when they made a spacewalk of five hours while installing additional panels on the space station's solar arrays.

Cosmos 1669 remained docked with Salyut 7 for more than a month, and was then undocked to free up a docking port (the station had two, and one was occupied by the crew's Soyuz T-13 craft) for the impending arrival of Soyuz T-14. Cosmos 1669 was de-orbited on August 30.

The Soyuz T-14 crew—commander Vladimir Vasyutin, flight engineer Georgi Grechko and research engineer Alexander Volkov—arrived at the station on September 17.

After about a week of joint operations, Dzhanibekov and Soyuz T-14 crew member Georgi Grechko prepared to leave Salyut 7 aboard Soyuz T-13; Savinykh would remain on the station with Vasyutin and Volkov as the next long-duration crew.

With his new crew mate Grechko, the veteran Dzhanibekov had one final exercise to perform before concluding his remarkable fifth mission; the two cosmonauts used their Soyuz T-13 for a lengthy series of rendezvous and docking tests before their return to Earth on September 26, 1985.

From launch to landing, Dzhanibekov spent 112 days, 3 hours and 51 minutes in space during his Salyut 7 repair flight. Grechko returned to Earth after a flight of 8 days, 21 hours and 13 minutes.

Aboard Salyut 7, meanwhile, Vasyutin, Savinykh and Volkov tried to settle into the routine of previous long-duration crews, which generally included scientific experiments, photography, medical and biological research, and a program of daily exercise aimed at lessening the losses in bone density and blood plasma volume that occur with long stays in space.

On October 2, 1985, the three cosmonauts welcomed the arrival of Cosmos 1686, an unmanned module filled with supplies and equipment. Like the Cosmos 1669 module that had been docked with the station in July, Cosmos 1686 effectively doubled the available space on Salyut 7.

Although the original plan for the flight apparently called for a stay of six months, a medical crisis intervened to shorten the mission. After about a month, the cosmonauts gradually became aware that commander Vladimir Vasyutin was ill, and that his condition was growing steadily worse. By mid-November he had become too sick to carry out his daily activities, and Savinykh was placed in command.

Unsure of the breadth of the crisis and perhaps not at that time certain of what it was that had made Vasyutin ill, Soviet mission controllers and medical personnel decided to end the flight early.

Vasyutin, Savinykh and Volkov returned to Earth aboard Soyuz T-14 on November 21, 1985. The three cosmonauts were initially placed in medical isolation, but an extensive battery of tests soon revealed Vasyutin's illness to be the result of an infection unrelated to his spaceflight, and not posing any threat to his fellow crew members—to the relief of all involved. Aggressively treated with antibiotics, Vasyutin recovered fully, although he was hospitalized for a month before returning to normal activities.

Savinykh had been in space for a total of 168 days, 2 hours and 51 minutes during his remarkable Salyut 7 salvage mission and abbreviated long-duration flight. Vasyutin and Volkov had each accumulated 64 days, 21 hours and 52 minutes in space during their trip to the space station.

Salyut 7 had hosted seven crews, failed in orbit, been revived by Dzhanibekov and Savinykh, and then been occupied again. With the safe return of the five cosmonauts involved in the Soyuz T-13 and Soyuz T-14 partial crew transfer and Vasyutin's subsequent complete recovery, the Soviet space program brought to a close a remarkable chapter in its history. The impending launch of the Mir space station (on February 20, 1986) would soon consume the program's resources and attention, but the achievements of the Soyuz T-13 crew in reviving Salyut 7 remain an outstanding example of expertise and dedication in overcoming the challenges of living and working in space.

SOYUZ MISSION 56: SOYUZ T-14

(September 17–November 21, 1985)

Soviet Manned Spaceflight

Launched on September 17, 1985 to deliver crew members to the Salyut 7 space station for a long-duration stay, Soyuz T-14 was commanded by Vladimir Vasyutin, who was accompanied on the flight by flight engineer Georgi Grechko (making his third spaceflight) and research engineer Alexander Volkov.

Vasyutin, Grechko and Volkov visited with Soyuz T-13/Salyut 7 crew members Vladimir Dzhanibekov and Viktor Savinykh, who had been living aboard Salyut 7 since June, and had been responsible for reviving the station after an electrical fault had caused it to fail in orbit in February.

Aleksandr Victorenko, Gennady Strekalov and Yevgeny Saley served as the backup crew for the Soyuz T-14 mission.

After about a week in which the Soyuz T-14 and Soyuz T-13 crews conducted joint operations, Grechko prepared to return to Earth with Dzhanibekov, while

Vasyutin and Volkov joined Savinykh as the station's next long-duration crew. The partial crew exchange was followed by a lengthy series of rendezvous and docking tests by Dzhanibekov and Grechko in Soyuz T-13, and they then used Soyuz T-13 to return to Earth, on September 26, 1985.

From launch to landing, Dzhanibekov had spent 112 days, 3 hours and 51 minutes in space during his Salyut 7 repair flight. Grechko returned to Earth after a flight of 8 days, 21 hours and 13 minutes.

After their fellow cosmonauts had left the station, Vasyutin, Savinykh and Volkov settled into the routine of previous long-duration crews, performing tasks that included scientific experiments, photography, medical and biological research, and daily exercise aimed at lessening the losses in bone density and blood plasma volume that occur with long stays in space.

On October 2, 1985, the unmanned Cosmos 1686 arrived at Salyut 7. Filled with supplies and equipment, Cosmos 1686 was similar to the Cosmos 1669 module that had been docked with the station in July. Like its predecessor, Cosmos 1686 effectively doubled the available space on Salyut 7.

The addition of the new module proceeded without any difficulty, but the mission aboard the space station was not going as planned. Although the Soyuz T-14/Salyut 7 stay had apparently been originally scheduled for 6 months, a medical crisis threw the original plan into serious doubt 30 days into the flight.

By mid-October the crew had gradually been forced to accept the fact that commander Vladimir Vasyutin was ill, and that his condition was growing steadily worse. A month later, by mid-November, Vasyutin became too sick to carry out his daily activities, and Savinykh was placed in command.

Unsure of the breadth of the crisis and perhaps at that time uncertain of what it was that had made Vasyutin sick, Soviet mission controllers and medical personnel decided to end the flight early.

Vasyutin, Savinykh and Volkov returned to Earth aboard Soyuz T-14 on November 21, 1985. The three cosmonauts were initially placed in medical isolation, but an extensive battery of tests soon revealed Vasyutin's illness to be the result of an infection unrelated to his spaceflight, and not posing any threat to his fellow crew members. Aggressively treated with antibiotics, Vasyutin recovered fully, although he was hospitalized for a month before he could return to normal activities.

Savinykh had been in space for a total of 168 days, 2 hours and 51 minutes during his remarkable Salyut 7 salvage mission and abbreviated long-duration flight. Vasyutin and Volkov each spent 64 days, 21 hours and 52 minutes in space during their trip to the space station.

SOYUZ MISSION 57: SOYUZ T-15

(March 13—July 16, 1986)

Soviet Manned Spaceflight

Launched on March 13, 1986, Soyuz T-15 commander Leonid Kizim and flight engineer Vladimir Solovyov successfully carried out a unique mission in which they became the first crew ever to visit two space stations during a single spaceflight. They also served as the first crew of the space station Mir, which had been launched unmanned on February 20, 1986.

As commander of Soyuz T-15, Kizim made his third flight into space; the mission was Solovyov's second spaceflight. The two cosmonauts had previously served, with their Soyuz T-10 crew mate Oleg Atkov, as the third long-duration crew aboard Salyut 7, in 1984. During that mission, they had spent a remarkable 236 days, 22 hours and 50 minutes in space and performed 6 spacewalks while repairing the propulsion system aboard Salyut 7.

Aleksandr Victorenko and Aleksandr Aleksandrov served as the backup crew for the Soyuz T-15 flight.

During the first part of their Soyuz T-15 mission, Kizim and Solovyov traveled to Mir, where they docked with the new station and began the initial on-orbit checking of its systems and equipment.

Mir differed from the Salyut series of space stations in that it was outfitted with multiple docking ports—a design that allowed the addition of multiple modules over an extended period of time. The first five Salyuts had been equipped with a single port, and Salyut 6 and 7 had each had two, which made long-duration missions possible because a crew could dock its Soyuz at one of the station's two ports and then receive supplies and fuel via unmanned, automated Progress cargo supply spacecraft. The Progress vehicles would be docked at the space station's second port for brief periods and then de-orbited.

The concept of expanding a space station in orbit by docking an unmanned expansion module to it had been successfully tested with the Salyut 6 and Salyut 7 space stations (and in fact, at the time of the Soyuz T-15 mission, the Cosmos 1686 module was docked with Salyut 7; the module remained attached to the station when it returned to Earth in February 1991 in a fiery re-entry).

During their initial stay aboard Mir, Kizim and Solovyov received supplies and equipment via two Progress flights, Progress 25 (which launched on March 19, 1986) and Progress 26 (launched on April 23).

On May 5 Kizim and Solovyov boarded Soyuz T-15 and left Mir, and traveled to Salyut 7, where they became the first cosmonauts to visit the station since the hurried exit of the Soyuz T-14 crew the previous November.

Because of an illness that turned out to be unrelated to his spaceflight and his scheduled long-duration stay aboard Salyut 7, Soyuz T-14 commander Vladimir Vasyutin had been forced to return to Earth early with his long-duration crew mates Viktor Savinykh and Alexander Volkov. Their emergency return cut short the program of experiments that had been planned for their mission. As a result, even though Mir had been launched in the intervening months and the main objective for Kizim and Solovyov was to visit the new space station and become its first crew, they were also given the task of visiting Salyut 7 to finish the work that Vasyutin, Savinykh and Volkov had been forced to abandon, and to collect experiment samples that had been left behind.

Kizim and Solovyov remained on Salyut 7 for 51 days. In one of their most important tasks, they performed two experimental EVAs to test orbital construction procedures. In their first spacewalk, a 3 hour and 50 minute excursion on May 28, 1986, they deployed a large truss structure in a test of assembly techniques, and then folded it and returned it to the station. They also retrieved experiment samples that had been mounted on the outside of Salyut 7.

They again tested the truss assembly procedure during their second spacewalk, on May 31, spending 4 hours and 40 minutes in EVA.

While Kizim and Solovyov worked aboard Salyut 7, the unmanned Soyuz TM-1 was flown to Mir and automatically docked with the station on May 21, 1986 in a test of the new version of Soyuz spacecraft. Its flight controlled by mission managers on the ground, Soyuz TM-1 remained linked to Mir for a little more than a week, and was then undocked and de-orbited on May 30.

Then, with their work aboard Salyut 7 complete, Kizim and Solovyov returned to Mir aboard Soyuz T-15 on June 25. Salyut 7 and its attached Cosmos 1686 module were boosted to a storage orbit for potential later use, but as things worked out, Kizim and Solovyov were the last cosmonauts to visit the station. With the Cosmos 1686 module attached, Salyut 7 plummeted back to Earth in an unplanned re-entry on February 7, 1991, mostly breaking up in the atmosphere but with some pieces surviving to rain down on parts of Argentina.

The second portion of their stay aboard Mir went smoothly for Kizim and Solovyov, who settled into the more usual routine of cosmonauts on long-duration missions, which normally included scientific experiments, medical and biological research, photography, and regular exercise to combat the effects of long stays in the microgravity environment. They remained aboard Mir until their return to Earth on Soyuz T-15 on July 16, 1986, after a total flight—to Mir, to Salyut 7 and then back to Mir, and then back to Earth—of 125 days, 2 hours and 1 minute.

In addition to the remarkable dual space station visits, the Soyuz T-15 flight was also notable for being the final use of the Soyuz T series of spacecraft. With the safe landing of Soyuz T-15 and the intervening successful test of the unmanned Soyuz TM-1, the Soyuz TM series replaced the Soyuz T.

SOYUZ MISSION 58: SOYUZ TM-1

(May 21–30, 1986)

Soviet Unmanned Spaceflight

An unmanned test of the new Soyuz TM series of spacecraft, Soyuz TM-1 launched on May 21, 1986 and was automatically docked with the space station Mir, which had previously been launched unmanned on February 20, 1986.

Prior to the Soyuz TM-1 test flight, Leonid Kizim and Vladimir Solovyov had flown to Mir aboard Soyuz T-15, the last of the Soyuz T series to fly in space. The cosmonauts left Mir aboard Soyuz T-15 on May 5 and traveled to the Salyut 7 space station, where they worked for nearly two months. The entire Soyuz TM-1 flight was completed before Kizim and Solovyov returned to Mir on June 25.

The Soyuz TM series was a major upgrade to the systems and equipment of the Soyuz T, reflecting the myriad of advances in electronic systems that had begun to emerge by the mid-1980s. Among the many upgrades in the TM series, the new generation of Soyuz included state-of-the-art computer systems and improvements in the spacecraft's rendezvous, guidance, electrical and hydraulic systems.

The basic Soyuz TM craft was used to ferry cosmonauts to and from the Mir space station from the mid-1980s, starting with Soyuz TM-2 in 1987, until the last visit to Mir in 2000, and was then used to transport crews to and from the International Space Station (ISS). The last of the TM series Soyuz craft was Soyuz TM-34, which visited the ISS in 2002.

Soyuz TM-1 spent about nine days in space; it was automatically undocked from Mir and de-orbited on May 30, 1986.

SOYUZ MISSION 59: SOYUZ TM-2

(February 5–December 29, 1987)

Soviet Manned Spaceflight

Launched on February 5, 1987, Soyuz TM-2 commander Yuri Romanenko and flight engineer Aleksandr Laveykin served as the second long-duration crew of the space station Mir. Romanenko had previously spent 96 days in space aboard Salyut 6, and had also flown aboard Soyuz 38 in September, 1980. Laveykin made his first flight in space during the Soyuz TM-2/Mir 2 mission.

Romanenko and Laveykin had originally been the backup crew, but the original prime crew of Vladimir Titov and Aleksandr Serebrov were forced off the flight when Serebrov suffered a medical problem.

Romanenko and Laveykin settled into a routine of scientific experiments, astronomical observations, photography, and the regular regimen of daily exercises necessary to combat the ill effects of weightlessness during long stays in space. They also unpacked supplies and equipment that had been delivered to Mir by the unmanned Progress 27 spacecraft, which had launched on January 16, 1987. A second supply and refueling ship, Progress 28, was launched on March 3, and the cosmonauts continued to receive regular resupply flights throughout their stay, including Progress 29 (launched April 21), Progress 30 (launched May 19), Progress 31 (launched August 4), Progress 32 (September 24) and Progress 33 (November 21).

Romanenko and Laveykin experienced the first excitement of their mission—and, as things turned out, more excitement than they expected—when the first Mir expansion module, Kvant 1, was launched on March 31, 1987. Designed to facilitate a program of astronomical observations and equipped with X-ray telescopes built by the European Space Agency (ESA), Great Britain, the Netherlands and West Germany, Kvant 1 was launched unmanned, and its flight was controlled by mission managers on Earth. Unlike the modules that had been attached to the earlier Salyut space stations, Kvant 1 had a docking port as well as a docking probe, which meant that it could accommodate visits from additional Soyuz or Progress spacecraft.

The controllers responsible for the mission tried to make an automated docking on April 5, but were unsuccessful; a second try, on April 9, seemed to work at first, but it soon became apparent that only a tenuous link had been established between Kvant 1 and Mir. The failure to make a hard docking (a secure link) between the two spacecraft made it necessary for Romanenko and Laveykin to make an unscheduled EVA on April 11. The cosmonauts spent 3 hours and 35 minutes in space while inspecting the Kvant 1 docking probe and the Mir docking port, and were shocked to find that the cause of the docking failure in orbit was a woefully humble Earthly object: a twisted piece of cloth (alternately reported as a plastic bag or simply an "extraneous white object"—in any case, a piece of trash) was wedged in Mir's rear docking port.

Further investigation revealed the source of the debris: the cloth had arrived at the station aboard the Progress 28 cargo supply craft in March, and had become enmeshed in the port when the Progress operations had been closed out at the end of its flight.

Despite the strangeness of the situation and the eerie sight of Earthly trash mucking up the expensive, ultra-high tech docking port, Romanenko and Laveykin were able to strip the material away from the station, clearing the way for Kvant 1 to be successfully docked with Mir on April 12.

Having met the challenge posed by the difficult Kvant 1 docking, the cosmonauts returned to their regular duties for several months. Their next spacewalking chore was far less eventful, and their EVA task of installing a third solar panel to the station to boost its electrical capacity generally went according to plan.

On June 12, 1987 they began the second EVA of their flight, spending 1 hour and 53 minutes outside of Mir while beginning the solar array installation. They finished the job on June 16, adding another 3 hours and 15 minutes to their spacewalking total. Romanenko and Laveykin spent 8 hours and 43 minutes in EVA during their three excursions outside the station.

After completing their spacewalking chores, the cosmonauts faced a difficult moment in early July, when mission controllers and Soviet medical officials monitoring their health began to detect indications that Laveykin had developed a heart ailment. Out of concern for his health, officials at mission control decided to replace Laveykin with a cosmonaut from the Soyuz TM-3 crew, which was scheduled to travel to Mir in July.

Commanded by Aleksandr Viktorenko, the Soyuz TM-3 crew also included flight engineer Aleksandr Aleksandrov and guest cosmonaut Muhammad Faris, who with the Soyuz TM-3 flight became the first citizen of Syria to fly in space.

It was necessary to make short flights to Mir during long-duration stays so a replacement Soyuz spacecraft could be switched with the Soyuz docked at the station before the 90-day safe operation limit of the docked Soyuz was exceeded. The systems and equipment within the Soyuz tended to degrade after 90 days in space, so regular flights were scheduled in which a visiting crew would travel to the station, stay a few days, and then leave their Soyuz at the station, effectively switching it with the docked Soyuz, which they would then use to return to Earth.

Beyond the technical necessity of the short flights, Soviet space officials recognized the Soyuz switching missions as a way to provide spaceflight opportunities to citizens from countries friendly to the Soviet Union. The resulting flights were collectively known as the manned portion of the Intercosmos Program—an expansion of the previously unmanned program, in which Soviet scientists and engineers had cooperated in the development of unmanned scientific satellites with their counterparts from other nations.

During the Soyuz TM-3 Intercosmos flight, Viktorenko, Aleksandrov and Faris visited with Romanenko and Laveykin for eight days, and when they left to return to Earth on July 30, they not only left their Soyuz TM-3 behind, but also transferred Aleksandrov to Mir, to serve out the rest of the long-duration mission while Laveykin returned to Earth for further medical evaluation.

Viktorenko, Faris and Laveykin landed in Soyuz TM-2 on July 30, 1987. Laveykin was subsequently found to be free of serious heart difficulties. He had spent 174 days, 3 hours and 26 minutes in space during his stay aboard Mir with Romanenko. Viktorenko and Faris had each been in space for 7 days, 23 hours and five minutes during their visit to Mir.

Romanenko and Aleksandrov continued the work of the long-duration mission for several more months, until they were visited by the Soyuz TM-4 crew in December. Launched on December 21, Soyuz TM-4 commander Vladimir Titov, flight engineer Musa Manarov and research engineer Anatoli Levchenko visited with Romaneko and Aleksandrov for eight days, and Titov and Manarov then took over as Mir's third long-duration crew. Romanenko, Aleksandrov and Levchenko returned to Earth aboard Soyuz TM-3 on December 29, 1987. The veteran Romanenko had added another 326 days, 11 hours and 38 minutes to his career total time in space during his remarkable Soyuz TM-2/Mir flight, while Aleksandrov had been in space for 160 days, 7 hours and 17 minutes from launch to landing. Levchenko accumulated 7 days, 21 hours and 58 minutes in space during his visit to Mir.

SOYUZ MISSION 60: SOYUZ TM-3

(July 22–30, 1987)

Soviet Manned Spaceflight

Aleksandr Viktorenko commanded the flight of Soyuz TM-3, which launched on July 22, 1987. He was accompanied by flight engineer Alexander Aleksandrov and Muhammed Faris, who with the Soyuz TM-3 flight became the first citizen of Syria to fly in space.

Anatoli Solovyev, Viktor Savinykh and Munir Habib served as the backup crew for the Soyuz TM-3 mission.

Soyuz TM-3 was part of the Soviet Intercosmos program, which made use of brief flights to the Salyut 6, Salyut 7 and Mir space stations to provide opportunities for countries allied with or friendly to the Soviet Union to send one of their citizens into space.

The manned Intercosmos flights had arisen from the fact that a Soyuz spacecraft could only safely remain in space for a period of about 90 days before its onboard systems would begin to deteriorate. With the advent of the Salyut 6 and later space stations which had more than one docking port and could therefore be resupplied and refueled, missions in which cosmonauts would be on the station for more than 90 days became possible. To ensure that the crew aboard the station had a reliable means of returning to Earth in the event of an emergency—and a spacecraft capable of bringing them safely home at the end of their mission—a replacement Soyuz had to be flown before the end of each 90-day interval, and switched with the Soyuz docked at the station. Soviet space officials had translated the necessity of the brief switching flights into an opportunity to provide spaceflight opportunities to citizens of nations friendly to the Soviet Union.

During a typical Intercosmos flight, the visiting crew would stay aboard the space station for about a week, and participate in a series of ceremonies and television broadcasts with the resident crew, touting the cooperative nature of the mission. The visiting cosmonaut also typically participated in scientific or symbolic activities, often designed by representatives of his home country.

During the Soyuz TM-3 Intercosmos flight, Viktorenko, Aleksandrov and Faris visited with the resident Mir crew, Yuri Romanenko and Aleksandr Laveykin, for about a week, and then made a partial crew transfer when they left the station to return to Earth on July 30. Because of medical concerns involving Laveykin, mission controllers had decided that he should return with Viktorenko and Faris. Aleksandrov remained with Romanenko aboard Mir, to serve out the rest of the long-duration mission that Romanenko and Laveykin had begun in February 1987.

In addition to leaving Aleksandrov at the station, the Soyuz TM-3 crew also left their spacecraft, switching it for Soyuz TM-2, which they used in their return to Earth.

Viktorenko, Faris and Laveykin landed in Soyuz TM-2 on July 30, 1987. Laveykin was subsequently found to be free of serious heart difficulties. He had spent 174 days, 3 hours and 26 minutes in space during his stay aboard Mir with Romanenko. Viktorenko and Faris had been in space 7 days, 23 hours and five minutes during their visit to the station.

After serving out the rest of the long-duration stay on Mir with Romanenko, Aleksandrov returned to Earth on December 29, 1987 aboard Soyuz TM-3. From launch to landing, he had spent 160 days, 7 hours and 17 minutes during his flight as a relief crew member.

SOYUZ MISSION 61: SOYUZ TM-4

(December 21, 1987–December 21, 1988)

Soviet Manned Spaceflight

Soyuz TM-4 lifted off on December 21, 1987, with commander Vladimir Titov, flight engineer Musa Manarov and research engineer Anatoli Levchenko, and traveled to the Mir space station, where Titov and Manarov would spend the next year in space during the third long-duration Mir mission.

Alexander Volkov, Alexander Kaleri and Aleksandr Shchukin served as the backup crew for Soyuz TM-4.

Titov, Manarov and Levchenko visited with the second long-duration Mir crew, Yuri Romaneko and Aleksandr Aleksandrov, for eight days, and Titov and Manarov then took over as the third long-duration crew. Romanenko, Aleksandrov and Levchenko returned to Earth aboard Soyuz TM-3 on December 29, 1987. Levchenko accumulated 7 days, 21 hours and 58 minutes in space during his visit to Mir.

Titov and Manarov settled into a routine of scientific experiments, astronomical observations, photography, and daily exercise to minimize the losses in bone density and blood plasma volume that occur with long exposure to the microgravity environment.

In January they received their first visit of an unmanned Progress supply and refueling spacecraft, Progress 34, which had launched on January 21. The cosmonauts received regular resupply flights throughout their long stay aboard the station, including Progress 35 (launched March 24), Progress 36 (May 13), Progress 37 (July 19) and Progress 38 (September 10).

On February 26, 1988, Titov and Manarov ventured outside the space station for the first time, making a spacewalk of 4 hours and 25 minutes while replacing one of Mir's solar arrays.

They received their first visitors at the station in June, when Soviets Viktor Savinykh and Anatoly Solovyov arrived at Mir with guest cosmonaut Aleksandr Aleksandrov of Bulgaria aboard Soyuz TM-5 on June 7.

The Soyuz TM-5 flight was part of the Soviet Intercosmos program, which made use of brief flights to the Salyut 6, Salyut 7 and Mir space stations to provide opportunities for countries allied with or friendly to the Soviet Union to send one of their citizens into space.

The manned Intercosmos flights had arisen from the fact that a Soyuz spacecraft could only safely remain in space for a period of about 90 days before its onboard systems would begin to deteriorate. With the advent of the Salyut 6 and later space stations which had more than one docking port and could therefore be resupplied and refueled, missions in which cosmonauts would be on the station for more than 90 days became possible. To ensure that the crew aboard the station had a reliable means of returning to Earth in the event of an emergency—and a spacecraft capable of bringing them safely home at the end of their mission—a replacement Soyuz had to be flown before the end of each 90-day interval, and switched with the Soyuz docked at the station. Soviet space officials had translated the necessity of the brief switching flights into a chance to offer spaceflight opportunities to cosmonauts from nations friendly to the Soviet Union.

During a typical Intercosmos flight, the visiting crew would stay aboard the space station for about a week, and participate in a series of ceremonies and television broadcasts with the resident crew, touting the cooperative nature of the mission. The visiting cosmonaut also typically carried out a scientific program or other activities devised by representatives from his home country.

Soyuz TM-5 generally followed the routine established by earlier Intercosmos flights, although it lasted slightly longer. Savinykh, Solovyov, and Aleksandrov visited with Titov and Manarov for a little more than a week, and then left their Soyuz TM-5 spacecraft docked to the station when they returned to Earth aboard Soyuz TM-4 on June 17, 1988.

Later in June, a problem arose with one of the X-ray instruments in the Kvant 1 module, imperiling future astronomical observations. Titov and Manarov made a spacewalk of 5 hours and 10 minutes on June 30 to repair the problem.

A second visiting crew arrived in August, aboard Soyuz TM-6, which was both an Intercosmos flight and a partial crew transfer mission. Launched on August 29, the Soyuz TM-6 crew—commander Vladimir Lyakhov, research engineer Valeri Polyakov and guest cosmonaut Abdul Ahad Mohmand of Afghanistan—visited with Titov and Manarov for a little more than a week.

The flight had been hurriedly prepared, against the background of the deteriorating Soviet invasion of Afghanistan. Mohmand was a pilot in the Afghan Air Force, which was supported by the Soviet regime. In sharp contrast to the bloody fighting on Earth, a highlight of the Soyuz TM-6 visit to Mir occurred when Mohmand read from the Koran.

At the end of the visiting flight, Polyakov transferred to Mir to begin a long-duration stay that began with Titov and Manarov and continued until April 1989, when he returned to Earth with Alexander Volkov and Sergei Krikalyov aboard Soyuz TM-7.

Lyakhov and Mohmand left their Soyuz TM-6 docked at Mir when they left on September 6 in Soyuz TM-5, in keeping with the standard procedure for a Soyuz switching flight. There was nothing routine about their return to Earth, however; a series of computer errors forced them to stay in space an extra day, and they then made a perilous—but safe—landing on September 7.

Titov, Manarov and Polyakov continued to serve aboard Mir.

On October 20, 1988 Titov and Manarov made another spacewalk to fix the still-malfunctioning Kvant 1 X-ray telescope, adding another 4 hours and 12 minutes to their EVA total. The cosmonauts also tested the new Orlan-DMA spacesuit, the latest iteration in what had originally been known as the Orlan-D suit, first introduced in 1977. An earlier upgrade, the Orlan-DM, had been in use since 1985.

A third visiting crew—which included the cosmonauts who would replace Titov and Manarov—arrived at Mir on November 26, aboard Soyuz TM-7. Commanded by Alexander Volkov, who would remain aboard Mir with fellow Soviet cosmonaut and Soyuz TM-7 crew mate Sergei Krikalyov and the already-in-residence Valeri Polyakov as the fourth long-duration Mir crew, Soyuz TM-7 also brought guest cosmonaut Jean-Loup J. M. Chrétien of France to the station.

Although they had nearly reached the end of their year-long stay on Mir, Titov and Manarov were still aboard the station on December 9, 1988 to witness an important milestone in space history, as Chrétien became the first person from a country other than the Soviet Union or the United States to make a spacewalk. He and Volkov spent 5 hours and 57 minutes in EVA; among their activities, they assembled a grid structure outside the station in a test of space construction procedures.

Chrétien's stay at Mir was known as the Aragatz mission—a joint project of the Soviet space program and the national space agency of France, Centre National d'Études Spatiales (CNES).

Their long stay complete, Titov and Manarov turned control of Mir over to Volkov, Krikalyov and Polyakov and returned to Earth with Chrétien on December 21, 1988, aboard Soyuz TM-6. Chrétien was in space for 24 days, 18 hours and 7 minutes during his historic Soyuz TM-7/Soyuz TM-6 flight.

During their remarkable Soyuz TM-4/Mir long-duration mission, Titov and Manarov became the first individuals to spend more than a year in space, as they accumulated a total of 365 days, 22 hours and 39 minutes, including 13 hours and 47 minutes in EVA, during the flight.

SOYUZ MISSION 62: SOYUZ TM-5

(June 7–17, 1988)

Soviet Manned Spaceflight

Launched on June 7, 1988, the Soyuz TM-5 flight to the space station Mir was commanded by Viktor Savinykh. He was accompanied by flight engineer Anatoly Solovyov and guest cosmonaut Aleksandr P. Aleksandrov of Bulgaria.

Vladimir Lyakhov, Andrey Zaytsev and Krasimir Stoyanov served as the backup crew for the Soyuz TM-5 mission.

Soyuz TM-5 was part of the Soviet Intercosmos program, which made use of brief flights to the Salyut 6, Salyut 7 and Mir space stations to provide opportunities for countries allied with or friendly to the Soviet Union to send their citizens into space.

The manned Intercosmos flights had arisen from the fact that a Soyuz spacecraft could only safely remain in space for a period of about 90 days before its onboard systems would begin to deteriorate. With the advent of the Salyut 6 and later space stations which had more than one docking port and could therefore be resupplied and refueled, missions in which cosmonauts would be on the station for more than 90 days became possible. To ensure that the crew aboard the station had a reliable means of returning to Earth in the event of an emergency—and a spacecraft capable of bringing them safely home at the end of their mission—a replacement Soyuz had to be flown before the end of each 90-day interval, and switched with the Soyuz docked at the station. Soviet space officials had translated the necessity of the brief switching flights into spaceflight opportunities for citizens of nations friendly to the Soviet Union.

During a typical Intercosmos flight, the visiting crew would stay aboard the space station for about a week, and the guest cosmonaut would typically carry out a scientific program or other activities devised by representatives of his home country. The visiting crew would also participate in a series of ceremonies and television broadcasts with the resident crew, touting the cooperative nature of the mission.

Soyuz TM-5 generally followed the routine established by earlier Intercosmos flights, although it lasted slightly longer. Savinykh, Solovyov, and Aleksandrov visited with Mir's third long-duration crew, Vladimir Titov and Musa Manarov, for a little more than a week, and then left their Soyuz TM-5 spacecraft docked to the station when they returned to Earth aboard Soyuz TM-4 on June 17, 1988, after a total flight of 9 days, 20 hours and 10 minutes in space.

SOYUZ MISSION 63: SOYUZ TM-6

(August 29–September 7, 1988)

Soviet Manned Spaceflight

As commander of the Soyuz TM-6 flight to the space station Mir, Vladimir Lyakhov made his third flight in space. He was accompanied by Soviet research cosmonaut Valeri Polyakov and guest cosmonaut Abdul Ahad Mohmand—who with the Soyuz TM-6 flight became the first citizen of Afghanistan to fly in space.

Anatoli Berezovoi, Gherman Arzamzov, and M. Dauran served as the backup crew for Soyuz TM-6.

Launched on August 29, 1988, the Soyuz TM-6 flight was both a part of the Soviet Intercosmos program and a partial crew transfer mission.

The Intercosmos program made use of brief flights to the Salyut 6, Salyut 7 and Mir space stations to provide opportunities for countries allied with or friendly to the Soviet Union to send their citizens into space.

The manned Intercosmos flights had arisen from the fact that a Soyuz spacecraft could only safely remain in space for a period of about 90 days before its onboard systems would begin to deteriorate. With the advent of the Salyut 6 and later space stations which had more than one docking port and could therefore be resupplied and refueled, missions in which cosmonauts would be on the station for more than 90 days became possible. To ensure that the crew aboard the station had a reliable means of returning to Earth in the event of an emergency—and a spacecraft capable of bringing them safely home at the end of their mission—a replacement Soyuz had to be flown before the end of each 90-day interval, and switched with the Soyuz docked at the station. Soviet space officials had translated the necessity of the brief switching flights into spaceflight opportunities for cosmonauts from nations friendly to the Soviet Union.

During a typical Intercosmos flight, the visiting crew would stay aboard the space station for about a week, and participate in a series of ceremonies and television broadcasts with the resident crew, touting the cooperative nature of the mission. The visiting cosmonaut would also typically carry out a program of activities designed by representatives of his home country.

In the case of Soyuz TM-6, the flight had been hurriedly prepared, against the background of the deteriorating Soviet invasion of Afghanistan. Mohmand was a pilot in the Afghan Air Force, which was supported by the Soviet regime. In sharp contrast to the bloody fighting on Earth, a highlight of the Soyuz TM-6 visit to Mir occurred when Mohmand read from the Koran, instilling a spiritual overtone in the tense circumstances under which the mission had been conceived and flown.

The partial crew transfer involved Polyakov remaining aboard Mir at the end of the brief Soyuz TM-7 visit. Polyakov began his own long-duration stay on Mir with the station's third long-duration crew, Vladimir Titov and Musa Manarov, and then remained on the station when they left in December 1987 to continue his stay with the next long-duration crew, Alexander Volkov and Sergei Krikalyov. Polyakov, Volkov, and Krikalyov would return to Earth in April 1989, aboard Soyuz TM-7.

Lyakhov and Mohmand left their Soyuz TM-6 docked at Mir when they left on September 6 in Soyuz TM-5, in keeping with the routine procedure for a Soyuz switching flight. There was nothing routine about their return to Earth, however.

On the return trip, a computer fault caused Lyakhov and Mohmand to miss their first opportunity for re-entry. A second attempt was marred by further complications, and neither Lyakhov or mission controllers on Earth were able to correct the situation.

Lyakhov and Mohmand spent a difficult twenty-four hours in orbit while awaiting a third attempt at returning to Earth. Ground controllers were finally able to overcome the software glitch that had caused the initial difficulty, and Lyakhov and Mohmand landed safely on September 7, 1988.

Polyakov returned to Earth on April 27, 1989 with Alexander Volkov and Sergei Krikalyov aboard Soyuz TM-7, after a remarkable 240 days, 22 hours and 35 minutes in space.

Lyakhov and Mohmand accumulated 8 days, 20 hours and 27 minutes in space during their Soyuz TM-6/Soyuz TM-5 mission.

SOYUZ MISSION 64: SOYUZ TM-7

(November 26, 1988–April 27, 1989)

Soviet Manned Spaceflight

Soyuz TM-7 lifted off on November 26, 1988, with commander Alexander Volkov, flight engineer Sergei Krikalyov, and guest cosmonaut Jean-Loup J. M. Chrétien of France. The Soyuz TM-7 crew traveled to the Mir space station, where Volkov and Krikalyov began their stay as the fourth long-duration Mir crew, and Chrétien participated in an historic spacewalk during a stay of slightly more than three weeks.

Aleksandr Victorenko, Aleksandr Serebrov and Michel Tognini served as the backup crew for the Soyuz TM-7 mission.

Soyuz TM-7 was both a crew replacement flight and a part of the Soviet Intercosmos program. Volkov and Krikalyov replaced the third long-duration Mir crew, Vladimir Titov and Musa Manarov, and joined Valeri Polyakov, who had arrived at Mir in August aboard Soyuz TM-6. They would serve with Polyakov until all three cosmonauts returned to Earth in April 1989.

Chrétien's flight aboard Soyuz TM-7—his second spaceflight—was part of the Intercosmos program, which made use of brief flights to the Salyut 6, Salyut 7 and Mir space stations to provide opportunities for countries allied with or friendly to the Soviet Union to send their citizens into space.

The manned Intercosmos flights had arisen from the fact that a Soyuz spacecraft could only safely remain in space for a period of about 90 days before its onboard systems would begin to deteriorate. With the advent of the Salyut 6 and later space stations which had more than one docking port and could therefore be resupplied and refueled, missions in which cosmonauts would be on the station for more than 90 days became possible. To ensure that the crew aboard the station had a reliable means of returning to Earth in the event of an emergency—and a spacecraft capable of bringing them safely home at the end of their mission—a replacement Soyuz had to be flown before the end of each 90-day interval, and switched with the Soyuz docked at the station. Soviet space officials had translated the necessity of the brief switching flights into an opportunity to provide spaceflight opportunities for citizens from nations friendly to the Soviet Union.

During a typical Intercosmos flight, the visiting crew would stay aboard the space station for about a week, and participate in a series of ceremonies and television broadcasts with the resident crew, touting the cooperative nature of the mission. The visiting international cosmonaut would also typically conduct a limited scientific program or other symbolic activities designed by representatives from his home country.

In the case of Soyuz TM-7, however, Chrétien's visit to Mir was part of a larger science mission jointly devised by the Soviet space program and the national space agency of France, Centre National d'Études Spatiales (CNES). Known as the Aragatz mission, the program called for Chrétien to stay some three weeks at the Soviet space station—in contrast to the usual international visits which generally lasted about a week—and to make an historic spacewalk.

Thus on December 9, 1988 Jean-Loup J. M. Chrétien became the first individual from a country other than the Soviet Union or the United States to make a spacewalk. He and Volkov worked outside Mir for 5 hours and 57 minutes. They installed a temporary grid on the outside of the station, and, although it failed to deploy properly at first, they managed to get the large frame to extend to its full length.

Chrétien also carried out his assigned program of scientific experiments and photography during his time aboard Mir. He returned to Earth with Titov and Manarov on December 21, 1988 aboard Soyuz TM-6, after a flight of 24 days, 18 hours and 7 minutes.

Volkov and Krikalyov remained aboard Mir, with the already-in-residence Valeri Polyakov, as the station's fourth long-duration crew. They returned to Earth on April 27, 1989 aboard Soyuz TM-7.

During his unique stay aboard Mir, which overlapped the Titov/Manarov and Volkov/Krikalyov missions, Polyakov spent a remarkable 240 days, 22 hours and 35 minutes in space.

Volkov and Krikalyov each spent 151 days, 12 hours and 8 minutes in space during their Soyuz TM-7/Mir long-duration flight, and Volkov spent 5 hours and 57 minutes in EVA during his spacewalk with Chrétien.

SOYUZ MISSION 65: SOYUZ TM-8

(September 6, 1989–February 19, 1990)

Soviet Manned Spaceflight

Launched on September 6, 1989, Soyuz TM-8 commander Aleksandr Viktorenko and flight engineer Aleksandr Serebrov served as the fifth long-duration crew of the space station Mir, and conducted the first test of the Sredstvo Peredvizheniy Kosmonavtov (SPK; translates to "Cosmonaut Maneuvering Equipment" in English) EVA backpack in space. They also were aboard Mir while the station received the Kvant 2 expansion module.

Anatoly Solovyov and Aleksandr Balandin served as the backup crew for the Soyuz TM-8/Mir mission.

During their first days aboard the station, Viktorenko and Serebrov unloaded supplies that had been delivered to Mir by the unmanned cargo spacecraft Progress M-1, which had launched on August 23, 1989 and been automatically docked with the station by flight controllers on Earth.

The first in the new "M" series of Progress resupply and refueling spacecraft, Progress M-1 featured a number of important improvements over the first generation Progress spaceships. Most prominently, it could carry more supplies and equipment, and it was also capable of collecting and storing its own power, thanks to its solar panels. The Progress M craft was also outfitted with an improved docking mechanism that enabled it to dock at the station's forward docking port, while the earlier version of the Progress craft could only be docked at Mir's rear port.

Viktorenko and Serebrov settled into a productive routine on Mir, conducting scientific studies, taking photographs and performing the daily exercise necessary to combat the losses in blood plasma and bone density that are common side effects of long stays in space.

After a series of frustrating delays, the Kvant 2 module was launched on November 26, 1989, aboard a D-1 rocket. Mission controllers were able to overcome a malfunction in the module's solar array mechanism, only to have a computer glitch aboard Mir force them to forego their first attempt to dock the unmanned Kvant 2 with the space station on December 2. The computer difficulty was also solved, after a good deal of analysis and cooperation between the crew and mission controllers, and Kvant 2 was successfully docked with Mir on December 6, 1989.

In addition to supplies and fuel, the next Progress craft to be flown to Mir also carried a commercial protein crystal growth experiment that was flown on behalf of the American company Payload Systems Inc. Launched aboard Progress M-2 on December 20, 1989, the experiment was the first in a series of flights the Soviets would deploy aboard Mir for Payload Systems.

On January 8, 1990, Viktorenko and Serebrov conducted the first of five EVAs they would make during their long stay at the station, spending 2 hours and 56 minutes outside of Mir while retrieving Mir Environmental Effects Payloads (MEEPs) packages from the outside of the station and installing equipment. They made a second, similar spacewalk on January 11, spending another 2 hours and 54 minutes in EVA. On January 26 they made a third spacewalk, spending 3 hours and 2 minutes in EVA while attaching hardware to the outside of the station that would later be used in the test of the SPK backpacks.

The SPK unit was basically equivalent to the U.S. Manned Maneuvering Unit (MMU); in both cases, the devices were designed to allow the wearer to propel himself around freely in space. Although they were designed to eliminate the necessity of a tether, a tether was used during the test as a safety precaution.

Serebrov had trained with the EVA backpack for several years before his flight to Mir, and on February 1, 1990 he put his long training to use in the first-ever test of the SPK in space. He successfully demonstrated the capabilities of the device during a 4 hour, 59 minute EVA, while Viktorenko documented the exercise on videotape. On February 5, 1990 the cosmonauts carried out a second test of the SPK, adding another 3 hours and 45 minutes to their EVA total.

Over the course of their 5 spacewalks together, Viktorenko and Serebrov accumulated a total of 17 hours and 36 minutes in EVA.

With their spacewalking duties successfully completed and Kvant-2 connected and operational, Viktorenko and Serebrov prepared to finish their remaining activities and leave the station. They received visitors on February 11, when their replacements, Anatoly Solovyov and Aleksandr Balandin, arrived. Then, after a little more than a week of joint operations, Viktorenko and Serebrov left Mir and returned to Earth in Soyuz TM-8 on February 19, 1990.

At the start of their return trip, they performed one last, unscheduled chore, as they carefully photographed the Soyuz TM-9 craft that had brought Solovyov and Balandin to Mir. Soyuz TM-9 had experienced a mishap at launch, when several pieces of the craft's insulation were partially torn away and were left hanging from the vehicle's Descent Module. The photos that Viktorenko and Serebrov took as they left Mir aboard Soyuz TM-8 later helped mission controllers to devise a solution to the Soyuz TM-9 insulation problem.

During their productive Soyuz TM-8/Mir 5 long-duration mission, Viktorenko and Serebrov accumulated 166 days, five hours and 58 minutes in space, including 17 hours and 36 minutes in EVA during their five spacewalks.

SOYUZ MISSION 66: SOYUZ TM-9

(February 11–August 9, 1990)

Soviet Manned Spaceflight

Launched on February 11, 1990, Soyuz TM-9 commander Anatoly Solovyov and flight engineer Aleksandr Balandin overcame a mishap at launch to become the sixth long-duration crew of the space station Mir, spending six months in space and making two spacewalks to perform repair duties in orbit.

Gennady Strekalov and Gennadi Manakov served as the backup crew for the Soyuz TM-9/Mir 6 mission.

At launch on February 11, three of the insulation blankets packed around the Soyuz TM-9 Descent Module were partially jarred loose by the force of the lift-off. The spacecraft was equipped with eight of the blankets to protect its heat shield, which in turn, would protect the craft during re-entry into the Earth's atmosphere. In addition to potentially exposing the heat shield to damage, the three loose blankets also threatened to interfere with the spacecraft's orientation procedures when the crew prepared to fire the craft's retrorockets when it came time to return to Earth.

Soyuz TM-8 crew members Aleksandr Viktorenko and Aleksandr Serebrov photographed Soyuz TM-9 as they left Mir at the end of their long-duration stay on February 19, to give mission controllers a close-up view of the problem with the blankets. After carefully analyzing the situation, Soviet space officials decided to have Solovyov and Balandin make a spacewalk to re-attach the loose blankets firmly.

Solovyov and Balandin carried out their less nerve-wracking assignments for most of their stay at the space station, since the equipment they would need to repair their Soyuz TM-9 Descent Module would not arrive at Mir until June, aboard the Kristall expansion module.

In the interim, the cosmonauts conducted scientific experiments—including a study of Japanese quail eggs called "Inkubator" that proved an embryo can be brought to full term in the microgravity environment—and participated in the daily exercise necessary to combat the ill-effects common to long-duration stays in space. They also had to focus on the task of installing new computers on the station.

Solovyov and Balandin received visits from two unmanned Progress resupply and refueling spacecraft, Progress M-3, which launched on March 1, 1990, and Progress 42—the last of the original, first-generation Progress cargo craft—which launched on May 6.

Launched unmanned on May 31, 1990, the Mir Kristall expansion module included a docking port that was to have accommodated the Soviet space shuttle Buran. (As things subsequently turned out, development of the shuttle program ended after the 1991 dissolution of the Soviet Union, and the one completed Soviet shuttle, the Buran, made only just one unmanned test flight in space. Buran never did visit Mir).

Like the earlier docking of the Kvant 2 module in December 1989, docking the Kristall module to Mir required two attempts. During the first try, on June 6, a malfunction in one of the module's booster rockets forced mission controllers to forego the attempt to dock. For the second attempt, on June 10, they used the module's backup propulsion system, and successfully docked Kristall to Mir.

On July 17, 1990, Solovyov and Balandin made a spacewalk of seven hours while examining their Soyuz TM-9 spacecraft. They carefully fastened several of the misplaced insulation blankets to the Descent Module, and searched for additional signs of damage—fortunately finding none. Reassured by both their inspection and their apparently successful repair work, they returned to Mir via the airlock on the new Kvant 2 module, exhausted after the long and tense EVA.

Adding to the tension of their repairs on the Soyuz craft, the cosmonauts had difficulty maneuvering around the exterior of the large Kvant 2 module, and their activities at the docked Soyuz were out of sight of the TV cameras mounted on Mir, so mission controllers on Earth could not view the cosmonauts while they worked. Solovyov and Balandin videotaped their repairs to the Soyuz TM-9 Descent Module for their support staff to evaluate after the conclusion of the long EVA.

Although the repair work went well and the seven hour spacewalk set a new record for the longest EVA of the Soviet space program, the cosmonauts faced another difficulty when they entered the airlock at the end of the EVA.

After the cosmonauts entered the airlock, they would normally seal and repressurize the compartment so they could pass from the vacuum of space back into the breathable atmosphere aboard the Kvant 2 module. But once inside the airlock, Solovyov and Balandin found that they could not close its outer hatch. After several attempts, they had to accept the fact that the hatch simply would not close—and as a result, they couldn't enter Kvant 2 without allowing the atmosphere inside the module to escape into space.

After carefully considering their predicament, Solovyov and Balandin improvised a novel solution. They purposely depressurized Kvant 2, opened the inner hatch of the airlock, entered Kvant 2, sealed the hatch between the airlock and Kvant 2 (leaving the airlock's recalcitrant outer hatch open), and then repressurized the Kvant 2 module—in effect using Kvant 2 itself as a very large airlock, and bypassing the actual, faulty airlock outside the module.

With admirable professionalism and grace under great pressure, Solovyov and Balandin had managed to overcome the airlock dilemma on-the-fly, even after their long, tense spacewalk. But even after they were safely back on board the space station, the problem with the airlock hatch remained; so, on July 26, 1990 they made another EVA, adding 3 hours and 31 minutes to their total spacewalking time while making repairs the balky hatch.

Solovyov and Balandin received visitors on August 1, when their replacements, Gennady Strekalov and Gennadi Manakov, arrived at Mir aboard Soyuz TM-10. The four cosmonauts spent a little more than a week together at the station, and Solovyov and Balandin then left in Soyuz TM-9 and returned to Earth on August 9, 1990.

During their long and occasionally tumultuous Soyuz TM-9/Mir 6 long-duration mission, Solovyov and Balandin accumulated 179 days, 2 hours and 19 minutes in space, including 10 hours and 31 minutes in EVA.

At 7 hours, their first EVA set a new record for the longest spacewalk of the Russian space program—a record that still stood nearly 20 years later, at the start of 2009.

SOYUZ MISSION 67: SOYUZ TM-10

(August 1–December 10, 1990)

Soviet Manned Spaceflight

Launched aboard Soyuz TM-10 on August 1, 1990, commander Gennadi Manakov and flight engineer Gennady Strekalov traveled to the Mir space station, where they served as the Mir 7 long-duration crew.

Viktor Afanasyev and Musa Manarov served as the backup crew for Soyuz TM-10/Mir 7.

When they arrived at the station Manakov and Strekalov visited with Mir 6 crew members Aleksandr Balandin and Anatoly Solovyov, who left Mir eight days later aboard Soyuz TM-9, ending their stay of more than 179 days in space.

Manakov and Strekalov conducted scientific work that included astronomical observations, biological and biotechnology experiments and materials processing investigations. They also performed space station maintenance tasks and participated in the regular physical exercise necessary to maintaining their own health during their long stay.

They received visits from the unmanned Progress M-4, which was launched with supplies and equipment on August 15, 1990, and Progress M-5, which brought additional cargo that the cosmonauts unpacked in October.

On October 29, 1990, Manakov and Strekalov made a spacewalk of 2 hours and 45 minutes to make repairs to the hatch on Mir's Kvant 2 module.

As they neared the end of their long flight, Manakov and Strekalov welcomed their replacements, Viktor Afanasyev and Musa Manarov, who arrived at Mir aboard Soyuz TM-11 with Japanese broadcast journalist Toyohiro Akiyama. The Tokyo Broadcasting System (TBS) paid an estimated total of $28 million to arrange the flight to Mir for Akiyama, who made regular daily broadcasts from the station during his eight days aboard Mir.

Then, having completed their long mission, Manakov and Strekalov returned to Earth on December 10, 1990 in Soyuz TM-10 with Akiyama. They had spent 130 days, 19 hours and 36 minutes in space during the Soyuz TM-10/Mir 7 mission. Akiyama accumulated 7 days, 21 hours and 55 minutes in space during his visit to Mir.

SOYUZ MISSION 68: SOYUZ TM-11

(December 2, 1990–May 26, 1991)

Soviet Manned Spaceflight

Viktor Afanasyev made his first trip into space as commander of the Soyuz TM-11/Mir 8 crew, which he began on December 2, 1990 with his long-duration crew mate and flight engineer Musa Manarov and visiting cosmonaut Toyohiro Akiyama, a Japanese broadcast journalist who spent a little over a week aboard Mir. With his trip aboard Soyuz TM-11, Akiyama became the first citizen of Japan to fly in space.

The backup crew for the Soyuz TM-11 flight was Anatoli Artsebarsky, Sergei Krikalyvov and Ryoko Kikuchi.

At Mir, the Soyuz TM-11 crew visited with Mir 7 cosmonauts Gennadi Manakov and Gennady Strekalov for about eight days. Akiyama made daily reports on life at the station during his stay, and he then returned to Earth with Manakov and Strekalov on December 10, 1990 in Soyuz TM-10.

Afanasyev and Manarov remained aboard Mir to begin their six month stay as the station's eighth long-duration crew.

During their long stay, they conducted scientific experiments, made astronomical observations, and performed the regular exercise necessary to combat losses in blood plasma volume and bone density that regularly occur during long stays in space.

On January 7, 1991, Afanasyev and Manarov made the first of 4 spacewalks they would make during their Mir 8 mission, venturing outside of the station for 5 hours and 18 minutes while continuing repairs on the hatch of Mir's Kvant 2 module, which Manakov and Strekalov had begun 2 months earlier.

Afanasyev and Manarov also made EVAs on January 23, January 26, and April 25, to complete a variety of construction tasks on the outside of the station, including the installation of supports on Mir's solar arrays. During their 4 Mir 8 spacewalks, they spent a total of 19 hours and 36 minutes in EVA.

On May 18, Anatoli Artsebarsky and Sergei Krikalyov—who would replace Afanasyev and Manarov aboard the station—arrived at Mir aboard Soyuz TM-12 with Helen Sharman—the first citizen of the United Kingdom to fly in space.

Afanasyev and Manarov returned to Earth with Sharman on May 26, 1991, after spending 175 days, 1 hour and 52 minutes in space.

SOYUZ MISSION 69: SOYUZ TM-12

(May 18–October 10, 1991)

Soviet Manned Spaceflight

Launched on May 18, 1991, commander Anatoli Artsebarsky and flight engineer Sergei Krikalyov traveled to the Mir space station with Helen Sharman—the first citizen of the United Kingdom to fly in space.

Alexander Kaleri, Alexander Volkov, and Timothy Mace served as the backup crew for the Soyuz TM-12 flight.

Sharman's trip to Mir was also known as the Juno Project. With Artsebarsky and Krikalyov, she visited with Mir 8 long-duration crew members Viktor Afanasyev and Musa Manarov for about eight days, and then she returned to Earth with Afanasyev and Manarov on May 26, 1991 while Artsebarsky and Krikalyov took over as the Mir 9 crew.

Artsebarsky and Krikalyov made six long spacewalks during their Mir 9 mission. They first ventured outside the station on June 24, to do some external maintenance work on Mir and to replace an antenna. On June 28, they deployed an experiment on the outside of the station during a spacewalk of 3 hours and

24 minutes. They then made four EVAs devoted to construction tasks. The first 3 of those EVAs, on July 15, July 19 and July 23, each lasted more than 5 hours, and their final spacewalk together, on July 27, 1991, took 6 hours and 49 minutes.

During the long flight, Artsebarsky and Krikalyov also conducted scientific experiments, performed space station maintenance and engineering tasks, and exercised regularly to combat the losses in blood plasma and bone density that regularly occur during long stays in space. They also unloaded supplies from the unmanned Progress-M 9 cargo spacecraft, which arrived at the station in late August.

While Artsebarsky and Krikalyov continued their flight in orbit, a series of turbulent events took place within the Soviet Union. An anti-reform group attempted a coup against Soviet leader Mikhail Gorbachov, which ultimately failed. The coup attempt and its aftermath led to the ultimate dissolution of the Soviet Union by the end of 1991.

The changes brought on by the crumbling of the totalitarian Soviet regime had a major impact on the nation's space program. Mission planners decided to accelerate the use of the Mir space station as a destination for "paying customers," as had been the case with Japanese journalist Toyohiro Akiyama, whose company had paid to have him visit the station during Soyuz TM-11 in December 1990.

For the Soyuz TM-13 flight, one Russian cosmonaut would visit Mir with two visitors—which would as result make it necessary for Krikalyov to extend his stay aboard the station, to make up one-half of the next long-duration crew.

Thus when Artsebarsky and Krikalyov welcomed cosmonaut Alexander Volkov, Franz Viehböck—the first citizen of Austria to fly in space—and Toktar Aubakirov, who with his Soyuz TM-13 flight became the first representative of the Kazakhstan Republic to fly in space, Krikalyov was in fact welcoming his fellow crew mate for the Mir 10 long-duration stay.

Artsebarsky bid farewell to Krikalyov on October 10, 1991 and returned to Earth aboard Soyuz TM-12 with Aubakirov and Viehböck, after a flight of 144 days, 14 hours and 22 minutes in space. Aubakirov and Viehböck each accumulated 7 days, 22 hours and 13 minutes during their visit to Mir.

Krikalyov remained aboard Mir as flight engineer for Mir 10, with Volkov serving as commander, until they returned to Earth in March 1992.

SOYUZ MISSION 70: SOYUZ TM-13

(October 2, 1991–March 25, 1992)

Russian Manned Spaceflight

With the Soyuz TM-13 flight, which launched on October 2, 1991, the Russian space program sent two visiting cosmonauts to the Mir space station with the Russian commander of the flight, Alexander Volkov. The visiting space travelers were Franz Viehböck—the first citizen of Austria to fly in space—and Toktar Aubakirov, who with his Soyuz TM-13 flight became the first representative of the Kazakhstan Republic to fly in space.

Aleksandr Viktorenko served as the backup commander of Soyuz TM-13, with fellow backup crew mates Clemens Lothaller of Austria and Talgat Musabayev of Kazakhstan.

Volkov, Viehböck and Aubakirov met Mir 9 commander Anatoli Artsebarsky and flight engineer Sergei Krikalyov at Mir, and after eight days at the station, the two visiting cosmonauts returned to Earth aboard Soyuz TM-12 with Artsebarsky, who ended a six month stay aboard Mir.

Volkov remained at the station to begin his long-duration flight as the commander of Mir 10, with Krikalyov—who had arrived at Mir with Artsebarsky on May 18, 1991—extending his long-duration stay to serve as Mir 10 flight engineer.

The necessity of Krikalyov serving two consecutive missions arose in part from the economic difficulties Russia faced as the Soviet Union collapsed. Mission planners decided to send the two visiting cosmonauts, whose flights were paid for by their respective countries, aboard Soyuz TM-13 with a single Russian cosmonaut, which left the station a cosmonaut short for the Mir 10 mission. Despite the potential difficulties involved in extending his flight after his already having had a long stay at the station, Krikalyov reportedly agreed to the arrangement without hesitation.

Volkov and Krikalyov continued the scientific work begun by Artsebarsky and Krikalyov, and on February 20, 1992 they made a spacewalk of 4 hours and 12 minutes to perform maintenance tasks outside the station. During the EVA, Volkov's spacesuit suffered a malfunction in its heat exchanger unit, which forced him to rely on the heating equipment of Mir's Kvant 2 module, and he had to stay near the module throughout the spacewalk.

On March 17, 1992, Mir 11 crew members Aleksandr Viktorenko and Alexander Kaleri arrived at the station aboard Soyuz TM-14, with guest cosmonaut Klaus-Dietrich Flade of Germany. They visited with Volkov and Krikalyov for a little more than a week, and Volkov, Krikalyov and Flade then returned to Earth aboard Soyuz TM-13 on March 25, 1992.

During Mir 10, Volkov added 175 days, 3 hours and 52 minutes to his career total time in space.

Krikalyov amassed a total stay in space of 311 days, 20 hours and 2 minutes during his consecutive Mir 9 and Mir 10 missions. He had lifted off from the Soviet Union on May 18, 1991; when he emerged from his Soyuz berth after landing on March 25, 1992, he took his first steps in the independent state of Kazakhstan—the Soviet Union having disintegrated entirely during his long stay in space.

SOYUZ MISSION 71: SOYUZ TM-14

(March 17–August 10, 1992)

Russian Manned Spaceflight

Alexander Kaleri and Aleksandr Viktorenko traveled to the Mir space station aboard Soyuz TM-14 with guest cosmonaut Klaus-Dietrich Flade of Germany on March 17,

1992. Kaleri and Viktorenko became the station's eleventh long-duration crew, while Flade remained aboard Mir for about eight days before returning to Earth with Mir 10 crew members Alexander Volkov and Sergei Krikalyov in Soyuz TM-13.

Russian cosmonauts Sergei Avdeyev and Anatoly Solovyov served as the backup crew for Soyuz TM-14/Mir 11, with Michel Tognini of France serving as backup to Klaus-Dietrich Flade.

Viktorenko and Kaleri performed scientific experiments, space station maintenance and engineering work, and participated in a program of regular exercise to maintain their health during their five month stay in space.

On July 8, 1992, they ventured outside of Mir for more than two hours while inspecting the station's Kvant module, which had suffered a failure in four of the six gyrodynes that were responsible for keeping the station in stable flight. Viktorenko and Kaleri cut through insulation to reach the devices, and then inspected and photographed them, transmitting the images back to Earth for later analysis.

Their replacements, Sergei Avdeyev and Anatoly Solovyov, arrived at Mir on July 27, 1992, aboard Soyuz TM-15 with Michel Tognini. After nearly two weeks of joint operations, Viktorenko and Kaleri turned over command of the station to Avdeyev and Solovyov, and returned to Earth aboard Soyuz TM-14 with Tognini, who logged 13 days, 18 hours and 57 minutes in space during his visit to Mir.

From launch to landing, Viktorenko and Kaleri had spent 145 days, 15 hours and 11 minutes in space.

SOYUZ MISSION 72: SOYUZ TM-15

(July 27, 1992–February 1, 1993)

Russian Manned Spaceflight

Launched on July 27, 1992, Soyuz TM-15 crew members Sergei Avdeyev and Anatoly Solovyov traveled to the Russian space station Mir to become the station's twelfth long-duration crew. They were joined on the flight by visiting cosmonaut Michel Tognini of France, who would spend nearly two weeks aboard Mir before returning to Earth with the Mir 11 crew.

Russian cosmonauts Gennadi Manakov and Alexander Poleschuk served as the backup crew for Soyuz TM-15/Mir 12; Jean-Pierre Haigneré of France served as Tognini's backup for the Soyuz TM-15 flight.

At Mir, Avdeyev, Solovyov, and Tognini met Mir 11 crew members Aleksandr Viktorenko and Alexander Kaleri. The five cosmonauts remained aboard the station for several weeks, and then Viktorenko, Kaleri, and Tognini returned to Earth aboard Soyuz TM-14 on August 10, 1992. Tognini spent 13 days, 18 hours and 57 minutes in space during his visit to Mir.

Avdeyev and Solovyov conducted a program of scientific experiments during their stay, and also kept the station in good repair and participated in a program of regular exercise to combat the losses of blood plasma and bone density that regularly occur during long stays in space.

In addition to their routine tasks, Avdeyev and Solovyov prepared for a series of four intense spacewalks in September, during which they would install a new thruster assembly on Mir's Sofora truss segment.

The new maneuvering rockets arrived at the station aboard the unmanned resupply spacecraft Progress-M 14 in mid-August, and were automatically jettisoned from the Progress craft on September 2. The next day, September 3, 1992, Avdeyev and Solovyov began their EVA work during a spacewalk of 3 hours and 56 minutes, when they added a bracket designed to hold the Sofora truss in position during the later installation process, which required the truss to be bent at a hinge about a third of the way along its total length.

In an EVA procedure similar to the use of the U.S. space shuttle Remote Manipulator System (RMS) robotic arm during shuttle spacewalks, the cosmonauts used the Mir Strela boom to position themselves at various spots around the station.

They made their second spacewalk on September 7, adding another 5 hours and 8 minutes while beginning the installation work, and they then finished the job on September 11 during another 5 hour, 44 minute excursion. The work was achieved more quickly than they'd planned, and as a result, they devoted their fourth Mir 12 EVA, on September 15, to maintenance tasks that included the re-positioning of an antenna to clear the way for future docking at the Mir Kristall module, and the retrieval of several pieces of equipment that had been left in the microgravity environment for extended periods to gauge the impact of the long exposure. On their fourth spacewalk, they added another 3 hours and 33 minutes to their EVA total for the flight.

Avdeyev and Solovyov spent a total of 18 hours and 21 minutes outside of Mir during their four Mir 12 spacewalks.

Mir 13 crew members Gennadi Manakov and Alexander Poleschuk arrived at the station in Soyuz TM-16 on January 24, 1993, and after a week of combined operations, Avdeyev and Solovyov brought their long flight to a close, returning to Earth aboard Soyuz TM-15 on February 1, 1993, after 188 days, 21 hours and 39 minutes in space.

SOYUZ MISSION 73: SOYUZ TM-16

(January 24–July 22, 1993)

Russian Manned Spaceflight

Launched aboard Soyuz TM-16 on January 24, 1993, Gennadi Manakov and Alexander Poleschuk traveled to the Mir space station, where they served as the station's Mir 13 long-duration crew.

Aleksander Serebrov and Vasili Tsibliyev served as the backup crew for Soyuz TM-16/Mir 13.

Manakov and Poleschuk lived and worked aboard Mir for six months, maintaining the station's equipment and systems and conducting scientific experiments, while also carrying out a program of regular exercise to maintain their health during the long stay.

On April 19, 1993, they began the first of two spacewalks to install solar array drives on the station. At the start of the 5 hour, 25 minute April 19 EVA, Polishchuck used the Mir Strela boom to position Manakov at Mir's Kvant module, where Manakov installed the electric drive for the solar array to a framework that had been previously installed on the Kvant module. In the midst of the work, one of the two handles on the Strela boom—which governed the device's movement—became unattached and drifted away. Manakov was not endangered by the incident, as he was still tethered to the station, but the loss of mobility severely curtailed the use of the Strela boom. As a result, mission managers decided to postpone the next-scheduled EVA, which had been planned for April 23, until a new handle could be delivered to the station aboard the next unmanned Progress resupply spacecraft.

Progress-M 19 arrived in May, with supplies and equipment as well as the new handle for the Strela control unit.

The cosmonauts next ventured outside of Mir on June 18, finishing the installation of the solar array electrical drives without further difficulty during a 4 hour, 33 minute spacewalk. Their two Mir 13 EVAs took a total of 9 hours and 58 minutes.

They welcomed visitors at Mir on July 1, 1993 when Mir 14 crew members Aleksandr Serebrov and Vasili Tsibliyev arrived aboard Soyuz TM-17 with Jean-Pierre Haigneré of France. The five cosmonauts spent a little over 3 weeks together on the station, and then, on July 22, 1993, Manakov and Poleschuk completed their 179 day, 1 hour and 45 minute Mir 13 mission when they returned to Earth in Soyuz TM-16 with Haigneré.

SOYUZ MISSION 74: SOYUZ TM-17

(July 1, 1993–January 14, 1994)

Russian Manned Spaceflight

Launched on July 1, 1993, Soyuz TM-17 carried Russian cosmonauts Vasili Tsibliyev and Aleksandr Serebrov and visiting cosmonaut Jean-Pierre Haigneré of France to the Mir space station. Serebrov and Tsibliyev became the station's fourteenth long-duration crew, and Haigneré spent several weeks at Mir before returning to Earth with the Mir 13 crew.

Victor Afanasyev and Yury Usachev served as the backup crew for the Soyuz TM-17/Mir 14 mission; Claudie Andre-Deshays (later Claudie Haigneré) of France served as the backup to Jean-Pierre Haigneré during Soyuz TM-17.

During their long stay aboard Mir, Tsibliyev and Serebrov performed scientific experiments, made astronomical observations, and carried out space station maintenance and engineering tasks. They also performed regular exercise to maintain their health in the microgravity environment.

On September 16, 1993, Tsibliyev and Serebrov made the first spacewalk of the Mir 14 mission, when they ventured outside the station for 4 hours and 18 minutes to prepare a truss, Rapana, to be installed onto the station. Contemporaneous evidence seemed to indicate that the work was related to the planned Mir-2 space station, which was reportedly envisioned as a larger, more advanced version of Mir, originally intended to have received visits from the Soviet space shuttle Buran. Subsequent events indicate that the forward progress of both Mir-2 and Buran ended with the 1991 fall of the Soviet Union, but both projects did remain theoretically alive for several years afterward, so it may well be true that the Rapana Truss work at Mir was related in one way or another to these projects.

The cosmonauts continued their work on September 20. They removed the 57.2-pound Rapana Truss from its container and attached it to Mir, where it was to have remained for nearly a year to examine the effect that the long stay in space might have on it. The second EVA took 3 hours and 13 minutes.

Tsibliyev and Serebrov took up a new task during their third spacewalk, a 1 hour, 52 minute excursion on September 28. Known as the Panorama survey, the cosmonauts' activity involved a detailed inspection of the station's exterior, with the intent of collecting evidence that the station was in good enough condition to be used for future cooperative missions with NASA and the European Space Agency (ESA). Mir had been damaged during the Perseid meteor storm a month earlier, and had been in space since 1986; American and European space officials—as well as the leaders of the Russian program themselves—sought reassurance that the station was still likely to be operated safely in the future, when the envisioned cooperative missions would be taking place.

The cosmonauts' work had to be cut short because of a malfunction in the cooling system of Tsibliyev's space suit. Serebrov was still able to capture video and still images of much of the exterior of the station, but the originally scheduled four hour exercise was curtailed before he could complete the full survey.

They attempted to finish the work on October 22, but that spacewalk also had to be cut short when Serebrov endured a fault in his Orlan-DMA spacesuit. The oxygen supply to the suit malfunctioned a little more than a half hour into the planned five hour EVA, and in keeping with the Russian buddy system of crew member support during spacewalking activities, Serebrov and Tsibliyev were forced to return to the safety of the space station's interior earlier than planned. Later investigation revealed that the spacecraft Serebrov wore during the EVA had been used on 13 previous spacewalks, and should have been discarded and replaced.

On October 29, Tsibliyev and Serebrov were finally able to complete the Panorama survey of the station's exterior during a 4 hour, 12 minute EVA. With the fifth spacewalk of the Mir 14 mission, Serebrov set a new record for the most career EVAs, at 10. He and Tsibliyev experienced an eerie moment at one point during their work, when a piece of metal suddenly, inexplicably appeared near their work site and drifted past them. Neither the spacewalking cosmonauts nor mission controllers tracking their progress from Earth could figure out where the unidentified space junk had come from.

At the end of the EVA, the cosmonauts indulged in a bit of mischievous good humor: gaining a small measure of revenge on the faulty spacesuits that had let them down during their previous spacewalks, they strapped the arm of the suit that had failed Serebrov to the neckpiece of the suit to create the illusion that the empty spacesuit was engaged in a salute, and then released the empty, worn-out piece of their wardrobe from the open hatch of Kvant 2.

Later analysis of the images that were gathered during the Panorama survey showed Mir to be in relatively good shape, despite damage from micrometeoroids and a grimy layer of contamination built up by firings of the station's thrusters—and damage from the impact of a variety of space junk, likely similar to the mysterious chunk of metal that had drifted past the cosmonauts during their fifth EVA—over the years.

On January 8, 1994, they welcomed their replacements, Viktor Afanasyev and Yury Usachev, who arrived at Mir aboard Soyuz TM-18 with Valeri Polyakov, who would spend more than a year in space aboard the station.

Slightly less than a week later, Serebrov and Tsibliyev returned to Earth in Soyuz TM-17, after completing a remarkable mission of 196 days, 16 hours and 57 minutes in space, including 14 hours and 13 minutes in EVA during their five Mir 14 spacewalks.

SOYUZ MISSION 75: SOYUZ TM-18

(January 8–July 9, 1994)

Russian Manned Spaceflight

The long-duration Mir 15 crew—commander Viktor Afanasyev and flight engineer Yury Usachev—launched on January 8, 1994 with Valeri Polyakov, who with his second flight in space would remain in orbit aboard the Mir space station for more than a year and two months, achieving the longest spaceflight in the history of space exploration.

Gherman Arzamazor, Yuri Malenchenko and Gennady Strekalov served as the backup crew for the Soyuz TM-18/Mir 15 mission.

Afanasyev and Usachev lived and worked aboard Mir for six months. The performed scientific experiments, astronomical observations, and space station maintenance and engineering tasks, and also participated in a regular program of exercise designed to combat the losses in bone density and blood plasma that regularly occur during long-duration stays in orbit.

Yury Usachev (seen here in 2001) traveled to the Mir space station in Soyuz TM-18 with Viktor Afanasyev and Valeri Polyakov. [NASA/courtesy of nasaimages.org]

They also received supplies and fuel from three unmanned Progress spacecraft: Progress-M 21, which was launched on January 26, 1994; Progress M-22, which followed in March; and Progress M-23, which lifted off on May 22.

Polyakov, a medical doctor who had been selected in March 1972 as part of the Medical Group Three selection of cosmonaut candidates, had previously spent 240 days, 22 hours and 35 minutes aboard Mir, from August 28, 1988, when he'd arrived on Soyuz

TM-6, to April 27, 1989, when he'd returned to Earth aboard Soyuz TM-7. As a doctor and researcher, he was given the job of evaluating the impact of long stays in space on a human test subject—himself—while also serving as an on-board general practitioner for the other cosmonauts who would live and work aboard Mir during his long stay.

For his second remarkable mission, Polyakov would arrive on Soyuz TM-18 in January 1994 and return to Earth aboard Soyuz TM-20 on March 22, 1995, accumulating a record 437 days, 17 hours and 59 minutes in space.

During his two long-duration visits to Mir, Polyakov amassed an amazing career total of 678 days, 16 hours and 34 minutes in space.

Soyuz TM-19, manned by Russian cosmonaut Yuri Malenchenko and Talgat Musabayev of Kazakhstan—who would replace Afanasyev and Usachev—arrived at Mir on July 1, 1994. Afanasyev and Usachev returned to Earth in Soyuz TM-18 on July 9, 1994, after 182 days in space.

SOYUZ MISSION 76: SOYUZ TM-19

(July 1–November 4, 1994)

Russian Manned Spaceflight

Launched on July 1, 1994, Soyuz TM-19 brought Russian cosmonaut Yuri Malenchenko and Talgat Musabayev of Kazakhstan to the Mir space station, where they served as the station's sixteenth long-duration crew.

Vladimir Dezhurov and Gennady Strekalov served as the backup crew for Soyuz TM-19/Mir 16.

At Mir, Malenchenko and Musabayev met Mir 15 crew members Viktor Afanasyev and Yury Usachev—who ended their long-duration stay aboard the station and returned to Earth on July 9—and Valeri Polyakov, who was in the midst of a record-setting long-duration mission of more than 437 days.

Malenchenko and Musabayev made two spacewalks during their stay aboard Mir. On September 9, 1994, they spent five hours and four minutes outside the station while inspecting and repairing damage that had been done during two previous docking maneuvers.

They began their inspection work at the Kristall module, where they found lighter-than-expected damage from the impact of Soyuz TM-17, which had hit the station during docking in January 1993. The impact spot revealed a small area where a piece of thermal insulating blanket had been torn away, along with minor damage to other insulation around the area. The cosmonauts quickly repaired the damage they had found at the first spot, and then moved on to the area where the unmanned Progress-M 24 resupply spacecraft had bumped into Mir in August 1994. They found no damage at the second site.

On September 14, Malenchenko and Musabayev added another six hours and one minute to their spacewalking total when they inspected the station's solar arrays. Their inspection was part of the preparation for the later moving of the arrays, in anticipation of the first docking at Mir of the American space shuttle Atlantis, which was planned for June 1995. The cosmonauts also retrieved experiments from the outside of the station, and installed an amateur radio antenna, which Polyakov tested from inside the space station.

On October 3, they welcomed their replacements, Mir 17 crew members Alexander Viktorenko and Yelena Kondakova, who arrived at Mir with European Space Agency (ESA) astronaut Ulf Merbold of Germany aboard Soyuz TM-20. In contrast to the usual short stay of visiting astronauts, Merbold's visit constituted the cooperative ESA-Russian EuroMir '94 mission, and as a result, he spent a month aboard Mir before returning to Earth with Malenchenko and Musabayev in Soyuz TM-19 on November 4, 1994.

During their long-duration Mir 16 mission, Malenchenko and Musabayev spent 125 days, 21 hours and 53 minutes in space, including 11 hours and 5 minutes in EVA during 2 spacewalks.

SOYUZ MISSION 77: SOYUZ TM-20

(October 3, 1994–March 22, 1995)

Russian Manned Spaceflight

Aleksander Viktorenko made his fourth flight in space as commander of the Soyuz TM-20/Mir 17 long-duration mission. Yelena Kondakova served as flight engineer for Mir 17, and she and Viktorenko were accompanied aboard Soyuz TM-20 by ESA astronaut Ulf Merbold of Germany, who worked aboard Mir for about a month during the joint Russian-European EuroMir '94 mission.

Nikolai Budarin and Anatoly Solovyov served as the backup crew for the Soyuz TM-20/Mir 17 mission; ESA astronaut Thomas Reiter of Germany served as Merbold's backup for EuroMir '94.

At Mir, Viktorenko, Kondakova and Merbold were greeted by Mir 16 crew members Yuri Malenchenko and Talgat Musabayev, and Valeri Polyakov, who was in the midst of a record-setting long-duration mission of more than 437 days.

Merbold conducted experiments designed by European scientists during his stay aboard the station, and he then returned to Earth aboard Soyuz TM-19 with Malenchenko and Musabayev on November 4, 1994. During EuroMir '94, Merbold spent 31 days, 12 hours and 35 minutes in space.

Viktorenko and Kondakova remained aboard Mir with Polyakov, and the three cosmonauts continued to conduct an ongoing program of scientific experiments, which included astronomical studies, medical research and physiological investigations. They also conducted a survey of the noise level in various parts of Mir's living quarters, and on November 13, they received a visit from the unmanned cargo supply spacecraft Progress M-25, which they subsequently unloaded.

On November 18, 1994, Mir passed a major milestone, as the original station (which had been launched on February 20, 1986 and subsequently augmented by additional modules) made its 50,000 orbit of the Earth.

Another milestone passed on January 9, 1995 when Valeri Polyakov marked day 366 of his long-duration visit to the station, breaking the previous record for longest stay in space, which had been set by Vladimir Titov and Musa Manarov during Mir 3 in 1987–1988.

In the first mission scheduled as part of the joint Russian and American cooperative human spaceflight program, the space shuttle Discovery launched on February 3, 1995 for STS-63—the first rendezvous and flyaround of the Russian Mir space station by an American space shuttle. The flight was the start of the two nations' development of the International Space Station (ISS) program, which in turn signified their joint entry into the modern era of routine international cooperation in space.

Mission controllers in the United States and in Russia were given an early opportunity to test their cooperative capabilities when Discovery developed a malfunction in its Reaction Control System (RCS) thrusters. The two Earth-based teams responded well to the problem, and the flight was able to proceed as planned.

The rendezvous was achieved at distances as close as 37 feet, and the shuttle and space station crews communicated with each other from within their vehicles. The STS-63 crew was commanded by James Wetherbee and included Russian cosmonaut Vladimir Titov; to mark the historic rendezvous, in which the two long-hostile nations achieved a major step forward in their cooperation in space, Wetherbee greeted the Mir cosmonauts with an appropriately stately reflection: "As we are bringing our spaceships closer together, we are bringing our nations closer together. The next time we approach, we will shake your hand and together we will lead our world into the next millennium."

Viktorenko, Kondakova and Polyakov continued their Mir stay through March. The unmanned resupply vehicle Progress-M 26 arrived on February 26, and the cosmonauts were able to celebrate one more historic moment during their flight when they marked the ninth anniversary of the station's launch.

The Mir 18 crew—Russian cosmonauts Vladimir Dezhurov and Gennadi Streakalov and NASA astronaut Norman Thagard—arrived at the station on March 14.

Viktorenko, Kondakova, and Polyakov turned the station over to the next crew, and then returned to Earth aboard Soyuz TM-20 on March 22, 1995. Viktorenko and Kondakov had spent 169 days, 5 hours and 21 minutes in space as the Mir 17 crew.

Polyakov returned from his long stay aboard Mir after accumulating a total of 437 days, 17 hours and 59 minutes in space—setting a new record for the longest spaceflight in the history of space exploration—and upping his career total time in space to 678 days, 16 hours and 34 minutes, during two remarkable long-duration missions.

SOYUZ MISSION 78: SOYUZ TM-21

(March 14–July 7, 1995)

Russian Manned Spaceflight

Launched on March 14, 1995, from the Baikonur Cosmodrome in Kazakhstan aboard Soyuz TM-21, Mir 18 commander Vladimir Dezhurov, flight engineer

Gennady Strekalov and cosmonaut researcher Norman Thagard of NASA traveled to the Mir space station, where they served a long-duration mission together as part of an ongoing cooperative initiative between the Russian and American space programs.

The backup crew for the Soyuz TM-21/Mir 18 mission included Russian cosmonauts Sergei Avdeyev and Yuri Gidzenko and NASA astronaut Bonnie Dunbar.

Dezhurov, Strekalov and Thagard visited briefly with Mir residents Aleksandr Viktorenko, Yelena Kondakova and Valeri Polyakov, who left the station about a week after the arrival of the Mir 18 crew.

Medical research was the primary goal of the first Russian-U.S. long-duration flight. Dezhurov, Strekalov, and Thagard were assigned a rigorous program of scientific research designed to measure various aspects of the experience of living in weightlessness for nearly four months.

Dezhurov and Strekalov also made five spacewalks during their Mir 18 stay at the station, primarily to reconfigure Mir for its impending addition by the installation of the Spektr module, and for its planned first docking with a U.S. space shuttle.

The first Mir 18 spacewalk began amid a swirl of minor controversy, after it became known that Gennady Strekalov had cut his hand while working inside the station. The wound became infected, giving rise to concerns that he might not be well enough to make the first scheduled EVA, but the cosmonauts emerged from Mir as scheduled on May 12, and by the end of their six hours and eight minutes outside the station, they had put to rest any concerns about Strekalov's health. During their first spacewalk, they prepared the Mir Kvant module's electrical system for the later transfer of solar arrays from the Kristall module.

They continued their work on May 17, accumulating an additional 6 hours and 52 minutes in EVA while moving the solar panel arrays from Kristall to Kvant. They were able to accomplish the move, but ran out of time before they could properly install the array, so they simply attached it to its new location and resigned themselves to finishing the job during a later EVA.

On May 22 they made the appropriate connection to activate the array, and Thagard successfully tested the array from within Mir. Dezhurov and Strekalov spent 5 hours and 15 minutes in EVA during their third spacewalk.

During their fourth spacewalk, on May 29, Dezhurov and Strekalov did not actually leave the space station. Their work was still considered an EVA, however, because they worked within Mir's transfer compartment—the chamber that linked Mir's core module to its Kristall expansion module—while it was depressurized. During the short EVA (about 21 minutes), they prepared the docking port on the core module for the later transfer of the Kristall module.

In the midst of all their preparations, the Spektr expansion module arrived at Mir on June 1, 1995, and was automatically (and temporarily) docked at the space station's front port.

On June 2, Dezhurov and Strekalov again made an "internal" spacewalk, when they depressurized the transfer compartment to realign the station for the final docking of the Spektr module in its appropriate port the following day. The cosmonauts added another 23 minutes to their EVA total during their final spacewalk of the flight.

For a time after Spektr was moved to its new docking position, it seemed that the fifth spacewalk might not be their last. A dramatic—and embarrassing—controversy erupted when the solar arrays on Spektr were tested for the first time on June 5. One array became stuck in a half-open position, reducing the amount of electricity by about 20 percent, and Russian mission controllers hurriedly announced their intention to have Dezhurov and Strekalov make a sixth EVA to rectify the problem before the scheduled first-ever docking between the space station Atlantis and Mir, which was scheduled to take place in late June.

After a good deal of discussion with the cosmonauts about how to approach the improvised EVA and the likely results it might achieve, Russian Mission Control announced first that the spacewalk would be put off for a few weeks, and then ultimately canceled the exercise altogether. Causing some consternation for Russian space officials as they hoped to present a worry-free environment for the impending joint operations with NASA, Strekalov vigorously expressed his opposition to the proposed EVA on the grounds that it presented an unacceptable safety risk. Dezhurov argued in favor of the exercise, and an unseemly rift developed between the two cosmonauts. Pressured both by his commander and by mission managers on the ground, the veteran Strekalov, who was on his fifth flight in space, finally let it be known that he would not carry out the proposed EVA if it was ordered.

To prevent the unpleasant situation from getting any further out of hand, Russian officials temporarily left the matter unresolved. A careful study of the station's electrical system indicated that there would be enough power for the scheduled shuttle docking in any case, and given the complexity and sensitivity of both the spacewalk controversy and the shuttle docking, it was deemed best to proceed without a sixth EVA exercise.

The controversy did carry a penalty for Dezhurov and Strekalov, however; when their long stay aboard Mir was finally completed, they were met on Earth with a fine that amounted to about 15% of their pay for the mission, which their home space agency levied for the failure to carry out the sixth EVA. Although the fine was likely viewed as an appropriate measure by the involved officials, who were responsible for ensuring the obedience of the cosmonauts in their charge, it seemed in retrospect a strange response to the safety concerns of a long-time veteran cosmonaut, no matter how those concerns might have been expressed.

On the other hand, the military origins and culture of the Soviet/Russian space program must have made Strekalov's refusal to participate in the proposed sixth EVA seem virtually mutinous to some officials within the program.

Whatever the merit or fault of the controversy over the unplanned EVA, Dezhurov and Strekalov did achieve the primary goals of their five planned spacewalks, in the process spending 18 hours and 59 minutes in EVA.

The STS-71 crew of the space shuttle Atlantis arrived at Mir in late June, and the first-ever docking of an American shuttle and the Russian space station took place without difficulty on June 29, 1995.

Aboard Atlantis for the historic STS-71 flight were commander Robert Gibson, pilot Charles Precourt, mission specialists Bonnie Dunbar, Ellen Baker, and Gregory Harbaugh, and Russian cosmonauts Anatoly Solovyov and Nikolai Budarin—who would transfer to Mir at the end of the period of docked operations, and replace Dezhurov, Strekalov and Thagard to begin a long-duration stay as the Mir 19 crew.

STS-71 commander Gibson piloted the shuttle during the final approach and docking to Mir. During the five days in which Atlantis and Mir were docked, they constituted the largest spacecraft flown in orbit up to that time, with a combined weight of half a million pounds.

Dezhurov, Strekalov and Thagard welcomed the STS-71 crew in a brief ceremony, and then relinquished command of the station to Solovyov and Budarin, who began their Mir 19 mission. The two crews transferred equipment and supplies to Mir, and removed a large number of medical samples that the Mir 18 crew had collected during their stay. The Mir 18 crew members also served as test subjects for medical investigations carried out by the STS-71 crew during the return trip to Earth.

After undocking from Mir on July 4, Atlantis returned to Earth with Dezhurov, Strekalov and Thagard, landing at the Kennedy Space Center (KSC) in Florida on July 7, 1995.

From their launch, aboard Soyuz TM-21 on March 14, 1995 to their landing aboard the shuttle Atlantis at the end of STS-71 on July 7, 1995, Mir 18 crew members Dezhurov, Strekalov and Thagard spent 115 days, 8 hours and 44 minutes in space.

During their time aboard the space station as the Mir 19 crew, commander Anatoliy Solovyev and flight engineer Nikolai Budarin made three spacewalks.

After the controversy surrounding the proposed EVA fix of the Spektr solar array during the previous long-duration mission, Solovyov and Budarin were able to effect the necessary repair during their first Mir 19 spacewalk, a 5 hour, 34 minute exercise on July 14, 1995.

They ventured outside the station again on July 19, spending three hours and eight minutes in EVA while making preparations to install an infrared spectrometer during their next excursion, and collecting a U.S.-built experiment from the exterior of the Kvant 2 module. The second EVA had originally been planned to last about two-and-a-half hours longer, but was curtailed when Solovyov's Orlan-DMA spacesuit suffered a malfunction in its cooling system.

Solovyov and Budarin successfully installed the Mir Infrared Atmospheric Spectrometer (MIRAS) during their third EVA, on July 21. They spent 5 hours and 35 minutes working outside of the Mir Spektr module during the installation, in an activity that French engineers characterized as the "most complex operation ever carried out during EVA." When mission controllers could not detect any data being transmitted by the instrument, Solovyov and Budarin made an on-the-spot repair, and MIRAS subsequently worked as planned.

Veteran cosmonaut Solovyov upped his career total EVA time to 41 hours and 49 minutes during the third Mir 19 spacewalk, giving him the record for the most career spacewalking time.

During their 3 spacewalks, Solovyov and Budarin accumulated a total of 14 hours and 17 minutes in EVA.

On September 3, 1995 their replacements—Mir 20 commander Yuri Gidzenko, flight engineer Sergei Avdeyev, and European Space Agency (ESA) cosmonaut researcher Thomas Reiter—arrived at Mir aboard Soyuz TM-22. Solovyov and Budarin left the station 8 days later, returning to Earth on September 11, 1995 in Soyuz TM-21, after a mission of 181 days and 41 minutes in space.

SOYUZ MISSION 79: SOYUZ TM-22

(September 3, 1995 –February 29, 1996)

Russian Manned Spaceflight

Launched aboard Soyuz TM-22 on September 3, 1995, commander Sergei Avdeyev, flight engineer Yuri Gidzenko and European Space Agency (ESA) astronaut Thomas Reiter of Germany traveled to the Mir space station, where they became the Mir 20 long-duration crew during a mission that was also known as "EuroMir '95," in honor of the ESA's participation in the flight.

Aleksander Kaleri, Valery Korzun, and ESA astronaut Pedro Duque of Spain served as the backup crew for Soyuz TM-22/Mir 20.

When Avdeyev, Gidzenko and Reiter arrived at Mir, they visited with long-duration crew members Anatoliy Solovyev and Nikolai Budarin for about eight days, before Solovyev and Budarin returned to Earth aboard Soyuz TM-21.

Avdeyev, Gidzenko and Reiter then settled into a productive routine aboard the space station, carrying out the maintenance tasks, engineering tests and regular physical exercise necessary to maintaining a safe environment and good health during a long stay in space. They also conducted a rigorous scientific program, and Reiter completed 40 experiments that had been developed for the flight at the ESA's European Astronaut Center.

On October 20, 1995, Avdeyev and Reiter ventured outside of Mir to do some external maintenance and to install the European Space Exposure Facility. The 5 hour, 11 minute excursion was the first-ever spacewalk by an ESA astronaut (Jean-Loup Chrétien of France had made the first spacewalk by a European in 1998, but did so as a representative of the French national space agency).

Avdeyev and Gidzenko made the second EVA of the mission on December 8, when they spent 37 minutes outside the station while transferring a docking cone from one port to another.

Gidzenko and Reiter made the final spacewalk of their stay on February 8, 1996 when they spent three hours and six minutes in EVA while retrieving the ESEF and repairing the antenna on the Mir Kvant 2 module.

On February 21 they received visitors, when their replacements, Yuri Onufrienko and Yury Usachev arrived at Mir aboard Soyuz -TM 23. Avdeyev, Gidzenko and Reiter returned to Earth 8 days later, landing on February 29, 1996, after a flight of 179 days, 1 hour and 41 minutes in space.

SOYUZ MISSION 80: SOYUZ TM-23

(February 21–September 2, 1996)
Russian Manned Spaceflight

Mir 21 commander Yuri Onufrienko and flight engineer Yury Usachev lifted off from the Baikonur Cosmodrome in Kazakhstan on February 21, 1996, aboard Soyuz TM-23. They traveled to the Mir space station, where they began a long-duration stay of nearly four months.

Aleksandr Lazutkin and Vasili Tsibliyev served as the backup crew for the Soyuz TM-23/Mir 12 flight.

Onufrienko and Usachev took over command of Mir from Yuri Gidzenko, Sergei Avdeyev and ESA cosmonaut Thomas Reiter, who ended their mission after 179 days in space and returned to Earth aboard Soyuz TM-22 on February 29, 1996.

During their long stay at Mir, Onufrienko and Usachev made a remarkable six long spacewalks. They began their EVA work on March 15, when they ventured outside of Mir to install a second Strela boom on the station. The 39.3-foot (12 meter) Strela devices were used by spacewalking cosmonauts to maneuver themselves to various worksites around the exterior of the station. During their first Mir 21 EVA, Onufrienko and Usachev spent 5 hours and 51 minutes working outside of Mir.

In the midst of its nine day STS-76 mission, the U.S. space shuttle Atlantis arrived at Mir on March 23. The STS-76 crew included commander Kevin Chilton, pilot Richard Searfoss, and mission specialists Shannon Lucid, Linda Godwin, Michael Clifford, and Ronald Sega.

Chilton directed Atlantis to the third docking of a U.S. shuttle and the Russian space station, and Lucid, on her fifth flight in space, transferred from Atlantis to Mir to begin an epic 188-day mission as the Mir 21 guest cosmonaut-researcher with Onufrienko and Usachev. Lucid's stay aboard Mir marked the beginning of a period of 907 days of continuous U.S. presence aboard the station, as part of a Russian-American program of cooperative living in space that represented the two nations' initial arrangements for their further cooperation in the International Space Station (ISS) program. Beginning with Lucid, a total of seven U.S. astronauts achieved consecutive long-duration missions aboard Mir with Russian cosmonauts; the program ended with the completion of Andrew Thomas's stay at Mir in June 1998.

Working together, the Mir 21 and STS-76 crews transferred equipment and supplies from the shuttle onto Mir, and moved materials no longer needed at the station onto Atlantis for return to Earth.

On March 27, 1996 STS-76 mission specialists Michael Clifford and Linda Godwin made the first-ever spacewalk by U.S. astronauts at the Mir space station. The historic EVA also marked the first time that a spacewalk was made from a docked shuttle. Clifford and Godwin wore the Simplified Aid for EVA Rescue

(SAFER) backpacks during the six hour, two minute spacewalk, during which they attached four experiment packages (the Mir Environmental Effects Payloads, or MEEPs) to the outside of the station, and tested boot restraints and tether hooks that were designed to accommodate both Russian- and U.S.-designed equipment in anticipation of their use on the ISS. Two of the MEEP packages were also related to ISS development, as they contained materials that engineers were considering for use on the international station, and were placed on Mir so the impact of the exposure to space might be figured into the designers' calculations during ISS planning. STS-76 mission specialist Ronald Sega monitored the spacewalkers' activities from inside Atlantis, using cameras mounted in the shuttle's payload bay.

Atlantis undocked from Mir on March 28, and the STS-76 crew returned to Earth three days later.

Onufrienko, Usachev and Lucid remained on the station to continue their work as the Mir 21 crew. They conducted a rigorous program of scientific work, which included life sciences and materials processing experiments and photography.

On April 23, Priroda—the last of the six Mir expansion modules—was launched from the Baikonur Cosmodrome. The module arrived at and docked with Mir on April 26, 1996 and the following day the crew moved it from its temporary docking port to its final position at the space station.

Another unmanned spacecraft, the Progress-M 31 cargo resupply craft, arrived at the station on May 7 and was automatically docked.

On May 21 Onufrienko and Usachev made their second spacewalk, to move the Mir Cooperative Solar Array (MCSA) from the station's Kristall module to the Kvant module.

At that point, they began a new set of chores that, depending on one's perspective, seemed utterly strange or oddly comforting, given the circumstances. The two Russian cosmonauts re-entered the airlock on Kvant 2 and began to unfold and assemble a nearly four-foot-tall structure that looked oddly like a very large Pepsi Cola can. When their pop art assemblage was complete, they took turns videotaping each other standing alongside it, and then took the faux can apart and stowed it for later return to Earth. The display was in fact intended to represent Pepsi in space, and the video record of the exercise was to be used later in a Pepsico Inc. advertising campaign.

The unique EVA took a total of 5 hours and 20 minutes to complete.

Onufrienko and Usachev returned to the MCSA installation on May 24, during a 5 hour, 34 minute third EVA. They deployed the 59-foot (18 meter) array on the Mir Kvant module. They linked the array's electrical cables to the station, but the lines only allowed the array to supply half of its intended power. Replacement cables were scheduled to be supplied by a later Progress cargo craft.

May also brought greetings from the space shuttle STS-77 crew, which established radio contact with the Mir 21 crew. The two crews also conducted a joint press conference from space for representatives of the news media in the United States and in Russia.

On May 30, Onufrienko and Usachev made their fourth spacewalk of the Mir 21 mission. The cosmonauts installed the Modular Optoelectrical Multispectral

Scanner (MOMS) on the outside of Mir's Priroda module during the 4 hour, 20 minute EVA. The MOMS instrument was a remote sensing camera designed to record data about the Earth's atmosphere and environment; it had previously flown aboard the U.S. space shuttle during STS-7 and STS-41B.

During their fifth EVA, which took place on June 6, the cosmonauts installed experiments on the outside of the Kvant 2 module, adding another 3 hours and 34 minutes to their spacewalking total for the flight.

They made their sixth and final spacewalk on June 13, 1996 when they attached a truss segment on the Mir Kvant module. The nearly 20-foot long truss segment was known generically as a Rapana structure; it was also referred to as the Strombus or Ferma 3 Truss. They also installed the Travers Synthetic Aperture Radar antenna on Priroda, and filmed additional footage for the Pepsi commercial they had begun during their second EVA. Their final EVA work took 5 hours and 42 minutes.

During their 6 Mir 21 spacewalks, Onufrienko and Usachev accumulated a total of 30 hours and 21 minutes in EVA.

In mid-July Lucid's scheduled stay aboard Mir was extended by virtue of the delayed launch of space shuttle mission STS-79, which had been scheduled to lift off on July 31, dock with Mir, and return Lucid to Earth at the end of its flight. A problem with the solid rocket boosters (SRBs) intended for use during the launch forced the scheduled shuttle launch to be delayed until mid-September.

The unmanned Progress-M 32 resupply and refueling spacecraft arrived at Mir on August 2, and the crew moved supplies from the vehicle onto the space station.

Soyuz TM-24 closely followed Progress-M 32, arriving on August 19 with Mir 22 crew members Alexander Kaleri and Valery Korzun, and Claudie Andre-Deshays (subsequently known by her married name, Claudie Haigneré) of France. Representing the French national space agency Centre National d'Études Spatiales (CNES), Andre-Deshays traveled to the space station to conduct the Cassiopee project of scientific research.

On September 2, 1996 Onufrienko and Usachev ended their long stay aboard Mir and returned to Earth aboard Soyuz TM-23 with Andre-Deshays, who had concluded her work aboard the station after 15 days, 18 hours, and 28 minutes.

During their Soyuz TM-23/Mir 21 long-duration flight, Onufrienko and Usachev spent 193 days, 19 hours and 7 minutes in space.

Lucid remained aboard Mir to await the STS-79 shuttle crew, who would visit the station in September and return her to Earth at that time, after delivering her replacement, NASA astronaut John Blaha.

SOYUZ MISSION 81: SOYUZ TM-24

(August 17, 1996–March 2, 1997)

Russian Manned Spaceflight

Mir 22 commander Valery Korzun and flight engineer Alexander Kaleri launched on August 17, 1996 aboard Soyuz TM-24 with guest cosmonaut Claudie

Andre-Deshays (subsequently known by her married name, Claudie Haigneré) of France, who with her flight aboard Soyuz TM-24 became the first French woman to fly in space. Representing the French national space agency Centre National d'Études Spatiales (CNES), Andre-Deshays traveled to Mir to conduct the Cassiopee project of scientific research.

Russian cosmonauts Aleksander Lazutkin and Vasili Tsibliyev were the backup crew for Soyuz TM-24/Mir 22, and Léopold Eyharts of CNES served as the backup crew member for Andre-Deshays for the Soyuz TM-24 flight.

The original Mir 22 crew had consisted of commander Gennadi Manakov and flight engineer Pavel Vinogradov, but they were removed from the assignment just days before the scheduled launch when a routine physical exam indicated that Manakov might have a heart condition.

When they arrived at the station on August 17, Korzun, Kaleri and Andre-Deshays were greeted by the Mir 21 crew—Russian cosmonauts Yuri Onufrienko and Yury Usachev, and Shannon Lucid of NASA. Onufrienko and Usachev left the station on September 2 with Andre-Deshays, after she completed the joint French-Russian research program; Lucid remained aboard Mir with Korzun and Kaleri until the space shuttle Atlantis arrived at the station during STS-79, on September 18.

During her Soyuz TM-24 trip to Mir, Andre-Deshays became the first French woman to fly in space. With Korzun and Kaleri, she conducted an intensive program of scientific research, during the fifth Russian-French cooperative human spaceflight. The French research included microgravity studies of the human cardiac and neurological systems, fluid dynamics investigations, analysis of structural dynamics in space, and studies of the role that gravity plays in the development of egg embryos. Andre-Deshays spent 15 days, 18 hours and 23 minutes in space during her Soyuz TM-24 visit to Mir.

After Onufrienko, Usachev and Andre-Deshays left the station on September 2, Korzun and Kaleri began their work as the Mir 22 crew. They conducted scientific research in fluids, life and materials sciences, performed station maintenance and engineering tasks, and regularly exercised to maintain their health during their long stay aboard Mir. They also played host to two visits from American astronauts aboard the space shuttle Atlantis, and made two spacewalks during the Mir 22 mission.

The space shuttle Atlantis arrived at Mir on September 18, during STS-79. Working together, the shuttle and space station crews transferred supplies and equipment onto the station, and U.S. astronaut John Blaha joined Korzun and Kaleri aboard Mir. Blaha replaced Shannon Lucid, who returned to Earth with the STS-79 crew after 188 days, 4 hours and 14 seconds in space—which set a new record for the longest space mission by a female astronaut up to that time, and the longest flight by a U.S. astronaut up to that time.

Korzun and Kaleri made their first spacewalk on December 2, 1996 when they ventured outside of Mir to install a cable on the Mir Cooperative Solar Array (MCSA). The MCSA had originally been installed by Mir 21 crew members Yuri Onufrienko and Yury Usachev in May, 1996, but the electrical cables supplies to Onufrienko and Usachev had only allowed the array to supply half of its intended power. The 75.5 foot (23 meter) replacement cable installed by Korzun and Kaleri brought the array's output up to 6 kilowatts—double the original amount. Korzun and Kaleri also re-positioned a Rapana girder on the outside of the Mir Kvant module during their 5 hour, 57 minute first EVA.

During their second Mir 22 spacewalk, on December 9, Korzun and Kaleri installed a radar antenna on Mir's docking module, for use during later rendezvous operations. Blaha followed their progress from within the station, and relayed information from Russian Mission Control to the spacewalkers during the 6 hour, 36 minute EVA.

Korzun and Kaleri spent a total of 12 hours and 33 minutes in EVA during their 2 Mir 22 spacewalks.

Blaha served aboard Mir with Korzun and Kaleri for four months, until Atlantis returned and docked with the station again on January 14, 1997 during STS-81. The space shuttle and space station crews transferred 6,000 pounds of supplies and equipment during that linking, and removed 2,400 pounds of experiment samples and other materials no longer needed on Mir, including wheat plants that were the first plants to complete an entire life cycle in space.

At the end of the STS-81/Mir docked operations, Blaha returned to Earth aboard Atlantis, after 118 days aboard Mir. He was replaced on the station by Jerry Linenger, who would remain aboard Mir for 4 months, until he was replaced by Michael Foale in May 1997.

Korzun, Kaleri and Linenger were aboard the station on February 10 when Mir 23 cosmonauts Vasili Tsibliyev and Aleksandr Lazutkin arrived in Soyuz TM-25 with visiting cosmonaut Reinhold Ewald of Germany. On February 24, 1997 an oxygen-generating device aboard the station caused a fire to break out in the Mir Kvant 1 module. The six crew members quickly equipped themselves with goggles and masks, and used three fire extinguishers to extinguish the fire. Heavy smoke filled the station for several minutes, and heat from the flames damaged equipment and cables.

Linenger, a medical doctor, examined his fellow crew members and gave them advice about how to protect themselves from the lingering effects of the smoke and the flames. Fortunately, none of the crew members were injured during the incident.

On March 2, 1997 Korzun and Kaleri completed their long, productive Mir 22 mission when they returned to Earth in Soyuz TM-24 with Ewald, while Tsibliyev, Lazutkin and Linenger remained aboard the station to continue their long-duration stay. Korzun and Kaleri had spent a total of 196 days, 16 hours and 26 minutes in space during their stay aboard Mir; Ewald spent 19 days, 16 hours and 35 minutes in space during his visit to the station.

SOYUZ MISSION 82: SOYUZ TM-25

(February 10–August 14, 1997)

Russian Manned Spaceflight

Launched aboard Soyuz TM-25 on February 10, 1997, commander Vasili Tsibliyev, flight engineer Aleksandr Lazutkin and guest cosmonaut Reinhold Ewald of

Germany traveled to the Mir space station, where Tsibliyev and Lazutkin served for more than six months as the Mir 23 long-duration crew. Ewald visited the station for about two weeks, and then returned to Earth with the Mir 22 crew on March 2, 1997.

Vladimir Dezhurov and Gennadi Padalka served as the backup crew for the Soyuz TM-25/Mir 23 mission; Hans Schlegel was Ewald's backup for Soyuz TM-25.

When they arrived at Mir, Tsibliyev, Lazutkin and Ewald were greeted by Mir 22 crew members Valery Korzun and Alexander Kaleri, and NASA astronaut Jerry Linenger, who was in the midst of a long-duration stay aboard the station as part of the joint Russian-American cooperative spaceflight program, which also served as preparation for the International Space Station (ISS) program.

For about two weeks, all six crew members participated in the chores of making the transition from one long-duration crew to the next, with the resident crew finishing its work before leaving and new crew members making themselves familiar with the environment they would call home for the next six months. Then, on February 24, calamity ensued when a fire broke out on the space station.

An oxygen-generating device malfunctioned and began emitting flames in the Mir Kvant 1 module, sending all six crew members scrambling to don protective goggles and masks. The crew used three fire extinguishers to extinguish the fire, and then dealt with the unpleasant results of the blaze, which included heavy smoke and damage to equipment and cables, for some time after the fire was out. Linenger, a medical doctor, helped his fellow crew members cope with the lingering after-effects. None of the crew members were injured during the incident.

Korzun, Kaleri and Ewald left Mir a short while later, as scheduled, and landed in Soyuz TM-24 on March 2, 1997. From launch to landing, Ewald accumulated 19 days, 16 hours and 35 minutes in space during his Soyuz TM-25 / Soyuz TM-24 flight.

Tsibliyev and Lazutkin remained aboard the station with Linenger, to start their long-duration stay as the Mir 23 crew.

In addition to the fire, the Mir 23 mission was hampered by a series of other mishaps that involved safety risks to the station, beginning with a leak in Mir's coolant system that loosed a dangerous level of ethylene glycol into the station's living area. Fortunately, the contamination proved to have no long-lasting ill effects on the crew; but Tsibliyev, Lazutkin and Linenger also had to endure an electrical outage and a failure of the station's attitude control system during their stay.

On April 29, 1997 Tsibliyev and Linenger made the first-ever spacewalk involving a Russian cosmonaut and an American astronaut. They spent 4 hours and 57 minutes outside of Mir, while testing the Russian Orlan-M spacesuit and retrieving and installing equipment and experiments. During their historic EVA, Tsibliyev and Linenger installed the Optical Properties Monitor on the station's Kristall module, retrieved several experiments from the exterior of the Kvant 2 module, and installed a radiation dosimeter.

On May 16 the space shuttle Atlantis arrived at Mir in the midst of STS-84—the sixth docking of a U.S. shuttle and the Russian space station. Charles

Precourt commanded STS-84, and the crew included pilot Eileen Collins, and mission specialists Jean-Francois Clervoy of the European Space Agency (ESA), Elena Kondakova of the Russian Federal Space Agency, and NASA astronauts Carlos Noriega, Edward Lu, and Michael Foale.

The STS-84 and Mir 23 crews transferred supplies and equipment from the shuttle to the station, and Foale transferred to the station to replace Linenger, who returned to Earth with the STS-84 crew after 123 days aboard Mir. The exchange of U.S. crew members furthered the continuous U.S. presence aboard Mir that was the centerpiece of the Russian-U.S. cooperative program of human spaceflight.

After the departure of the space shuttle, Tsibliyev, Lazutkin and Foale turned their attention to scientific research. Foale's program of scientific work included 35 experiments.

June brought further misadventure to Mir, when an unmanned Progress supply spacecraft collided with the station's Spektr module on June 25. The Spektr module was damaged during the incident, and its atmosphere escaped—putting the crew at immediate risk of having the entire station depressurized. Tsibliyev, Lazutkin and Foale worked quickly to shut the hatch to the leaking module, thereby insuring their safety but raising the uncomfortable prospect of having to make a spacewalk within the station if they were to assess and repair the damage to Spektr.

They were training for just such an exercise when a further mishap caused the station to briefly lose electrical power. Concerned officials in both the Russian space agency and NASA decided to put off the internal EVA until the arrival of Mir 24 cosmonauts Anatoly Solovyov—a veteran commander and spacewalker—and Pavel Vinogradov.

Solovyov and Vinogradov arrived at Mir on August 5 aboard Soyuz TM-26. For a little more than a week, they conferred with Tsibliyev, Lazutkin and Foale about the accident and the station's condition, and Tsibliyev and Lazutkin then ended their stay at the station on August 14, 1997, when they returned to Earth aboard Soyuz TM-25.

During their tumultuous Mir 23 mission, Vasili Tsibliyev and Aleksandr Lazutkin spent 185 days, 1 hour and 41 minutes in space. Michael Foale remained aboard the station after they left, serving with Solovyov and Vinogradov until he was replaced by David Wolf in September and returned to Earth aboard the space shuttle Atlantis at the end of STS-86.

SOYUZ MISSION 83: SOYUZ TM-26

(August 5, 1997–February 19, 1998)

Russian Manned Spaceflight

Soyuz TM-26/Mir 24 commander Anatoly Solovyov made his fifth flight in space—and began his fourth long-duration mission—on August 5, 1997 when he lifted off with flight engineer Pavel Vinogradov to travel to the Mir space station. During their long stay at Mir, Solovyov and Vinogradov would make a series of long spacewalks

Responding to the dilemmas of their pending visit to the Mir space station in September, 1997, NASA astronauts (left to right) Scott Parazynski, David Wolf, and Wendy Lawrence sport name tags detailing their prospects for fitting into the Russian Orlan spacesuit: Parazynski's tag reads "Too Tall Parazynski," while Lawrence's label identifies her as "Too Short." The just-right Wolf, meanwhile, wears an unaltered tag. [NASA/courtesy of nasaimages.org]

to inspect and repair the station's Spektr module, which had been damaged during a collision with an unmanned Progress supply spacecraft on June 25, 1997.

Gennady Padalka and Sergei Avdeyev served as the backup crew for Soyuz TM-26/Mir 24.

At Mir, Solovyov and Vinogradov were greeted by Mir 23 crew members Vasili Tsibliyev and Aleksandr Lazutkin, and NASA astronaut Michael Foale, who was in the midst of a long-duration stay aboard the station as part of the joint Russian-American program of cooperative human spaceflight that also served as the first stage of the two nations' development of the International Space Station (ISS) program.

The crew members spent about a week conferring about the status of the Spektr module and the general state of the station, and Tsibliyev and Lazutkin then left Mir and returned to Earth on August 14, 1997.

Solovyov and Vinogradov began their stay with Foale as mission managers on Earth devised repairs for the station. The Spektr module had been depressurized during the June collision, and had been sealed off from the rest of Mir. As a result, crew members entering the module would have to be fully suited in pressurized spacesuits and equipped with life support systems—in essence making an internal spacewalk to inspect the inside of the depressurized module to try to locate the area or areas in which it had been punctured during the collision.

Thus, the veteran Solovyov and rookie Vinogradov (who was making his first flight in space) arrived at Mir with a full program of planned EVA work and the intense responsibility of ensuring the station's safe operation for themselves, and for future inhabitants.

In response to the seriousness of the situation, NASA also changed its plans for future U.S. activities on Mir. Although the American space agency was committed to continuing its cooperative program of living and working in space with Russian cosmonauts aboard the Russian space station, NASA officials decided to replace Wendy Lawrence, the astronaut originally scheduled to replace Foale at the end of his stay, with her backup, David Wolf. Lawrence was smaller than the minimum size specified to wear the Russian Orlan spacesuit, which would be required for the internal spacewalk, so Wolf, who had been scheduled for a later flight, was moved up so he could serve as a backup for the spacewalks that were planned to fix the station over the following months.

Solovyov and Vinogradov made the first of their Mir 24 EVAs, the internal spacewalk, on August 22. They carefully inspected the damaged Spektr module, and connected power cables to the three Spektr solar arrays that were not damaged during the June collision. They spent 3 hours and 16 minutes in EVA during the work.

During Solovyov's second Mir 24 spacewalk, on September 6, he and Foale ventured outside of the Spektr module to try to locate the point where the craft's hull had been punctured. They also repositioned the solar arrays to increase the amount of energy available to the module during their six hour EVA.

The space shuttle Atlantis arrived at Mir on September 27, during STS-86, and achieved the seventh docking of the U.S. shuttle and the Russian space station. The shuttle and station crews worked together to transfer 10,400 pounds of supplies and equipment onto Mir, and removed experiment samples and other material for return to Earth aboard the shuttle.

On October 1, STS-86 mission specialists Scott Parazynski of NASA and Russian cosmonaut Vladimir Titov made the first joint Russian-American spacewalk from a space shuttle. They spent five hours and one minute in EVA while they moved a cap for the Spektr solar array to the module for later use, retrieved several experiments from the outside of the space station, and tested NASA's Simplified Aid for EVA Rescue (SAFER) jet packs.

After much deliberation by NASA, including an in-depth study of conditions aboard Mir and two independent assessments of the station's ability to continue to support safe operations, the planned long-duration Mir stay of U.S. astronaut David Wolf had been approved in the months leading up to STS-86. As a result, during the period in which Atlantis and Mir were docked, Wolf transferred to the station, exchanging places with Michael Foale, who left Mir after a stay of 134 days. Foale returned to Earth aboard Atlantis with the STS-86 crew on October 6.

Solovyov and Vinogradov returned to their Spektr repair duties during three long, intense spacewalks in October and early November. On October 20 they spent 6 hours and 38 minutes inside the damaged module, searching for damage and making repairs; and on November 3 and November 6 they made spacewalks of more than six hours each while finishing the refurbishing of the Spektr solar arrays.

They spent another three hours and six minutes in EVA in their final Mir 24 spacewalk together, on January 9, 1998, while making hatch repairs.

Wolf, meanwhile, settled into a productive routine. Like his NASA predecessors on Mir, he worked on an extensive scientific program and participated in station maintenance tasks with his Russian crew mates.

On January 14, 1998, Wolf got a first-hand look at the damage to Spektr, when he donned a Russian Orlan spacesuit and accompanied Solovyov during a trip outside the station to continue to look for holes in the module's outer skin. Solovyov and Wolf spent 3 hours and 52 minutes in EVA during the exercise.

The STS-89 crew of the U.S. space shuttle Endeavour achieved the eighth docking of a shuttle and the Mir space station on January 24, 1998. The shuttle's launch had been delayed twice at the request of Russian space officials, who required more time for the Mir 24 crew to finish its work on the station before the arrival of the STS-89 shuttle crew.

In keeping with the routine established during earlier shuttle-Mir dockings, the STS-89 crew worked with the Mir 24 crew members to transfer equipment and supplies to the station and to remove materials no longer needed on Mir.

David Wolf ended his stay aboard the station after 119 days, and was replaced by NASA astronaut Andrew Thomas in the fifth and final switch of U.S. astronauts aboard Mir, as the joint Russian-American first phase of preparation for the ISS neared its conclusion. The last of 7 U.S. astronauts to stay aboard Mir during a string of 907 consecutive days of continuous U.S. presence aboard the station, Thomas began his stay with Solovyov and Vinogradov, who were replaced a week later by the Mir 25 crew.

The Mir 25 crew—Talgat Musabayev and Nikolai Budarin—arrived at Mir on January 29 with guest cosmonaut Léopold Eyharts of France, aboard Soyuz TM-27.

With their replacements in place, Solovyov and Vinogradov closed out their Mir 24 stay and returned to Earth aboard Soyuz TM-26 on February 19, 1998 with Eyharts, who logged 20 days, 19 hours and 36 minutes in space during his visit to Mir.

During their exceptional Soyuz TM-26/Mir 24 long-duration repair mission, Solovyov and Vinogradov had been in space for 197 days, 17 hours and 34 minutes.

Vinogradov spent 25 hours and 16 minutes in EVA during his five Mir 24 spacewalks.

Solovyov's spacewalking total for Mir 24 was a remarkable 34 hours and 54 minutes, during 7 EVAs. With his Mir 24 EVA heroics, he became the most prolific spacewalker in history, with more EVAs, 16, and more career time in EVA, 78 hours and 3 minutes, than any other individual.

SOYUZ MISSION 84: SOYUZ TM-27

(January 29–August 25, 1998)

Russian Manned Spaceflight

Launched aboard Soyuz TM-27 on January 29, 1998, commander Talgat Musabayev of Kazakhstan, Russian cosmonaut flight engineer Nikolai Budarin, and guest cosmonaut Léopold Eyharts of France traveled to the Mir space station, where Musabayev and Budarin served as the station's twenty-fifth long-duration crew.

Viktor Afanasyev and Sergei Treschev served as the backup crew for the Soyuz TM-27/Mir 25 mission, and Jean-Pierre Haigneré served as Eyharts's backup for the Soyuz TM-27 flight.

At Mir, Musabayev, Budarin and Eyharts were greeted by Mir 24 crew members Anatoly Solovyov and Pavel Vinogradov, and NASA astronaut Andrew Thomas, who had arrived at the station just five days earlier, on January 24, aboard the U.S. space shuttle Endeavour. Thomas was the seventh and last U.S. astronaut to serve aboard Mir as part of the joint Russian-American first phase of preparation for the International Space Station (ISS) program.

Solovyov, Vinogradov and Eyharts left the station and returned to Earth aboard Soyuz TM-26 on February 18. Thomas remained aboard Mir to serve with Musabayev and Budarin.

From his launch aboard Soyuz TM-27 with Musabayev and Budarin to his landing aboard Soyuz TM-26 with Solovyov and Vinogradov, Eyharts accumulated a total of 20 days, 19 hours and 36 minutes in space.

Thomas conducted an intensive program of scientific research, and worked with Musabayev and Budarin to help maintain the station.

During their long stay aboard Mir, Musabayev and Budarin made a series of long spacewalks to make repairs to the station. Their first two EVAs—a 6 hour, 40 minute excursion on April 1 and a 4 hour, 23 minute spacewalk on April 6—continued repairs to the station's solar array panels, which had also been a focus of earlier spacewalkers on previous missions.

They then made three long spacewalks to repair the engine that propelled Mir's thruster system, beginning with a 6 hour, 25 minute EVA on April 11. They continued the work during 6 hours and 33 minutes outside the station on April 17, and then finished the job on April 22, when they added another 6 hours and 21 minutes to their total EVA time for the flight.

On June 4, 1998, the U.S. space shuttle Discovery arrived at Mir during STS-91 for the ninth and final docking of an American shuttle and the Russian space station. The shuttle-Mir program and the program of continuous U.S. presence aboard Mir were remarkable steps forward in spaceflight cooperation for the former Cold War rivals.

The cooperative program of living and working together had also served as the first phase of the U.S. and Russian development of the International Space Station (ISS) program. In addition to hundreds of U.S. experiments carried out aboard Mir, flights aboard the shuttle by Russian cosmonauts, and the goodwill it engendered between representatives of the two countries' space programs, the cooperative program also resulted in 9 shuttle-Mir dockings by 3 different space shuttles (Atlantis and Endeavour had preceded Discovery), and a series of 7 consecutive long-duration stays at Mir by NASA astronauts for a total of 907 days of continuous U.S. presence aboard the station.

Charles Precourt commanded Discovery for the STS-91 docking with the station. He was accompanied by pilot Dominic Gorie and mission specialists Wendy Lawrence, Franklin Chang-Diaz, Janet Kavandi, and Russian cosmonaut Valeri Ryumin. The shuttle and space station crews transferred equipment and supplies to Mir, and moved the last of the U.S. science experiments and their related equipment onto Discovery for return to Earth.

After four days of docked operations, the STS-91 crew left Mir with Andrew Thomas, who completed his long-duration stay at the station after 130 days; he landed in Discovery on June 12, 1998.

Musabayev and Budarin remained aboard Mir for several more months, carrying out their assigned scientific research and performing the necessary maintenance and engineering tasks to keep the station operating safely.

Mir 26 commander Gennady Padalka and flight engineer Sergei Avdeyev arrived at Mir on August 13 aboard Soyuz TM-28 with Yuri Baturin, who made a short visit to the station before returning to Earth with Musabayev and Budarin.

After nearly two weeks of joint operations, Musabayev and Budarin completed their Mir 25 mission and returned to Earth in Soyuz TM-27 with Baturin. Musabayev and Budarin each logged a remarkable 207 days, 12 hours and 51 minutes during their long-duration stay, while Baturin accumulated 11 days, 22 hours and 41 minutes in space during his Soyuz TM-28 visit to Mir.

SOYUZ MISSION 85: SOYUZ TM-28

(August 13, 1998 –February 28, 1999)

Russian Manned Spaceflight

Launched on August 13, 1998, Soyuz TM-28 carried Mir 26 commander Gennady Padalka and flight engineer Sergei Avdeyev to the Mir space station with their Soyuz TM-28 crew mate Yuri Baturin, who would visit the station for about two weeks and then return to Earth with the Mir 25 long-duration crew. Baturin was an adviser to Russian President Boris Yeltsin; during his time at the station, he carried out a program of scientific experiments.

Sergei Zalyotin and Alexander Kaleri served as the backup crew for Padalka and Avdeyev for the Soyuz TM-28/Mir 26 mission.

At Mir, Padalka, Avdeyev and Baturin met Mir 25 commander Talgat Musabayev and flight engineer Nikolai Budarin, who were in the midst of closing out their stay of over 207 days on the station. On August 25 Musabayev, Budarin and Baturin returned to Earth in Soyuz TM-27.

From his launch with Padalka and Avdeyev aboard Soyuz TM-28 to his landing with Musabayev and Budarin in Soyuz TM-27, Baturin accumulated a total of 11 days, 22 hours and 41 minutes in space.

Padalka and Avdeyev settled into a productive routine as they began their six month stay aboard Mir. They conducted scientific research, performed the space station maintenance and engineering tasks necessary to keeping Mir operating safely, and performed the regular exercise necessary to combat losses in bone density and blood plasma that occur during long stays in space.

They also unloaded supplies from the unmanned cargo supply spacecraft Progress M-40, which launched on October 25, 1998 and remained docked with Mir until February 5, 1999.

Padalka and Avdeyev made two spacewalks during their Mir 26 mission. The first was a 30 minute excursion on September 15, 1998, during which they repaired a motor for one of the Spektr module's solar arrays. They made the EVA inside the module, which had been damaged in a collision with an unmanned Progress supply spacecraft on June 25, 1997 that had left the module depressurized and shut off from the rest of the station.

During their second EVA, on November 10, Padalka and Avdeyev deployed a satellite and attached experiments to the outside of the station. They accumulated an additional 5 hours and 54 minutes in EVA, bringing their total spacewalking time for the flight to 6 hours and 24 minutes.

On February 20, Mir 27 crew members Viktor Afanasyev and Jean-Pierre Haigneré of France arrived at Mir with guest cosmonaut Ivan Bella of Slovakia aboard Soyuz TM-29. The five cosmonauts spent about a week together at the station, and Padalka and Bella then returned to Earth in Soyuz TM-28 on February 28, 1999.

During his Mir 26 long-duration mission, Padalka spent 198 days, 16 hours and 31 minutes in space. Bella returned to Earth after 7 days, 21 hours and 56 minutes in space.

Avdeyev remained aboard Mir to serve with Afanasyev and Haigneré.

SOYUZ MISSION 86: SOYUZ TM-29

(February 20–August 28, 1999)

Russian Manned Spaceflight

Launched on February 20, 1999 with Russian commander Viktor Afanasyev, Jean-Pierre Haigneré of France, and Ivan Bella—the first citizen of Slovakia to fly in space—Soyuz TM-29 was designed as the final flight to the Mir space station (although the subsequent attempt to commercialize Mir as a tourist destination resulted in one additional flight, Soyuz TM-30, in 2000).

During Soyuz TM-29, the international crew visited with the station's resident crew members Gennady Padalka and Sergei Avdeyev for about a week. Afanasyev and Haigneré then remained aboard the station with Avdeyev to become the Mir 27 long-duration crew, and Padalka and Bella returned to Earth aboard Soyuz TM-28 on February 29, 1999. Padalka had spent 198 days, 16 hours and 31 minutes in space during his Mir 26 long-duration mission.

During his historic flight to Mir as the first citizen of Slovakia to fly in space, Bella logged 7 days, 21 hours and 56 minutes in space.

Afanasyev, Haigneré and Avdeyev spent more than six months aboard Mir, making good use of the last official mission to conduct scientific research and related experiments. They also participated in 3 spacewalks during their stay, beginning with a 6 hour, 19 minute excursion by Afanasyev and Haigneré on April 16, 1999 during which they deployed experiments on the station's exterior, and Haigneré deployed a small satellite.

Afanasyev and Avdeyev made the two final Mir 27 EVAs, to install a communications antenna on the station, on July 23 and July 28. The first exercise took 6 hours and 7 minutes, and the cosmonaut spent 5 hours and 22 minutes outside the station during their second spacewalk.

At the end of their long stay, Afanasyev, Avdeyev and Haigneré closed out the last planned mission to the station, entered their Soyuz TM-29 spacecraft, and returned to Earth on August 28, 1999.

Afanasyev and Haigneré logged 188 days, 20 hours and 16 minutes in space during their Mir 27 mission, which included 17 hours and 48 minutes in EVA for Afanasyev and 6 hours and 19 minutes of spacewalking by Haigneré.

Avdeyev, who had begun his long-duration stay at Mir with Gennady Padalka on August 13, 1998, spent a remarkable 379 days, 14 hours and 51 minutes in space during his Mir 26/Mir 27 mission. He made two spacewalks with Padalka and two with Afanasyev, accumulating a total of nearly 18 hours in EVA during his long stay at Mir.

With the safe return of Afanasyev, Avdeyev and Haigneré on August 28, 1999, Russian space officials assumed that they had launched and recovered the final crew to occupy Mir, and began finalizing their plans to de-orbit the station later that year or in the early months of 2000.

However, private sector interest in acquiring the station and operating it as a tourist destination, which was embodied in the commercial endeavors of MirCorp, resulted in the legendary space station remaining in orbit until March 2001.

SOYUZ MISSION 87: SOYUZ TM-30

(April 4–June 16, 2000)

Russian Manned Spaceflight

Soyuz TM-30 commander Sergei Zalyotin and flight engineer Alexander Kaleri launched on April 4, 2000 and traveled to the Mir space station, where they carried out the first commercially financed long-duration space mission.

The Russian government had announced plans to de-orbit Mir in the months following the end of the Mir 27 long-duration mission in August 1999. Faced with financial constraints and no doubt pleased with the station's remarkable performance in the 14 years since its February 20, 1986 launch, Russian space officials were, in 1999, in the midst of preparing the personnel, systems and equipment they would use in the development and occupation of the International Space Station (ISS).

Private investors in the West, however, saw a potentially lucrative opportunity in keeping the aging space station in orbit. Hoping to exploit the legendary Mir station for a variety of commercial uses, including space tourism, an international conglomerate of private investors based in Amsterdam struck a deal with RKK Energia to lease Mir to companies or individuals interested in using the station for a variety of commercial ventures. The arrangement gave RKK Energia a 60% stake in MirCorp.

In the lead-up to the launch of Soyuz TM-30, MirCorp provided a reported $21 million to fund the Mir 28 mission. By some estimates, the company spent as much as $40 million in its effort to keep Mir in orbit, before the costs involved out-ran MirCorp's funding capability.

Zalyotin and Kaleri arrived at Mir after the station had been vacant for seven months. Their main duties during the flight were to inspect the station and its equipment and systems, in order to provide a first-hand evaluation of Mir's condition and its sustainability for future flights.

In addition to their work inside the station, they also made a five hour, three minute spacewalk on May 12, 2000 to inspect the outside of the station.

Zalyotin and Kaleri stayed at Mir for nearly two-and-a-half months during their Soyuz TM-30/Mir 28 mission. They returned to Earth in Soyuz TM-30 on June 16, 2000, after a total flight of 72 days, 19 hours and 46 minutes in space.

Although MirCorp officials had announced in April that they intended to finance a second mission to Mir, the company's plans failed to materialize and it was unable to raise the necessary funds required to pursue any further involvement with the space station.

On March 23, 2001, the Mir space station was de-orbited, directed by Russian mission controllers to a fiery demise during a controlled re-entry into the Earth's atmosphere. The station had been in orbit for just over 15 years.

SOYUZ MISSION 88: SOYUZ TM-31

(October 31, 2000 –March 21, 2001)

Russian Manned Spaceflight

Launched from the Baikonur Cosmodrome in Kazakhstan on October 31, 2000 aboard Soyuz TM-31, NASA astronaut William Shepherd and Rosaviakosmos cosmonauts Yuri Gidzenko and Sergei Krikalyov traveled to the International Space Station (ISS), where they served for six months as the station's first long-duration crew.

The backup crew for the Soyuz TM-31/ISS Expedition 1 crew included Frank Culbertson of NASA and Vladimir Dezhurov and Mikhail Tyurin of Rosaviakosmos.

As the first resident crew aboard the ISS, Shepherd, Gidzenko and Krikalyov were the first to test out the station's equipment and systems, and its next-generation orbital living conditions. They also set a productive standard for scientific work aboard the station, which would increase steadily during subsequent missions.

Their scientific work included Earth observations and the first protein crystal growth experiment to be carried out on the station, and they also deployed an experiment known as SEEDS—the Space Exposed Experiment Developed for Students.

During their ISS Expedition 1 stay, Shepherd, Krikalyov and Gidzenko were visited by three U.S. space shuttle crews.

On December 2, 2000 the space shuttle Endeavour docked with the ISS during STS-97 to begin seven days of joint operations with the Expedition 1 crew. The STS-97 crew, which was commanded by Brent Jett and also included pilot Michael Bloomfield and mission specialists Carlos Noriega, and Joseph Tanner of NASA, and Marc Garneau of the Canadian Space Agency (CSA), delivered supplies and equipment to the ISS and also delivered and installed the station's P6 Truss segment, which included the first U.S. solar arrays for the ISS.

The STS-97 flight was the fifth space shuttle mission devoted to ISS construction tasks; the four earlier flights had occurred before the station's first crew had arrived.

STS-97 mission specialists Noriega and Tanner made three spacewalks to install and activate the 240-foot long P-6 Truss segment. After a brief visit with Shepherd, Krikalyov and Gidzenko near the end of the docked period, the Endeavour crew left the station and returned to Earth on December 11, 2000.

The space shuttle Atlantis arrived at the ISS on February 9, 2001 during STS-98. The STS-98 crew, which delivered the 16-ton U.S. Destiny Laboratory module to the ISS, included commander Kenneth Cockrell, pilot Mark Polansky, and mission specialists Robert Curbeam, Marsha Ivins, and Thomas Jones.

Curbeam and Jones made three long spacewalks to connect and activate the ISS Destiny module—the primary U.S. science facility at the station. With the addition of the Destiny Lab, the ISS surpassed the Russian Mir station as the largest space station ever flown.

The first test of the ISS control moment gyroscopes, which were designed to provide the station with an electrically-powered method of controlling its attitude in space, was successfully carried out during the period of docked operations, on February 13. Mission managers on Earth were glad to hear that the gyroscopes had performed as expected, as the devices were intended to conserve fuel by electrically orienting the ISS, as opposed to maintaining its orientation by firings of its propulsion system.

Atlantis undocked from the ISS on February 15, after the STS-98 and space station crews had transferred 3,000 pounds of equipment and supplies from the shuttle to the ISS and removed 850 pounds of materials no longer needed aboard the station. The shuttle returned to Earth on February 20, 2001.

Shepherd, Gidzenko and Krikalyov also welcomed the STS-102 crew of the space shuttle Discovery—which included their replacements, the crew of ISS Expedition 2—during their long stay at the ISS. Discovery docked with the station in the early morning hours of March 10, 2001, bringing STS-102 commander James Wetherbee, pilot James Kelly, mission specialists Paul Richards and Andrew Thomas, and ISS Expedition 2 crew members James Voss and Susan Helms of NASA, and Rosaviakosmos cosmonaut Yury Usachev to the station.

James Voss and Susan Helms began their time at the ISS with a record-setting spacewalk of 8 hours and 56 minutes—which set new records for the longest EVA in the history of the space shuttle program and the longest spacewalk by a female astronaut—to prepare the ISS Unity module to receive the Leonardo Multi-Purpose Logistics Module (MPLM), which was used for the first time during STS-102.

The first of three modules designed to ease the transfer of supplies and equipment from the shuttle to the ISS, Leonardo was manufactured by Italian aerospace company Alenia Aerospazio under the auspices of the Italian space agency Agenzia Spaziale Italiana (ASI). Subsequent MPLMs were named Raffaello and Donatello.

Using the Leonardo module to move equipment and supplies onto the ISS, the shuttle and space station crews transferred nearly five tons of material to the station and removed about one ton of refuse for the return trip to Earth.

STS-102 mission specialists Richards and Thomas also made an EVA, to attach a platform to the outside of the ISS and to connect cables to the Destiny module.

At the end of the docked operations, Voss, Helms and Usachev transferred to the space station to begin their stay as the ISS Expedition 2 crew, and Shepherd, Krikalyov and Gidzenko ended their long-duration mission at the station and returned to Earth aboard Discovery with the STS-102 crew.

Shepherd, Krikalyov and Gidzenko spent 136 days, 17 hours and 9 minutes aboard the ISS.

On April 21, 2001 the Expedition 2 crew received its first visitors at the station, with the arrival of the space shuttle Endeavour and the STS-100 crew.

Commanded by Kent Rominger, the STS-100 crew included pilot Jeffrey Ashby and mission specialists Scott Parazynski and John Phillips of NASA, Chris Hadfield of the Canadian Space Agency (CSA), European Space Agency (ESA) astronaut Umberto Guidoni of Italy, and Russian cosmonaut Yuri Lonchakov.

Hadfield and Parazynski made two spacewalks during their STS-100 visit to the ISS, to install the Canadarm2 robotic arm on the station.

Designed and built in Canada as a follow-up to the successful space shuttle Remote Manipulator System (RMS) robotic arms, the Canadarm2 device for the space station was the first of three parts of the Space Station Mobile Servicer System (SSMSS), which also included the Mobile Base System (MBS) work platform and the Special Purpose Dexterous Manipulator (SPDM) or "Canada Hand." The Canadarm2 portion of the system weighed 3,968 pounds, and was designed with a larger range of motion than a human arm.

During the period of docked operations between Endeavour and the ISS, a computer problem caused the space station to lose contact with Mission Control. In a unique solution to the problem, ISS Expedition 2 flight engineer Susan Helms was able to communicate with ground controllers via Endeavour's communications link. Working with engineers on Earth, she was able to solve the computer difficulty and reestablish communications between the ISS and Mission Control.

The STS-100 crew also delivered 6,000 pounds of equipment and supplies to the ISS, using the Raffaello Multi-Purpose Logistics Module (MPLM).

Endeavour undocked from the space station on April 29; the next day, April 30, 2001, cosmonauts Talgat Musabayev and Yuri Baturin and American businessman Dennis Tito—the first space tourist, who paid the Russian space agency $20 million to arrange his flight to the ISS—arrived aboard Soyuz TM-32 for a brief visit. After a week's stay, Musabayev, Baturin and Tito returned to Earth aboard Soyuz TM-31 on May 6, 2001.

SOYUZ MISSION 89: SOYUZ TM-32

(April 28–May 6, 2001)

Russian Manned Spaceflight

Soyuz TM-32 commander Talgat Musabayev and flight engineer Yuri Baturin lifted off aboard Soyuz TM-32 on April 28, 2001 to travel to the International Space Station (ISS) with a unique passenger: American businessman Dennis Tito, who arranged for his flight by paying the Russian space agency $20 million, thus becoming the first space tourist.

Russian cosmonauts Viktor Afanasyev and Konstantin Kozeyev served as the Soyuz TM-32 backup crew for Musabayev and Baturin.

In addition to delivering Tito to the ISS for his brief stay, the Soyuz TM-32 flight was also intended as a Soyuz switching flight, in that the Soyuz TM-32 spacecraft was left at the ISS at the end of the short flight to replace the Soyuz TM-31 vehicle that Musabayev, Baturin and Tito used to return to Earth. The Soyuz TM-and Soyuz TMA-vehicles can be safely operated in space for about six months before their on-board systems begin to deteriorate, and thus must be exchanged within that time if they are to be available for long-duration crews at the space station.

At the ISS, the Soyuz TM-32 crew was met by the ISS Expedition 2 crew—commander Yury Usachev of Rosaviakosmos, and flight engineers James Voss and Susan Helms of NASA.

The two crews visited for about a week, and Musabayev, Baturin and Tito then returned to Earth aboard Soyuz TM-31 on May 6, 2001 after a total flight of 7 days, 22 hours and 4 minutes.

SOYUZ MISSION 90: SOYUZ TM-33

(October 21–31, 2001)

Russian Manned Spaceflight

Soyuz TM-33 commander Viktor Afanasyev and flight engineer Konstantin Kozeyev lifted off aboard Soyuz TM-33 on October 21, 2001 to travel to the International Space Station (ISS) with European Space Agency (ESA) astronaut Claudie Haigneré of France (who earlier flew in space as Claudie Andre-Deshays, before her marriage to fellow French astronaut Jean-Pierre Haigneré). Haigneré's trip to the ISS was known as the ESA Andromeda mission.

Sergei Zalyotin and Nadeshda Kushelnaja served as the backup crew for the Soyuz TM-33 flight.

At the ISS, Afanasyev, Kozeyev and Haigneré were greeted by the ISS Expedition 3 crew—commander Frank Culbertson of NASA, and Russian cosmonauts Vladimir Dezhurov and Mikhail Tyurin of Rosaviakosmos.

In addition to delivering Haigneré to the ISS for the Andromeda program of scientific research, the Soyuz TM-33 flight was also intended as a Soyuz switching flight. Because the Soyuz TM-and Soyuz TMA-vehicles can be safely operated in space for about six months, they must be exchanged within that time if they are to be available for long-duration crews at the space station. In keeping with the need to exchange the Soyuz at the station, Afanasyev, Kozeyev and Haigneré left Soyuz TM-33 at the ISS when they departed from the station after a brief stay, and returned to Earth aboard Soyuz TM-32.

The first in a series of ISS science missions designed by the ESA, the Andromeda mission was arranged by the French space agency Centre National d'Études

Soyuz TM-32 undocks from the International Space Station (ISS) in October, 2001. [NASA/courtesy of nasaimages.org]

Spatiales (CNES) and Rosaviakosmos. The Andromeda research included experiments in a variety of fields, including biology and life and materials sciences, and also incorporated educational and Earth observation activities.

Afanasyev, Kozeyev and Haigneré landed in Soyuz TM-32 on October 31, 2001, after a flight of 9 days, 19 hours and 59 minutes in space.

SOYUZ MISSION 91: SOYUZ TM-34

(April 25–May 5, 2002)

Russian Manned Spaceflight

Soyuz TM-34 commander Yuri Gidzenko of Rosaviakosmos, European Space Agency (ESA) flight engineer Roberto Vittori of Italy, and Mark Shuttleworth—the first citizen of South Africa to fly in space, and second space tourist—lifted off aboard Soyuz TM-34 on April 25, 2002 to travel to the International Space Station (ISS).

Gennady Padalka and Oleg Kononenko served as the backup crew for the Soyuz TM-34 flight.

Vittori's visit to the ISS, which included a program of scientific research designed by ESA scientists, was also known as the ESA Marco Polo mission.

Shuttleworth had arranged his flight to the space station by paying $20 million to the Russian space agency Rosaviakosmos.

At the ISS, Gidzenko, Vittori and Shuttleworth were greeted by the ISS Expedition 4 crew—commander Yuri Onufrienko of Rosaviakosmos, and NASA flight engineers Daniel Bursch and Carl Walz.

Vittori's work during the Marco Polo project included tests of a nonintrusive blood pressure monitor, a microgravity study, an investigation into the impact that cosmic particles have on the human body during spaceflight, and an experiment involving a test of clothing materials.

Soyuz TM-34 was also designed as a Soyuz switching flight, in which one Soyuz vehicle was exchanged for another craft that had been in space for a period approaching the six month safe operations limit of the Soyuz TM-and Soyuz TMA-series. In keeping with the need to leave a fresh Soyuz at the station as a safeguard in case the ISS crew might need to make an emergency return to Earth, Gidzenko, Vittori and Shuttleworth left Soyuz TM-34 docked to the ISS when they left the station. They returned to Earth in Soyuz TM-33 on May 5, 2002 after a total flight of 9 days, 21 hours and 25 minutes.

Onufrienko, Bursch and Walz remained aboard the ISS until June, 2002, when they completed their Expedition 4 mission and returned to Earth on the space shuttle Endeavour during STS-111, after being replaced aboard the station by the ISS Expedition 5 crew—commander Valery Korzun and flight engineer Sergei Treschev of Rosaviakosmos, and flight engineer Peggy Whitson of NASA.

SOYUZ MISSION 92: SOYUZ TMA-1

(October 30–November 10, 2002)

Russian Manned Spaceflight

The first Soyuz TMA-spacecraft—an upgrade to the Soyuz TM-series to take into account the vehicle's expanded role in ferrying cosmonauts and astronauts from many nations—lifted off from the Baikonur Cosmodrome on October 30, 2002 to travel to the International Space Station (ISS) with commander Sergei Zalyotin and flight engineer Yuri Lonchakov of Rosaviakosmos and European Space Agency (ESA) flight engineer Frank De Winne of Belgium.

De Winne's flight to the ISS, which included a program of scientific research, was known as the ESA Odissea mission.

Yuri Lonchakov was originally a member of the backup crew for the Soyuz TMA-1 flight, with fellow Russian cosmonaut Aleksandr Lazutkin.

For a time it seemed likely that Zalyotin and De Winne would be accompanied by a paying passenger, as the American entertainer Lance Bass of the musical group 'N Sync announced plans to become the third space tourist by paying Rosaviakosmos $20 million. The arrangement failed to materialize, however, and as a result, Lonchakov moved up from the backup crew to become the third member of the primary Soyuz TMA-1 crew.

At the ISS, Zalyotin, De Winne and Lonchakov were greeted by the ISS Expedition 5 crew—commander Valery Korzun and flight engineer Sergei Treschev of Rosaviakosmos, and flight engineer Peggy Whitson of NASA.

De Winne's Odissea program of scientific research included 23 experiments, 4 of which involved the use of the ESA-developed Microgravity Science Glovebox (MSG).

Soyuz TMA-1 was designed as a Soyuz switching flight, in that it involved the exchange of Soyuz spacecraft at the ISS. Because the Soyuz TMA-series of spacecraft have a six month safe operations limit, a vehicle docked at the space station must be exchanged for a new Soyuz before six months pass. With that in mind, Zalyotin, De Winne and Lonchakov left Soyuz TMA-1 at the ISS when they departed from the station.

They returned to Earth aboard Soyuz TM-34—the last of the Soyuz TM-series—on November 10, 2002. During their Soyuz TMA-1/Odissea mission, Zalyotin, De Winne and Lonchakov spent 10 days, 20 hours and 53 minutes in space.

Korzun, Treschev and Whitson completed their six month stay as the ISS Expedition 5 crew when the U.S. shuttle Endeavour arrived at the station on November 25, 2002 during STS-113. The STS-113 crew installed the ISS P1 Truss segment, and also delivered the ISS Expedition 6 crew—commander Kenneth Bowersox and flight engineer Don Pettit of NASA, and flight engineer Nikolai Budarin of Rosaviakosmos—to the station.

Bowersox, Budarin and Pettit lived and worked aboard the ISS until May 2003. They performed an extensive program of scientific research with an emphasis on health and medical observations, including studies of the effects of EVA on pulmonary function, the impact of radiation absorbed by astronauts during EVA, and the spaceflight-induced reactivation of latent Epstein-Barr virus. They also did materials processing work, and performed protein crystal growth experiments.

Bowersox and Pettit made two spacewalks during their ISS Expedition 6 mission, accumulating a total of 13 hours and 17 minutes in EVA.

On April 26, 2003, ISS Expedition 7 commander Yuri Malenchenko of Rosaviakosmos and flight engineer Edward Lu of NASA arrived at the station in Soyuz TMA-2. A week later, Bowersox, Budarin and Pettit completed their long-duration mission and left the station in Soyuz TMA-1, to return to Earth on May 3, 2003. They endured a difficult landing when a computer error caused them to make a ballistic re-entry and touch down nearly 300 miles from their intended landing site. Fortunately, they were not injured during the incident.

During their ISS Expedition 6 mission, Bowersox, Budarin and Pettit were in space for 161 days, 1 hour and 17 minutes.

SOYUZ MISSION 93: SOYUZ TMA-2

(April 26–October 28, 2003)

Russian Manned Spaceflight

The International Space Station (ISS) Expedition 7 crew—commander Yuri Malenchenko of Rosaviakosmos and Flight engineer Edward Lu of NASA—lifted

Russian Mission Control, April 30, 2003. [NASA/courtesy of nasaimages.org]

off from the Baikonur Cosmodrome aboard Soyuz TMA-2 on April 26, 2003 to travel to the ISS, where they would live and work for more than six months.

Alexander Kaleri of Rosaviakosmos and Michael Foale of NASA served as the backup crew for the Soyuz TMA-2/ISS Expedition 7 mission.

Malenchenko and Lu had previously visited the ISS during STS-106 in September 2000, prior to the arrival of the station's first long-duration crew. During that flight, the STS-106 shuttle crew prepared the ISS for the Expedition One crew, and Malenchenko and Lu made a spacewalk of 6 hours and 14 minutes to attach the station's magnetometer and to link power, data and communications cables between the station and its Zvezda Service Module.

During their six month Expediton 7 stay at the ISS, Malenchenko and Lu conducted a vigorous program of scientific research that included experiments in biotechnology, fluid dynamics, medical research, and protein crystal growth, among others.

Another highlight of their flight occurred on August 10, 2003 when Malenchenko became the first person to be married while in space. The cosmonaut and his bride, Ekaterina Dmitriev, exchanged vows in a long-distance service, with her at the NASA Johnson Space Center in Houston, Texas, and him in orbit aboard the ISS 240 miles above New Zealand.

On October 18, 2003 Malenchenko and Lu welcomed the ISS Expedition 8 crew—commander Michael Foale and flight engineer Alexander Kaleri—who arrived at the station aboard Soyuz TMA-3 with European Space Agency (ESA) astronaut Pedro Duque of Spain, who was traveling to the ISS for a brief visit to conduct the ESA Cervantes program of scientific research.

Primarily sponsored by the Spanish Ministry of Science and Technology, the Cervantes program included 22 experiments in a variety of fields, including biology, human physiology and physical science, and also featured educational activities and technology demonstrations.

Duque also gave interviews to media representatives of Spain and Germany during his stay at the ISS, and communicated with the student winners of the Habla ISS competition via amateur radio.

Malenchenko and Lu turned over the station's operations to Foale and Kaleri, and then returned to Earth with Duque aboard Soyuz TMA-2 on October 28, 2003. During their ISS Expedition 7 mission, Malenchenko and Lu spent 184 days, 22 hours and 47 minutes in space.

SOYUZ MISSION 94: SOYUZ TMA-3

(October 18, 2003–April 30, 2004)

Russian Manned Spaceflight

The International Space Station (ISS) Expedition 8 crew—commander Michael Foale of NASA and flight engineer Alexander Kaleri of Russia's Federal space agency—launched from the Baikonur Cosmodrome in Kazakhstan aboard Soyuz TMA-3 on October 18, 2003 with European Space Agency (ESA) astronaut Pedro Duque of Spain to travel to the ISS.

Russian cosmonaut Valery Tokarev and NASA astronaut William McArthur, Jr. served as the backup crew for the long-duration mission, and European Space Agency (ESA) astronaut André Kuipers of The Netherlands served as Pedro Duque's backup for the Soyuz TMA-3 flight.

Duque's visit to the ISS was also known as the ESA Cervantes mission of scientific research.

When they arrived at the station, Foale, Kaleri and Duque were met by ISS Expedition 7 crew members Yuri Malenchenko of Rosaviakosmos and Edward Lu of NASA.

During his short visit to the station, Duque conducted 22 experiments in carrying out the Cervantes research program, which was primarily sponsored by the Spanish Ministry of Science and Technology. The Cervantes experiments covered a variety of fields, including biology, human physiology and physical science, and also featured educational activities and technology demonstrations.

Media representatives from Spain and Germany interviewed Duque during his stay in orbit, and he also communicated with the student winners of the Habla ISS competition via amateur radio.

After the five crew members visited for more than a week and Duque finished his science work, Malenchenko and Lu officially turned over command of the ISS to Foale and Kaleri, and then left the station to return to Earth with Duque aboard Soyuz TMA-2 on October 28, 2003. Malenchenko and Lu accumulated a total of 184 days, 22 hours and 46 minutes in space during their ISS Expedition 7 mission; from launch to landing, Duque spent a total of 9 days, 21 hours and 3 minutes in space during his Cervantes visit to the ISS aboard Soyuz TMA-3.

Foale and Kaleri settled into a productive routine aboard the ISS, conducting a vigorous program of scientific research and performing the maintenance and engineering tasks necessary to keep the station operating smoothly. They also exercised regularly, to combat the losses in blood plasma volume and bone density that occur during long stays in space.

Among their many scientific tests, Foale and Kaleri conducted a variety of innovative experiments that included the use of an advanced diagnostic ultrasound in microgravity, study of chromosomal aberrations in the astronauts' blood lymphocytes, an investigation of the reactions of nematode worms in the weightless environment, a space soldering experiment, and several protein crystal growth exercises.

The ISS Expedition 8 flight also featured a spacewalk by Foale and Kaleri, when they left the station late on February 26, 2004 to attach new experiments to the outside of the ISS and to retrieve experiments deployed on earlier missions. They spent 3 hours and 55 minutes in EVA during the exercise, and returned to the interior of the station in the early hours of February 27.

On April 19, 2004, Foale and Kaleri welcomed their replacements, ISS Expedition 9 commander Gennady Padalka and flight engineer Mike Fincke, who arrived at the station aboard Soyuz TMA-4 with European Space Agency (ESA) astronaut André Kuipers of The Netherlands, who traveled to the ISS for a short visit to conduct the ESA DELTA program of scientific research.

Foale and Kaleri returned to Earth with Kuipers on April 30, 2004 in Soyuz TMA-3 after spending 194 days, 18 hours and 34 minutes in space during their ISS Expedition 8 mission. Kuipers spent 10 days, 20 hours and 52 minutes in space during his DELTA visit to the ISS.

SOYUZ MISSION 95: SOYUZ TMA-4

(April 19–October 24, 2004)

Russian Manned Spaceflight

The International Space Station (ISS) Expedition 9 crew—commander Gennady Padalka of Russia's Federal Space Agency and flight engineer Michael Fincke of NASA—launched from the Baikonur Cosmodrome in Kazakhstan on April 19, 2004 to travel to the space station aboard Soyuz TMA-4, with European Space Agency (ESA) astronaut André Kuipers of The Netherlands.

Kuipers's visit to the ISS was also known as the ESA DELTA program of scientific research.

Russian cosmonaut Saliszan Sharipov and NASA astronaut Leroy Chiao served as the backup crew for the Soyuz TMA-4/ISS Expedition 9 mission, and ESA astronaut Gerhard Thiele of Germany backed up Kuipers for the Soyuz TMA-4/DELTA mission.

When they arrived at the station, Padalka, Fincke and Kuipers were met by ISS Expedition 8 crew members Michael Foale and Alexander Kaleri.

Kuipers, a medical doctor, conducted 21 experiments during his DELTA research program at the ISS, covering areas that included biology, human physiology and microbiology. Specific experiments included studies of how the human cardiovascular system adapts to the microgravity environment, an investigation into nematode worm genetics, and observations of microbial activity on the space station. The program also featured three technology demonstrations, which included a test of an astronaut orientation system, and educational activities that involved school children in The Netherlands and throughout Europe.

One of Kuipers's most innovative assignments was a study of sprites, the meteorological phenomena that result when atmospheric electrical discharges take place above cumulonimbus clouds during a thunderstorm. He also performed space photography chores, capturing an image of his home nation that provided a stunning view of the Dutch coast and the Wadden Islands, and the Afsluitdiijk dam between the Zuiderzee inlet and Lake IJssel.

The DELTA mission was sponsored by the Dutch ministries of Economic Affairs and Education, Culture and Science. Kuipers was able to complete about 85 percent of the scientific program during his stay at the ISS, and in one case—a test of a heat transfer apparatus known as the HEAT experiment—the experiment was completed long after he left the station, when Fincke and Padalka tried the experiment again in September 2004.

Kuipers left the ISS with Foale and Kaleri on April 30, 2004, returning to Earth aboard Soyuz TMA-3. The two ISS Expedition 8 crew members had been in space for 194 days, 18 hours and 34 minutes; Kuipers accumulated 10 days, 20 hours and 52 minutes in space during his DELTA visit to the ISS.

Padalka and Fincke remained aboard the station to begin their long-duration stay. They settled into a productive routine, carrying out an intensive program of science work that included ongoing experiments, fluid dynamics investigations, and four spacewalks to maintain and upgrade the station's exterior.

They first ventured outside of the ISS on June 24, but ran into an immediate problem when the oxygen tank on Fincke's spacesuit lost pressure, forcing them to rush back to the safety of the station's interior just 13 minutes into the planned five hour-plus EVA exercise.

On June 30 they tried again, and fared far better. Starting late in the day and working into the early hours of July 1, Padalka and Fincke were able to complete their assigned task of servicing one of the station's gyroscopes in 5 hours and 40 minutes.

In their final two EVAs, on August 3 and September 3, they installed antennas and made repairs to laser reflecting devices on the outside of the station. They added four-and-a-half hours to their spacewalking time during the August 3 EVA, and the excursion on September 3 added another 5 hours and 20 minutes.

On October 14, Padalka and Fincke welcomed their replacements, Salizhan Sharipov and Leroy Chiao, who arrived at the ISS aboard Soyuz TMA-5 with Russian cosmonaut Yuri Shargin.

Padalka and Fincke completed their ISS Expedition 9 stay 10 days later, when they left the station and returned to Earth in Soyuz TMA-4 on October 24, 2004 with Shargin, who spent 9 days, 21 hours and 30 minutes in space during his Soyuz TMA-5 visit to the ISS. Padalka and Fincke accumulated 187 days, 21 hours and 17 minutes in space, including 15 hours and 43 minutes in EVA, during their ISS Expedition 9 mission.

SOYUZ MISSION 96: SOYUZ TMA-5

(October 14, 2004–April 24, 2005)

Russian Manned Spaceflight

The International Space Station (ISS) Expedition 10 crew—commander Leroy Chiao of NASA and flight engineer Salizhan Sharipov of Russia's Federal Space Agency—lifted off in Soyuz TMA-5 from Baikonur Cosmodrome in Kazakhstan on October 14, 2004 to travel to the ISS with Yuri Shargin, a test cosmonaut representing the Russian Space Forces.

William McArthur of NASA and Russian cosmonaut Valeri Tokarev served as the backup crew for the ISS Expedition 10 mission.

Chiao, Sharipov and Shargin met ISS Expedition 9 crew members Gennadi Padalka and Mike Fincke at the ISS, and after a brief period of joint operations, Padalka and Fincke turned over command of the station to Chiao and Sharipov, and then returned to Earth in Soyuz TMA-4 with Shargin on October 24, 2004.

During his brief visit to the ISS, Shargin spent a total of 9 days, 21 hours and 30 minutes in space.

Chiao and Sharipov lived and worked aboard the ISS for the next six months. During their stay as the station's tenth long-duration crew, they conducted a program of 23 experiments and performed space station maintenance and engineering tasks necessary to keep the ISS in good working order. They also exercised regularly, to combat the losses in blood plasma volume and bone density that occur during long stays in space, and made two spacewalks.

They worked on the outside of the ISS Zvezda module during their first spacewalk, on January 26, 2005, spending 5 hours and 28 minutes in EVA. On their second excursion, on March 28, they spent another five hours and six minutes outside the ISS while deploying a small Russian satellite and preparing the station for the planned first rendezvous by the European Space Agency (ESA) Automated Transfer Vehicle (ATV), the 20.7 ton cargo spacecraft designed as the ESA's first-ever spacecraft capable of rendezvous and docking.

In April, Chiao and Sharipov welcomed their replacements, ISS Expedition 11 crew members Sergei Krikalyov and John Phillips, who arrived at the station aboard Soyuz TMA-6 with ESA astronaut Roberto Vittori of Italy. Vittori conducted the ESA Eneide program of scientific research during his brief stay at the ISS, and then returned to Earth with Chiao and Sharipov in Soyuz TMA-5 with Chiao and Sharipov.

During their long-duration stay at the ISS as the Expedition 10 crew, Chiao and Sharipov accumulated 192 days, 19 hours and 2 minutes in space, including

10 hours and 34 minutes in EVA. Vittori spent 9 days, 21 hours and 22 minutes in space during the Eneide mission.

SOYUZ MISSION 97: SOYUZ TMA-6

(April 15, 2005–October 11, 2005)

Russian Manned Spaceflight

The International Space Station (ISS) Expedition 11 crew—commander Sergei Krikalyov of the Russian Federal Space Agency and flight engineer John Phillips of NASA—launched from Baikonur Cosmodrome in Kazakhstan on April 15, 2005 to travel to the station aboard Soyuz TMA-6 with European Space Agency (ESA) astronaut Roberto Vittori of Italy.

Vittori's visit to the ISS was also known as the ESA Eneide program of scientific research.

Russian cosmonaut Mikhail Tyurin and NASA astronaut Daniel Tani served as the backup crew for the ISS Expedition 11 mission, and Canadian Space Agency (CSA) astronaut Robert Thirsk served as Vittori's backup for the Soyuz TMA-6 flight.

At the ISS, Krikalyov, Phillips and Vittori were met by ISS Expedition 10 crew members Leroy Chiao and Salizhan Sharipov.

Vittori successfully completed all major objectives of the Eneide research program, which included 22 experiments designed by scientists in Italy, Denmark, Germany, Russia, Switzerland, and the United States. The work included biological studies, educational activities, investigations in human physiology, and technology tests.

The primary technology experiment involved a test of the reliability of the signal emitted by the European Geostationary Navigation Overlay Service (EGNOS), ESA's satellite navigation service.

ESA mission controllers managed the Eneide mission at the agency's new Columbus Control Center at the German Aerospace Center at Oberpfaffenhofen, Germany.

In addition to his scientific work, Vittori also communicated with several groups of his fellow citizens during his time at the ISS. He spoke with officials from the Italian government, members of the media, and with school children, and also talked with representatives of the Italian Ministry of Defense and the Region of Lazio, which had co-sponsored the Eneide mission.

An agricultural experiment, the Agrospace Experiment Suite, consisted of a student exercise called Space Beans for Students and a scientific experiment known as Seedlings. In each case, seeds were grown in sealed plastic bags to determine the impact of the space environment on their ability to sprout and grow.

School children in Rome, Civitavecchia and Pisa duplicated Vittori's work in their classrooms for comparison to the results he achieved aboard the ISS.

After completing the Eneide research program, Vittori returned to Earth with Expedition 10 crew members Chiao and Sharipov in Soyuz TMA-5 on April 24, 2005.

Vittori spent 9 days, 21 hours and 22 minutes in space during the Eneide mission.

Krikalyov and Phillips lived and worked aboard the ISS for nearly six months. They continued the ongoing scientific research of previous crews and carried out the maintenance and engineering tasks necessary to keep the station in good working order.

During their Expedition 11 stay, Krikalyov and Phillips also welcomed the STS-114 crew of the space shuttle Discovery, which docked with the station on July 28, 2005. The first shuttle flight after the loss of the shuttle Columbia in February, 2003, the STS-114 mission had gotten off to a shaky start when foam insulation from its External Tank came loose during launch—an incident similar to the one that had doomed Columbia. Fortunately, the insulation that broke off during the STS-114 launch did not hit Discovery, but the failure to secure the foam caused NASA officials to ground all future shuttle flights until a better solution could be found.

Prior to Discovery's docking with the ISS, STS-114 commander Eileen Collins piloted the shuttle through the first-ever Rendezvous Pitch Maneuver, which caused the shuttle to slowly revolve end over end at a rate of three quarters of a degree per second. The maneuver allowed Krikalyov and Phillips to get a clear view of the bottom of the shuttle—where the majority of the vehicle's heat-resistant tiles are located—from their vantage point within the ISS, at a distance of about 600 feet. They carefully photographed the area with digital cameras, and then transmitted the images they captured to Mission Control, where a team of 200 specialists closely examined the pictures for any sign of damage to Discovery's vulnerable underside.

As things worked out, the shuttle had not been damaged during launch, and the STS-114 crew safely returned to Earth on August 9, 2005 after delivering equipment and supplies to the ISS and helping out with maintenance and repair tasks at the station.

On August 18, 2005, Krikalyov and Phillips ventured outside of the station to make a spacewalk of 4 hours and 58 minutes, which stretched into the early morning hours of the following day. They installed a television camera on the exterior of the ISS, and retrieved experiments.

The ISS Expedition 12 crew—commander Valeri Tokarev and flight engineer William McArthur, Jr.—arrived at the station on October 3 aboard Soyuz TMA-7 with space tourist Gregory Hammond Olsen. Krikalyov and Phillips ended their Expedition 11 stay on the ISS and landed in Soyuz TMA-6 with Olsen on October 11, 2005.

From launch to landing, Olsen spent 9 days, 21 hours and 15 minutes in space during his Soyuz TMA-7/Soyuz TMA-6 flight.

Krikalyov and Phillips accumulated 179 days and 23 minutes in space during their ISS Expedition 11 mission.

SOYUZ MISSION 98: SOYUZ TMA-7

(October 1, 2005–April 8, 2006)

Russian Manned Spaceflight

The International Space Station (ISS) Expedition 12 crew—commander William McArthur, Jr. of NASA and Flight engineer Valery Tokarev of the Russian Federal

Space Agency Roscosmos—launched from the Baikonur Cosmodrome aboard Soyuz TMA-7 on October 1, 2005 to fly to the station with American space tourist Gregory Hammond Olsen.

Jeffrey Williams of NASA and Russian cosmonaut Mikhail Tyurin served as the backup crew for the ISS Expedition 12 mission, and Russian space tourist Sergei Kostenko was in line to make the flight if Olsen had dropped out.

Olsen arranged his visit to the ISS by paying $20 million to Roscosmos. After a short visit to the station, he returned to Earth with ISS Expedition 11 crew members Sergei Krikalyov and John Phillips on October 11, 2005, aboard Soyuz TMA-6. From launch to landing, he spent 9 days, 21 hours and 15 minutes in space.

McArthur and Tokarev lived and worked aboard the ISS for six months. They conducted scientific experiments, performed health, nutrition and safety activities, and made two spacewalks during their long stay.

On November 7, 2005 they spent 5 hours and 22 minutes outside the ISS while installing a camera and removing and replacing a variety of parts in various fixtures on the outside of the station, including the installation of a replacement controller module on the station's Mobile Transporter system.

They made a second spacewalk on February 3, 2006. During 5 hours and 43 minutes in EVA, McArthur and Tokarev deployed SuitSat-1—a spacesuit fitted with a radio transmitter—and worked on several experiments.

During their two ISS Expedition 12 spacewalks, McArthur and Tokarev spent 11 hours and 5 minutes in EVA.

At the end of their long stay, McArthur and Tokarev welcomed their replacements, ISS Expedition 13 commander Pavel Vinogradov of Roscosmos and flight engineer Jeffrey Williams of NASA, who launched on March 29, 2006 in Soyuz TMA-8 with Marcos Pontes—the first citizen of Brazil to fly in space.

Pontes's flight to the ISS was also known as the Centennial mission, in honor of Brazil's pioneering pilot Alberto Santos Dumont.

Michael Fincke of NASA and Russian cosmonaut Fyodor Yurchikhin served as the backup crew for the ISS Expedition 13 mission.

McArthur, Tokarev and Pontes returned to Earth aboard Soyuz TMA-7 on April 8, 2006. McArthur and Tokarev spent 189 days, 19 hours and 53 minutes in space during their ISS Expedition 12 mission.

From his launch, aboard Soyuz TMA-8 with Vinogradov and Williams, to his landing in Soyuz TMA-7 with McArthur and Tokarev, Pontes spent 9 days, 21 hours and 18 minutes in space during his historic spaceflight.

SOYUZ MISSION 99: SOYUZ TMA-8

(March 30, 2006 –September 29, 2006)

Russian Manned Spaceflight

Soyuz TMA-8 launched from the Baikonur Cosmodrome on March 30, 2006. Pavel Vinogradov, representing the Russian Federal Space Agency Roscosmos, commanded Soyuz TMA-8. He was joined by flight engineer Jeffrey Williams of NASA, and Marcos Cesar Pontes, the first citizen of Brazil to fly in space.

The Soyuz TMA-crew traveled to the International Space Station (ISS), where Vinogradov became commander of the ISS Expedition 13 crew, with Williams serving as flight engineer for the long-duration flight. Vinogradov and Williams replaced the Expedition 12 crew of William McArthur and Valery Tokarev.

Although Pontes' flight to the ISS was arranged via a contractual arrangement with Roscosmos in a fashion similar to that of individuals who flew as space tourists, the arrangement was a national project that served as a point of pride for Brazil and all its citizens. The flight was celebrated in Brazil as the Centennial mission, in honor of the nation's pioneering pilot Alberto Santos Dumont.

After a short stay at the ISS, Pontes returned to Earth aboard Soyuz TMA-7 on April 8, 2006 with McArthur and Tokarev. From launch to landing, he spent 9 days, 21 hours and 18 minutes in space during his historic spaceflight.

Vinogradov and Williams conducted a productive program of scientific research during their Expedition 13 stay at the ISS, and on June 1, 2006 they ventured outside the station for a spacewalk of 6 hours and 31 minutes while performing repair work and retrieving experiments.

On July 6, 2006 they welcomed a new Expedition 13 crew member when European Space Agency (ESA) astronaut Thomas Reiter arrived aboard the space shuttle Discovery during STS-121. Reiter served as the second flight engineer for Expedition 13.

Reiter and Williams made the second spacewalk of the ISS Expedition 13 mission on August 3 when they installed replacement hardware outside the ISS during an EVA of just under six hours.

At the end of their Expedition 13 mission, Vinogradov and Williams returned to Earth in Soyuz TMA-8 on September 29, 2006, after a flight of 182 days, 23 hours and 44 minutes. Reiter remained aboard the ISS and became a member of the Expedition 14 crew.

SOYUZ MISSION 100: SOYUZ TMA-9

(September 18, 2006–April 21, 2007)

Russian Manned Spaceflight

Launched from the Baikonur Cosmodrome on September 18, 2006, Soyuz TMA-9 was commanded by Mikhail Tyurin of the Russian Federal Space Agency Roscosmos, who lifted off with NASA astronaut Michael Lopez-Alegria and Anousheh Ansari, the first native of Iran to fly in space.

For Tyurin and Lopez-Alegria, the flight to the International Space Station (ISS) represented the start of their long-duration mission as the ISS Expedition 14 crew—for which they would switch roles, with Lopez-Alegria serving as the commander of ISS Expedition 14, and Tyurin serving as the Expedition 14 flight engineer.

For Ansari, the Soyuz TMA-9 flight marked the realization of a lifelong dream to fly in space.

As a driving force behind the sponsorship of the $10 million Ansari X Prize competition for commercial spacecraft, an extremely successful entrepreneur and a life member of the Association of Space Explorers and the advisory board of the Teacher in Space project, Ansari was a well-known figure in the space industry even before her flight aboard Soyuz TMA-9.

Her rise to prominence in the corporate world and as an advocate of commercial space travel have provided her with a unique perspective, and she has used her fame to promote the virtues of space exploration as a means of furthering the cause of peace among the disparate nations of the world.

Ansari was originally a backup for Daisuke Enomoto, who was originally scheduled to fly aboard Soyuz TMA-9 but was forced off of the flight about a month before launch due to a medical issue, clearing the way for Ansari to move up to the prime crew.

During her eight-day stay at the ISS, Ansari performed several experiments for the European Space Agency (ESA), and visited with resident Expedition 13 crew members Pavel Vinogradov and Jeffrey Williams, with whom she returned to Earth aboard Soyuz TMA-8 on September 29, 2006. From launch to landing, Anousheh Ansari accumulated 10 days, 21 hours and five minutes in space during her first spaceflight.

At the ISS, Lopez-Alegria and Tyurin joined ESA astronaut Thomas Reiter, who had arrived at the station in July 2006 aboard the STS-121 flight of the space shuttle Discovery. Reiter had served with the Expedition 13 crew since that time, and became a second flight engineer on Expedition 14 when Lopez-Alegria and Tyurin officially assumed their duties as the station's fourteenth crew.

On November 23, 2006, Lopez-Alegria and Tyurin made an EVA of more than five-and-a-half hours while tending to a number of experiments for the Russian Federal Space Agency, Roscosmos. They also experienced some light-hearted moments when Tyurin carried out a bit of business for a paying customer: at the behest of a Canadian golf equipment company that paid for the privilege, Tyurin drove a golf ball off into space from the exterior of the ISS Pirs airlock.

In December, NASA astronaut Sunita Williams arrived with the STS-116 crew of the space shuttle Discovery and remained at the ISS as a member of the Expedition 14 crew with Lopez-Alegria and Tyurin, while Thomas Reiter returned to Earth aboard Discovery.

Before the space shuttle departed from the station, Williams participated in an EVA of seven-and-a-half hours with STS-116 mission specialist Robert Curbeam, during which they made electrical repairs to the ISS.

During the remainder of their stay at the station as the Expedition 14 crew, Lopez-Alegria and Williams made three more spacewalks while working on the station's cooling system. The first of their EVAs took place on January 31, 2007 when they spent nearly eight hours in EVA, and they added an additional seven-plus hours on February 4 and over six-and-a-half hours more to their mission EVA total on February 8.

Lopez-Alegria and Tyurin made the final spacewalk of the Expedition 14 mission when they repaired a damaged antenna on the unmanned ISS Progress 24 cargo spacecraft on February 22, 2007. During that EVA, they spent more than six hours outside the space station.

With the conclusion of the Expedition 14 mission, Lopez-Alegria and Tyurin returned to Earth on April 21, 2007, after a total flight of 215 days, 8 hours and 23 minutes. Sunita Williams remained on board the space station as a flight engineer for the ISS Expedition 15 crew. She returned to Earth with the STS-117 crew of the space shuttle Atlantis, after setting a new record for the longest space mission by a female astronaut, at 194 days, 18 hours and 58 minutes. During her long stay at the ISS, Williams also set a new record for most career EVA time by a female astronaut when she accumulated a total of 29 hours and 17 minutes in EVA during 4 spacewalks.

SOYUZ MISSION 101: SOYUZ TMA-10

(April 7–October 21, 2007)

Russian Manned Spaceflight

Launched from the Baikonur Cosmodrome on April 7, 2007 to travel to the International Space Station (ISS), Soyuz TMA-10 featured two cosmonauts representing the Russian National Space Agency Roscosmos: Oleg Kotov, who commanded Soyuz TMA-10 and served aboard the ISS as Expedition 15 flight engineer, and Fyodor Yurchikhin, who served as Soyuz TMA-10 flight engineer and ISS Expedition 15 commander.

Accompanying the cosmonauts aboard Soyuz TMA-10 was Charles Simonyi, the fifth space tourist, whose flight was arranged via a contractual agreement with Roscosmos. A successful software entrepreneur who was born in Budapest, Hungary, Simonyi was the first native of Hungary to fly in space since Bertalan Farkas lifted off in Soyuz 36 in the fifth manned flight of the Soviet Union's Intercosmos program, on May 26, 1980.

At the ISS, Yurchikhin, Kotov and Simonyi visited with ISS Expedition 14 crew members Michael Lopez-Alegria, Mikhail Tyurin and Sunita Williams for several days, and Yurchikhin and Kotov officially assumed their responsibilities as the station's next long-duration crew. Williams remained aboard the ISS to become the second flight engineer for

Fyodor Yurchikhin traveled to the International Space Station (ISS) aboard Soyuz TMA-10 to serve as commander of the ISS Expedition 15 long-duration crew. [NASA/courtesy of nasaimages.org]

Expedition 15, and Lopez-Alegria and Tyurin returned to Earth with Simonyi on April 21, 2007, aboard Soyuz TMA-9.

During his flight to the ISS, Charles Simonyi spent 13 days, 18 hours and 59 minutes in space.

Yurchikhin and Kotov conducted two spacewalks together during their Expedition 15 stay, working outside the ISS for about five-and-a-half hours on May 30–31, 2007 and slightly longer on June 6.

With the arrival of the STS-117 flight of the space shuttle Atlantis later in June, Williams completed her stay at the space station and returned to Earth, having set a new long-duration record for the longest space mission by a female astronaut, at 194 days, 18 hours and 58 minutes. NASA astronaut Clayton Anderson replaced her at the ISS, joining Yurchikhin and Kotov as a member of the Expedition 15 crew.

Williams also set a new record for most career EVA time by a female astronaut, accumulating a total of 29 hours and 17 minutes in EVA during 4 spacewalks during her time at the ISS.

Yurchikhin and Anderson made the final spacewalk of the Expedition 15 mission on July 23, 2007, to make repairs and perform maintenance on the station's exterior. They worked outside the station for more than seven-and-a-half hours.

At the end of their long stay at the ISS, Yurchikhin and Kotov returned to Earth in Soyuz TMA-10 on October 21, 2007, after a flight of more than 196 days in space. They endured a difficult landing, with their spacecraft making a ballistic re-entry and landing far from its intended landing site, but they emerged from the ordeal unharmed.

After seeing off his Expedition 15 crew mates, Clayton Anderson remained at the ISS to serve as a flight engineer for the station's next long-duration crew. He returned to Earth with the STS-120 crew of the space shuttle Discovery, in November 2007.

SOYUZ MISSION 102: SOYUZ TMA-11

(October 10, 2007–April 19, 2008)

Russian Manned Spaceflight

When she arrived at the International Space Station (ISS) aboard Soyuz TMA-11 in October 2007 to begin her second long-duration stay at the station, NASA astronaut Peggy Whitson became the first female commander of an ISS crew.

Yuri Malenchenko, a veteran Russian cosmonaut making his fourth flight into space, accompanied Whitson on the flight and served as commander of Soyuz TMA-11. At the ISS he served as flight engineer for the station's Expedition 16 long-duration mission. Whitson and Malenchenko were accompanied aboard Soyuz TMA-11 by Sheikh Muszaphar Shukor, who became the first citizen of Malaysia to fly in space.

Whitson, Malenchenko and Shukor were greeted by the resident ISS Expedition 15 crew: Fyodor Yurchikhin, Oleg Kotov and Clayton Anderson.

During his stay at the ISS, Shukor carried out several health-related experiments and a protein crystal experiment. A devout Muslim, he also celebrated Ramadan, serving as an example for religious individuals on Earth who are committed to practicing their faith regardless of their environment or circumstances.

After a visit of nearly 10 days, Shukor left the station and returned to Earth on October 21, 2007 in Soyuz TMA-10 with Yurchikhin and Kotov. Whitson and Malenchenko remained at the ISS with Anderson, who joined them as a member of the station's sixteenth long-duration crew.

A spacecraft malfunction forced Soyuz TMA-10 into a ballistic re-entry, and the spacecraft landed more than 300 kilometers from its intended landing site in Kazakhstan. Despite the difficult trip, Yurchikhin, Kotov and Shukor were uninjured. Shukor's historic flight lasted 10 days, 21 hours and 14 minutes, and he made 171 orbits of the Earth.

The STS-120 flight of the space shuttle Discovery arrived at the station shortly afterward, to deliver and install the ISS Harmony Node 2 module. NASA astronaut Daniel Tani arrived with the STS-120 crew to begin his mission as a flight engineer for the ISS Expedition 16 crew. He replaced Clayton Anderson, who returned to Earth aboard Discovery on November 7 after completing a long-duration stay of more than 151 days.

Whitson and Malenchenko made the first spacewalk of the Expedition 16 mission on November 9, 2007 when they spent just under seven hours outside the station while carrying out a variety of maintenance tasks.

Tani then joined Whitson for the remaining four spacewalks of the ISS Expedition 16 flight. On November 20, they spent over seven hours in EVA while configuring the Harmony Node, a job they completed four days later with a second EVA of seven-plus hours.

On December 18, Whitson and Tani ventured outside the station to inspect components of the ISS S4 starboard solar array, adding nearly seven more hours to their EVA total. They made the final spacewalk of the Expedition 16 mission on January 30, 2008, when they made a repair to the outside of the station over the course of more than seven hours in EVA.

Making his second flight in space, European Space Agency (ESA) astronaut Léopold Eyharts of France traveled to the ISS aboard space shuttle Atlantis during STS-122 in February 2008. He joined the Expedition 16 crew as a flight engineer, replacing Daniel Tani, who returned to Earth with the STS-122 crew. Tani had spent four months at the station, from October 2007 to February 2008, and accumulated nearly 35 hours in EVA.

In March 2008, NASA astronaut Garrett Reisman arrived at the ISS with the STS-123 crew of the space shuttle Endeavour. He joined the Expedition 16 crew as a flight engineer, replacing Léopold Eyharts, who returned to Earth with the STS-123 crew on March 26, 2008. Eyharts spent more than 48 days in space during his Expedition 16 stay at the ISS.

At the end of their long stay at the space station, Whitson and Malenchenko returned to Earth in Soyuz TMA-11 with So-yeon Yi, the first citizen of Korea to fly in space, on April 19, 2008. They endured a harsh ballistic re-entry and landed far from their targeted landing site in Kazakhstan, but emerged safely from the ordeal. Their remarkable odyssey as the ISS Expedition 16 crew served as both a separate and collective career highlight; for Whitson, the honor of being the first female ISS Expedition commander neatly bookended a brightly distinguished career that included a six month stay at the station as flight engineer for the Expedition 5 crew in 2002, while Malenchenko adorned his already exceptional resume as a cosmonaut by adding the successful Expedition 16 to his previous missions as commander of ISS Expedition 7, commander of Mir 16, and the STS-106 flight of the space shuttle Atlantis in September 2000.

During their stay at the International Space Station as the Expedition 16 crew, Peggy Whitson and Yuri Malenchenko spent more than 191 days in space, and made 3,028 orbits of the Earth.

Expedition 16 flight engineer Garrett Reisman remained aboard the ISS at the conclusion of that mission, and became flight engineer for the ISS Expedition 17 crew until he returned to Earth on June 14, 2008 with the STS-124 Discovery crew.

SOYUZ MISSION 103: SOYUZ TMA-12

(April 8, 2008–October 24, 2008)

Russian Manned Spaceflight

With the launch of Soyuz TMA-12 from the Baikonur Cosmodrome on April 8, 2008, So-yeon Yi became the first citizen of South Korea to fly in space.

Russian cosmonaut Sergei Volkov commanded Soyuz TMA-12, accompanied by flight engineer Oleg Kononenko, who like Volkov was also representing the Russian Federal Space Agency Roscosmos. Together with Yi, they traveled to the International Space Station (ISS), where Volkov became commander of the station's Expedition 17 long-duration crew, with Kononenko serving as Expedition 17 flight engineer.

They were met at the station by the Expedition 16 crew: NASA's Peggy Whitson and Garrett Reisman, and Yuri Malenchenko of Roscosmos.

Yi carried out a program of scientific experiments during a stay of nine days at the station, and she then returned to Earth with Whitson and Malenchenko in Soyuz TMA-11 on April 19, 2008. They survived a harsh ballistic re-entry and were located by recovery forces despite landing far from their intended landing site in Kazakhstan.

During her historic flight to the ISS, Yi spent 10 days, 21 hours and 13 minutes in space.

Garrett Reisman remained at the space station, and joined Volkov and Kononenko as a member of the Expedition 17 crew.

In June 2008, NASA astronaut Gregory Chamitoff arrived at the ISS with the STS-124 crew of the space shuttle Discovery. He replaced Garrett Reisman as flight engineer for Expedition 17, and Reisman returned to Earth with the STS-124 crew after more than 95 days in space.

Volkov and Kononenko began the first spacewalk of the Expedition 17 mission on July 10, 2008 when they ventured outside the ISS for a little over six hours.

They also spent just under six hours in EVA on July 15, when they worked on the station's Zvezda service module and performed other maintenance tasks around the exterior of the station.

At the end of their stay aboard the ISS, Volkov and Kononenko returned to Earth in Soyuz TMA-12 with U.S. space tourist Richard Garriott on October 24, 2008, after a mission of more than 198 days.

The return flight had a particular poignancy for longtime space buffs, as it paired the first second generation cosmonaut (Sergei Volkov is the son of veteran cosmonaut Alexandr A. Volkov) with the son of a pioneering NASA astronaut (Richard Garriott's father is Skylab 3 veteran Owen Garriott, who also flew aboard the space shuttle Columbia during his long career at NASA).

Gregory Chamitoff remained aboard the ISS to serve as flight engineer and NASA science officer for the Expedition 18 long-duration crew.

SOYUZ MISSION 104: SOYUZ TMA-13

(October 12, 2008– April 8, 2009)

Russian Manned Spaceflight

Launched aboard Soyuz TMA-13 on October 12, 2008, Michael Fincke of NASA and Yuri Lonchakov of the Russian Federal Space Agency traveled to the International Space Station (ISS) to become the station's eighteenth long-duration crew. For the flight to the station, Lonchakov served as commander of Soyuz TMA-13 and Fincke served as flight engineer; at the ISS they switched roles, with Fincke commanding the long-duration ISS Expedition 18 and Lonchakov serving as flight engineer.

Traveling with them aboard Soyuz TMA-13 was video game entrepreneur Richard Garriott, whose flight was arranged via a commercial spaceflight agreement with the Russian Federal Space Agency Roscosmos.

At the ISS, Fincke, Lonchakov and Garriott visited with the station's Expedition 17 crew: commander Sergei Volkov, flight engineer Oleg Kononenko and flight engineer Gregory Chamitoff. After a little more than a week, Volkov and Kononenko returned to Earth with U.S. space tourist Richard Garriott aboard Soyuz TMA-12 on October 24, 2008. Chamitoff remained at the ISS to become part of the Expedition 18 crew with Fincke and Lonchakov.

NASA astronaut Sandra Magnus and Japan Aerospace Exploration Agency (JAXA) astronaut Koichi Wakata were also scheduled to participate in the ISS Expedition 18 mission as flight engineers, with Magnus scheduled to replace Chamitoff at the end of his stay aboard the station, and Wakata scheduled to replace Magnus when she concluded her mission.

SPACE SHUTTLE PROGRAM

(1981–)

U.S. Manned Spaceflight Program

Officially known as the Space Transportation System, the U.S. space shuttle program was given life by President Richard Nixon in January 1972. NASA

Space shuttle Columbia lifts off from the Kennedy Space Center, April 12, 1981. [NASA/courtesy of nasaimages.org]

received $100 million in shuttle-related funding that year, as it sought to define the future course of the U.S. space program as the Apollo lunar landing program neared its completion.

A variety of options had been considered by NASA planners for the post-Apollo period, including a space station in Earth orbit—which was realized in the Skylab program—and even a manned lunar base or the long-term development of a manned landing on Mars. In the interest of providing frequent access to Earth orbit in support of such 'real world' practical applications as satellite deployments, military studies and space science initiatives—all at a cost that promised to be far lower than further Moon missions or a trip to Mars—the U.S. Congress and the Nixon Administration encouraged NASA to pursue the shuttle program, and the agency responded with intense enthusiasm for the project.

Optimistic predictions of a completed shuttle and a first test flight by 1978 went hand-in-hand with discussions of shuttle launch schedules expressed in terms of weeks rather than months; but the complexity of the strategic planning and the logistical preparations necessary for a typical shuttle mission made such plans painfully unrealistic. The fiscal and political realities of the time also mitigated against even NASA's most practicable plans for the shuttle.

To the credit of all those involved in its development, the shuttle proceeded fairly rapidly from an expensive engineering concept to a remarkable spacecraft that its proponents credibly declared "the most complex machine ever built."

Construction of the shuttle itself was assigned to Rockwell International Corp. (which at the time was known as North American Rockwell Corp.), while the contract to build the massive external fuel tank was awarded to the Martin Marietta Corporation, and the Thiokol Corporation was given the job of developing and building the shuttle's Solid Rocket Boosters (SRBs).

In anticipation of the first shuttle flight, a prototype shuttle was built in the late 1970s for use as a test vehicle. Dubbed "Enterprise" in honor of the fictional spacecraft in the 1960s television series "Star Trek," the prototype vehicle was never flown in space, but it played a central role in the proving flights that were collectively known as the Approach and Landing Test (ALT) program.

Enterprise was built at Rockwell International's manufacturing plant in Palmdale, California, and was completed in September 1976. Beginning in February 1977, it was used in a series of test flights designed to test the basic systems and aerodynamic capabilities of the shuttle vehicles. Because the space shuttles were designed as reusable spacecraft, and the operational shuttles that would follow Enterprise were intended to be flown both in space and in the Earth's atmosphere at the end of each spaceflight, the vehicles had to function well and maintain the integrity of their structure and systems in both environments. Given the revolutionary concept of creating a single vehicle that could be launched into space, maneuvered in orbit as a spacecraft, and then be flown to a runway landing so it could subsequently be used again for additional flights, the engineers at NASA and at each of the major shuttle contractors required enormous amounts of test data to ensure that each element of the complex shuttle system would work as planned, and that all of the elements would properly work together.

The ALT flights ensured that the shuttle could be flown like an airplane during the final stage of a shuttle mission, when the vehicle had re-entered the atmosphere and would need to be piloted to a landing on a runway. The ALT program also helped shuttle system designers to understand the ways in which many of the spacecraft's systems were likely to function during actual flights.

Beginning in February 1977, a total of 13 ALT flights were carried out at the Dryden Flight Research Center at Edwards Air Force Base in California. Of the 13 test flights, 8 were crewed. Five were flown by astronauts Fred Haise and Gordon Fullerton, and astronauts Joe Engle and Richard Truly served as the crew for three of the flights.

At the start of each test, Enterprise began its flight atop an airplane that had been specially adapted by the Boeing Corporation to serve as a transport. Known as the Shuttle Carrier Aircraft (SCA), the plane demonstrated that the shuttle could be ferried safely from one location to another on Earth in between flights into space.

The ALT program was completed in October 1977, and the Enterprise shuttle—which some at NASA had hoped might ultimately be adapted for flights in space—was the centerpiece of a NASA goodwill tour before it was officially retired. Enterprise was subsequently acquired by the Smithsonian Institution.

In March 1979, the space shuttle Columbia became the first fully operational shuttle to be completed and delivered to NASA. Although the basic design of the spacecraft would undergo a far-reaching evolution during the program's first two decades, the first iteration of the shuttle vehicle (which was referred to within NASA as an orbiter) featured a crew compartment area, a large payload bay, and the vehicle's three Space Shuttle Main Engines (SSMEs).

Each shuttle was roughly the size of a DC-9 airplane: 122 feet (37 meters) long, with a 60-foot (18.3 meter) payload bay initially capable of carrying payloads with a combined weight of approximately 50,000 pounds (22,680 kilometers). From one wing tip to the other, the shuttle measured 78 feet (24 meters); and when first delivered and flown, the vehicle weighed about 200,000 pounds (90,718 kilograms) when empty of payloads.

An elaborate combination of protective materials was used on the outside of the shuttle to ensure that the vehicle could survive re-entry. The Rockwell Corporation assigned the development of the heat protection system to the Lockheed Missiles and Space Company, and in cooperation with both NASA and Rockwell, the engineers at Lockheed devised an innovative array of heat-resistant solutions. Materials offering different levels of protection were deployed on different areas of the shuttle, according to the degree of heat exposure in each area.

Central to the system was the use of a variety of silica fiber tiles along the bottom of the vehicle, and reinforced carbon was deployed in high-heat areas such as the nose and the leading edges of the wings.

Crew members entering the new space shuttle for the first time found a crew area located in the front (or nose end) of the vehicle containing 2,325 cubic feet (65.8 cubic meters) of workspace divided between the cockpit, the astronauts' living area and a laboratory station for conducting scientific research. On the upper portion of the split-level crew module, the Flight Deck featured controls and displays for piloting the shuttle and for controlling the deployment and operation of payloads carried in the payload bay. The Flight Deck instrument panels featured over 200,000 controls and displays, and virtually the entire front end of the Flight Deck was wrapped with specially designed windshields—six around the cockpit area, plus two windows overhead and two more windows that allowed the astronauts to see into the payload bay.

Sleep stations, supplies and storage space were located in the shuttle's Middeck, along with the astronauts' dining area, personal hygiene station and waste management system. The Middeck accommodations were capable of supporting as many as seven crew members in the event of an emergency.

The forward fuselage of the shuttle also provided space for astronaut researchers to carry out scientific work in a pressurized Spacehab module, and an airlock that could be moved and mounted in different locations within the crew area to support a variety of extravehicular activities (EVAs).

In the center of the shuttle, the payload bay and its supporting systems and equipment were used to carry satellites, space station components and supplies, and a variety of other payloads into orbit. Perhaps the most remarkable feature of the payload bay, the large doors of the compartment, ran nearly the entire length of the vehicle. As a result, when its payload doors were fully open, with the large doors extending outward from the vehicle, the shuttle appeared almost twice its actual size.

Closed off from the rest of the shuttle by a structure known as the forward bulkhead, the tail end of the aft fuselage contained the vehicle's Orbital Maneuvering System (OMS) for steering left and right in orbit, and the shuttle's three SSMEs and their supporting systems and hardware.

Designed to be used for as many as 50 shuttle missions before requiring major maintenance, the SSMEs built and delivered to NASA by the Rockwell Corporation's Rocketdyne Division were designed to provide a maximum thrust of 375,000 pounds (179,097 kilograms) at launch and 470,000 pounds (213,188 kilograms) in space. A single SSME measured 14 feet (4.3 meters) long, with a diameter of 8 feet (2.4 meters) at its widest point, and weighed about 7,000 pounds (3,175 kilograms). Each shuttle would be equipped with three of the engines for each flight.

During launch, the three SSMEs helped to provide the thrust necessary for lift-off, along with the thrust provided by the shuttle's Solid Rocket Boosters (SRBs). Once the shuttle lifted off the ground, the SSMEs continued to fire for a total of 8 minutes and 30 seconds during the mission phase known as the shuttle's powered flight. Consuming some 500,000 gallons of liquid propellant (a mix of liquid hydrogen and liquid oxygen supplied by the shuttle's External Tank), the SSMEs sped the shuttle toward orbit at 17,000 miles per hour (27,358 kilometers per hour) during the powered flight phase.

Following nearly a decade of development, the large External Tank delivered by the Martin Marietta Corporation stood 154 feet (47 meters) tall and measured 27.5 feet (8.4 meters) in diameter, with an initial weight (when filled prior to launch) of about 1,700,000 pounds (760,220 kilograms).

Within the shell of each External Tank there were separate internal tanks for liquid hydrogen and liquid oxygen, and a third tank that connected the first two and further served as a connection point between the External Tank and the SRBs.

A one inch (2.5 centimeter) layer of polyisocyanurate foam, applied as a spray, coated the outside of each External Tank to control the temperature of the liquids inside the structure and to protect the External Tank during its ride aboard the shuttle, during which it would be carried to an altitude of about 70 miles (113 kilometers).

Unlike the shuttle itself and the SRBs, each External Tank was designed to be used only once. After providing the 500,000-plus gallons of fuel necessary to propel the shuttle to near-orbital velocity during the first eight-and-a-half minutes of a flight, the External Tank would be detached from the shuttle and would then largely be destroyed as it descended through the atmosphere. Any surviving portions of the discarded tank were carefully plotted on a trajectory designed to send them safely into the ocean.

Working in conjunction with the SSMEs during the initial phase of a shuttle launch, the 149-foot (45.5 meter) long, 12-foot (3.7 meter) round SRBs developed by the Thiokol Corporation were designed to provide 80 percent of the thrust necessary for launch. Containing the largest solid propellant motor ever used to launch a spacecraft, each SRB utilized over one million pounds (450,000 kilograms) of solid fuel, which consisted of powdered aluminum and oxygen derived from ammonium perchlorate.

The SRBs also played an important role in guiding the shuttle onto the proper path into orbit, as they contained several important electronic subsystems that helped to control various aspects of the early phase of the launch.

Each pair of SRBs used in a shuttle launch were designed to generate a total thrust of 5,300,000 pounds. The SRBs were the first of the three primary launch components (the shuttle, the External Tank, and the SRBs) to finish their work; during a launch, they would separate from the shuttle and the large fuel tank at an altitude of 28 miles (45 kilometers) and then each return to Earth beneath a parachute, splashing down in the Atlantic Ocean. Recovery ships then retrieved the spent SRBs, which would be refurbished and used again for subsequent launches.

Depending on the particular requirements of a given mission, shuttle flights were generally conducted in Earth orbits ranging from 115 miles (185 kilometers) to 400 miles (643 kilometers) high.

In addition to the commercial advantages of frequent manned launches with vehicles large enough to deploy modern satellites, the shuttle program also vastly increased opportunities for scientific research in space. Crews with specialized training conducted a wide array of experiments in disciplines as diverse as life sciences, astronomy, materials processing, technology and a host of other fields.

A particularly engaging aspect of space shuttle research was embodied in the Getaway Special program, which provided opportunities for paying customers to place small, self-contained experiments aboard a shuttle to be carried during a given flight. The program gave a broad spectrum of diverse clients the chance to carry out experiments that, given the very high costs and stringent requirements of other alternatives, would not have otherwise flown in space.

Similarly, the Shuttle Student Involvement Program allowed students a stake in the activities of the astronauts in space, as shuttle crew members carried out experiments designed by students.

As the sole launch facility for the manned U.S. space program, the Kennedy Space Center (KSC) at Cape Canaveral in Florida was naturally chosen as the launch site for shuttle missions, with shuttle flights managed by teams of engineers at the Johnson Space Center (JSC) in Houston, Texas. With a single exception (STS-3), all shuttle landings from the program's start in 1981 through mid-2009 were made at KSC or at Edwards Air Force Base in California.

As was the case with NASA's earlier human spaceflight programs, astronauts chosen for the shuttle program were officially assigned to the agency's Astronaut Office at the JSC, while administration of the space shuttle program was overseen by officials at NASA headquarters in Washington, D.C.

The very first space shuttle mission, STS-1, took place in April 1981. Longtime veteran astronaut and Apollo 16 Moon explorer John Young served as commander of the flight, accompanied by Robert Crippen. Young and Crippen spent a little over two days in orbit during STS-1, which was the first in a series of test flights designed to prove the shuttle's capabilities as the world's first reusable spacecraft, and to evaluate the craft's systems and equipment during actual spaceflight.

NASA's initial plans for very frequent shuttle flights, coupled with the high degree of complexity involved in planning and preparing individual flights, had made it clear from the program's earliest development that an entire fleet of space shuttles would be required for a true space transportation system. With that in mind, the agency worked closely with its suppliers to ensure the construction and delivery of additional vehicles.

In July 1982, NASA took delivery of the second space shuttle, Challenger; the shuttle Discovery was completed and delivered in November 1983; and the shuttle Atlantis completed the original fleet in April 1985.

After 24 successful flights, at a time when the space shuttle program had clearly achieved the effect of making spaceflight appear to most ordinary citizens to be virtually routine, the program and the nation suffered a shocking loss with the January 28, 1986 explosion of the space shuttle Challenger. The seven crew members aboard the shuttle for the STS-51L mission were killed in the accident, which was subsequently found to have been caused by a fault in an O-ring—the gasket used to seal the joint between sections of an SRB. The O-ring that led to the Challenger accident had become brittle in the cold weather leading up to the launch and had failed as a result, allowing the SRB to leak fuel onto the External Tank, which in turn ignited and exploded slightly more than a minute after the launch.

A long, difficult recovery followed the Challenger accident, and the nation mourned the lost crew members with great sadness. Concerns about the future of the U.S. space program were addressed by President Ronald Reagan, who vowed that the program would recover after a thorough investigation of the accident and the amelioration of its causes.

In the wake of the accident, the U.S. Congress approved construction of the shuttle Endeavour, which was completed in May 1991.

In the meantime, NASA returned to spaceflight with the STS-26 mission, from September 29 to October 3, 1988. More than a decade of often spectacular missions followed; an early example of how fully the agency had recovered from its post-Challenger trauma was the first flight flown by the new shuttle Endeavour, STS-49, in May 1992. During the landmark mission, the STS-49 crew retrieved and repaired the Intelsat VI satellite in orbit. In doing so, they set a new record for the longest spacewalk, and utilized the skills of three crew members during a single EVA for the first time in history.

The shuttle fleet also played a central role in the immediate post-Soviet era, when the United States and Russia worked together to develop a cooperative human spaceflight project. With the mid-1990s Shuttle-Mir program, the United States and Russia each provided access to the "crown jewel" of their respective space programs—the United States for the first time counting Russian cosmonauts among the members of its shuttle crews and using the shuttle to deliver crew members and supplies and equipment to Mir; and Russia making room aboard its legendary space station for a succession of NASA astronauts, which resulted in an unprecedented U.S. presence aboard Mir for a period of 907 days.

In the modern era of routine international cooperation in space, the shuttle's capacity for ferrying large payloads into space became a central asset in the multinational development of the International Space Station (ISS).

Once again casting a pall over the future of the shuttle program, however, the loss of the STS-107 crew and the shuttle Columbia on February 1, 2003 gave rise to perhaps the gravest challenges the U.S. space program has ever faced. In some ways reminiscent of the Challenger accident 17 years earlier, the loss of the shuttle Columbia, which had been damaged during launch and then broke into pieces over the southern United States just minutes before she was scheduled to land, the loss of the STS-107 crew occurred in an America vastly different from the nation that had witnessed the terrible images of the Challenger exploding in 1986.

Tempered by the terrorist attacks of September 2001 and enveloped in fear and uncertainty about the country's future course—as well as their own personal security and well-being—many Americans appeared to accept the Columbia accident with a certain fatalistic detachment. Perhaps due to the lack of dramatic video footage (as had been the case with Challenger) or perhaps due to a certain deconditioning to shock engendered by the 9/11 attacks, the Columbia accident did not occasion a national moment of self-analysis and resolve in the manner that the Challenger accident had nearly two decades earlier.

After an exhaustive reconstruction of the events that had led to the tragedy, the Columbia Accident Investigation Board (CAIB) attributed the accident to a piece of insulating material that had broken off the External Tank at launch and struck the leading edge of one of Columbia's wings, impairing the vehicle's ability to maintain its heat protection system and its aerodynamic stability during the final minutes of its flight.

In response to the board's recommendations, NASA created an elaborate system of cameras and electronic sensors to document the launch of the first shuttle mission following the Columbia accident. As a result, the STS-114 return-to-flight mission featured the most carefully scrutinized launch in the history of the shuttle program. The new monitoring system worked exactly as planned, and did in fact find evidence of loose insulation hitting the shuttle Discovery shortly after the STS-114 crew left the launch pad on July 26, 2005.

While the STS-114 flight was in progress, NASA officials grounded all planned future shuttle flights in response to the obvious conclusion that the problem with the foam insulation had not been fixed despite the agency's best efforts during the

18 months between the Columbia accident and the launch of the return to flight mission.

Fear for the safety of the crew increased as the nation awaited Discovery's return to Earth. For their part, however, the STS-114 astronauts performed remarkably well, and enjoyed a successful flight despite the tension surrounding their impending return. They performed their tasks with great skill and courage, and as events unfolded, they landed safely on August 9, 2005.

Another long delay followed before the next mission, STS-121, in July, 2006, while NASA staffers continued their efforts to alleviate the potential problems caused by the foam insulation shedding off the External Tank.

By the launch of STS-121, the insulation issue appeared to have been sufficiently enough addressed to, at least, enable the crew—and the U.S. public and media—to go about their business without the tension that surrounded the first return to flight mission. The subsequent STS-115 in September 2006 also proceeded relatively smoothly, offering further promise that the shuttle program might once again achieve a routine of relatively worry-free trips into space—or at least a routine as free of worry as possible, considering the obvious dangers inherent in launching into space and safely returning to Earth the most complex machine ever built.

SPACE SHUTTLE MISSION 1: STS-1

(April 12–14, 1981)

U.S. Manned Spaceflight

The first mission of America's space shuttle program lifted off from launch pad 39A at the Kennedy Space Center in Florida on April 12, 1981. The first shuttle flight marked the end of nearly six years during which the United States had not flown a single manned space mission.

Commanded by long-time veteran astronaut John Young—who also flew during the Gemini and Apollo programs, and landed on and explored the Moon during the flight of Apollo 16—STS-1 was intended to prove that the space shuttle could be successfully flown into space, safely operated in orbit, and returned safely to Earth.

The first space shuttle crew: STS-1 commander John Young (left) and pilot Robert Crippen. [NASA/courtesy of nasaimages.org]

Robert Crippen was pilot for the STS-1 flight, which took place aboard the space shuttle Columbia. The main payload for the mission was a group of instruments designed to collect information about the shuttle during launch, ascent, orbital flight, re-entry and landing.

The short proving mission proceeded smoothly; Young and Crippen traveled over one million miles in 2 days, 6 hours, 20 minutes and 53 seconds prior to landing at Edwards Air Force Base in California on April 14. Columbia performed well and landed safely even though it lost 16 of the special heat shield tiles designed to protect the spacecraft during its flight through the Earth's atmosphere, and suffered damage to 148 additional tiles. The mission was deemed a major success overall, and a good start for the shuttle program.

SPACE SHUTTLE MISSION 2: STS-2

(November 12–14, 1981)

U.S. Manned Spaceflight

Joe Engle was commander for shuttle mission STS-2, with Richard Truly serving as pilot for the flight, November 12–14, 1981, aboard the shuttle Columbia.

The launch of STS-2 was delayed four separate times, having been rescheduled on October 9 and November 4 due to technical difficulties and then experiencing shorter delays twice on launch day, November 12. The successful launch proved the value of changes that NASA technicians had made to the launch platform since the first shuttle launch, as damage caused to Columbia's heat shield tiles during the launch of STS-1 by the shuttle's solid rocket boosters (SRBs) was avoided during STS-2. Post-flight inspection showed that no tiles were lost during STS-2 (although 12 were damaged).

The primary payload for the second shuttle mission was a group of Earth observation experiments designed by the Office of Space and Terrestrial Applications, which were collectively known as OSTA-1.

The experiments were largely successful and returned useful data, and Engle and Truly were also able to test the shuttle's Remote Manipulator System (RMS, also known as "Canadarm" in honor of its nation of origin) for the first time.

The mission, however, had to be cut short by three days when one of the shuttle's three fuel cells failed, endangering the crew's supply of electricity and drinking water. Columbia landed safely at Edwards Air Force Base in California, after a flight of 2 days, 6 hours, 13 minutes and 12 seconds.

SPACE SHUTTLE MISSION 3: STS-3

(March 22–30, 1982)

U.S. Manned Spaceflight

Commander Jack Lousma and pilot Gordon Fullerton made the first diverted landing of the shuttle program at the end of STS-3, and their flight was extended by one day due to high winds at their eventual landing site in White Sands, New Mexico.

Lousma had previously served aboard the Skylab space station, and veteran test pilot Fullerton had flown in the shuttle's approach and landing test program aboard the prototype shuttle Enterprise.

STS-3 began with the launch of the shuttle Columbia on March 22, 1982, following a one hour delay.

During the flight, Lousma and Fullerton completed tests of the shuttle's remote manipulator system (RMS) and placed the shuttle in various orbital positions to measure its interaction with the Sun.

The primary scientific payload for STS-3 was a group of experiments for the Office of Space Science (OSS-1) that included a Monodisperse Latex Reactor (MLR), an Electrophoresis Equipment Verification Test (EEVT) and a Heflex Bioengineering Test (HBT). The experiments included the measurement of gas and dust contaminants propelled into the space environment near the Earth by the launch and flight of the shuttle. STS-3 also carried the first experiment designed by students, as part of the Shuttle Student Involvement Program (SSIP).

A series of minor problems occurred during the flight. The crew suffered space sickness and had difficulty sleeping because of an unnerving static, and also had other sporadic problems with the shuttle's communications system.

Wet weather in California forced NASA mission controllers to change the landing to the Northrup landing sight in White Sands, New Mexico for the first time, rather than using the rain-soaked dry lake bed at Edwards Air Force Base in California. The White Sands landing was itself put off when high winds forced the extension of the flight by one day. After 8 days, 4 minutes and 46 seconds, Lousma and Fullerton returned to Earth on March 30, 1982 during a dust storm, in a difficult landing that caused damage to the shuttle's brakes and covered the spacecraft in dust. Fortunately, they emerged from the shuttle unharmed.

SPACE SHUTTLE MISSION 4: STS-4

(June 27–July 4, 1982)

U.S. Manned Spaceflight

Apollo 16 veteran Thomas Mattingly returned to space as commander of shuttle mission STS-4, launched on June 27, 1982 with pilot Henry Hartsfield, Jr. on the shuttle Columbia. In addition to proving the shuttle safe for future flights, which would carry larger crews and stay in space for longer and more complex missions, the STS-4 flight was devoted to a classified payload for the U.S. Department of Defense. It also carried several scientific experiments, including a Continuous Flow Electrophoresis System (CFES), experiments designed by students in the Shuttle Student Involvement Program (SSIP), and projects designed by researchers at Utah State University.

The crew also made medical observations to gather data for NASA health services personnel, and used the shuttle's robotic arm to manipulate the Induced Environment Contamination Monitor, which was designed to measure the amount of contamination the shuttle produces in the near-Earth space environment at launch and during flight. Mattingly and Hartsfield also captured images of lightning activity in the Earth's atmosphere.

After launch, mission controllers evaluating the shuttle's performance found that rainwater had seeped into the protective coating of several of Columbia's heat shield tiles, exposing the shuttle to potential damage in orbit, when the water would freeze and could damage the tiles. The problem was fixed, however, by turning the waterlogged tiles toward the Sun, which evaporated the water.

After a flight of 113 orbits, in 7 days, 1 hour, 9 minutes and 31 seconds, Mattingly and Hartsfield landed at Edwards Air Force Base in California on July 4, 1982.

SPACE SHUTTLE MISSION 5: STS-5

(November 11–16, 1982)

U.S. Manned Spaceflight

The first fully operational shuttle mission (the first flight that was not envisioned primarily as a test of the spacecraft and its systems), shuttle mission STS-5 resulted in the successful deployment of two commercial satellites.

Vance Brand, who had previously flown in space during the Apollo-Soyuz Test Project (ASTP) in 1975, was commander of STS-5, aboard the shuttle Columbia. The pilot for the flight was Robert Overmyer, and also aboard were mission specialists Joseph Allen and William Lenoir. The STS-5 crew launched on November 11, 1982.

The ANIK C-3 Canadian communications satellite and the Satellite Business Systems' SBS-C satellite were placed in orbit during the five-day mission. They were propelled into orbit by Payload Assist Module-D (PAM-D) motors, in the first use of the PAM-D devices during the shuttle program.

The crew successfully accomplished most of their primary goals by deploying the commercial payloads, but Allen and Lenoir were frustrated by faults in the spacesuits they were to have worn during the shuttle program's first spacewalk, and the EVA had to be canceled.

At the end of the STS-5 mission, having spent 5 days, 2 hours, 14 minutes and 26 seconds in flight, Columbia and its crew touched down at Edwards Air Force in California on November 16, 1982.

SPACE SHUTTLE MISSION 6: STS-6

(April 4–9, 1983)

U.S. Manned Spaceflight

Shuttle mission STS-6 was the space shuttle Challenger's first flight in space. It also featured the first space walk of the shuttle program.

Paul Weitz was commander of STS-6; he flew into space with pilot Karol Bobko and mission specialists Donald Peterson and Story Musgrave.

The original launch, scheduled for January 20, had to be rescheduled after NASA technicians found cracks in one of Challenger's three Space Shuttle Main

Engines (SSMEs), and severe weather then caused further delay. After repairs to all three SSMEs and extensive testing, STS-6 launched on April 4, 1983.

During the flight, Donald Peterson and Story Musgrave achieved the first spacewalk of the shuttle era. They spent 4 hours and 17 minutes in EVA.

The primary payload of Challenger's maiden flight was the first Tracking and Data Relay Satellite (TDRS-1). A fault in the rocket designed to propel the satellite into orbit caused TDRS-1 to be placed in the wrong orbit, but NASA ground controllers were eventually able to position the satellite properly by executing additional maneuvers over a period of months, long after the end of STS-6.

Other payloads included a Continuous Flow Electrophoresis System (CFES), a Monodisperse Latex Reactor (MLR), and instruments to monitor radiation and to survey lightning activity in the Earth's atmosphere.

The STS-6 crew landed at Edwards Air Force Base in California on April 9, 1983, after 81 orbits in 5 days, 23 minutes and 42 seconds.

SPACE SHUTTLE MISSION 7: STS-7

(June 18–24, 1983)

U.S. Manned Spaceflight

Launched aboard the shuttle Challenger on June 18, 1983, STS-7 deployed two satellites and featured an historic first for the U.S. space program, as Sally Ride became the first American woman to fly in space.

Sally Ride, the first American woman to fly in space, with STS-7 crew mates (front, left to right): Robert Crippen, Frederick Hauck, (back, left to right): John Fabian and Norman Thagard. [NASA/courtesy of nasaimages.org]

Robert Crippen was commander of STS-7. Frederick Hauck served as pilot, and John Fabian, Norman Thagard, and Sally Ride were mission specialists for the flight.

The two satellites successfully deployed during STS-7 were the Canadian Anik C-2 communications satellite and the PALAPA-B1 Indonesian satellite, both of which were propelled into the correct orbit by Payload Assist Module-D (PAM-D) motors.

Scientific experiments carried out during the mission included studies in the formation of metal alloys under microgravity conditions and the use of a remote sensing scanner mounted on the Shuttle Pallet Satellite SPAS-01; the deployment of an experimental payload for the Office of Space and Terrestrial Applications; a study of the cause and effects of space sickness; a series of Getaway Special projects, including one designed to examine the impact of space on the social behavior of an ant colony; and an experiment carried out as part of the Shuttle Student Involvement Program (SSIP).

Challenger's STS-7 flight was originally scheduled to end with a landing at the Kennedy Space Center after 96 orbits, but poor weather in Florida caused the flight to be extended for 2 additional orbits. The diverted landing occurred at Edwards Air Force Base in California, where the shuttle touched down safely on June 24, 1983, after a mission of 6 days, 2 hours, 23 minutes and 59 seconds.

SPACE SHUTTLE MISSION 8: STS-8

(August 30–September 5, 1983)

U.S. Manned Spaceflight

During STS-8, Guion Bluford became the first African American to fly in space.

Richard Truly was commander of STS-8; Daniel Brandenstein served as pilot; and Dale Gardner, Guion Bluford, and William Thornton served as mission specialists.

Space shuttle Challenger lifted off in inclement weather on August 30, 1983 in the first nighttime launch of the shuttle program.

The crew deployed the Indian INSAT-1B multipurpose satellite, using the Payload Assist Module-D (PAM-D) motor to propel the satellite into its proper orbit.

In an experiment designed to test the effect that extremely cold conditions might have on the flight deck area, the front end of the shuttle was positioned away from the Sun for 14 hours; and in another test, the spacecraft was flown at an altitude of 139 miles in an effort to determine the source of the glow that surrounds part of the shuttle at night.

Other exercises included the use of the Remote Manipulator System (RMS), observations of the effect of spaceflight on rats housed in the shuttle's "animal enclosure module," an Incubator-Cell Attachment Test (ICAT), the operation of radiation monitoring equipment, an experiment for the Shuttle Student

Involvement Program (SSIP), operation of a Continuous Flow Electrophoresis System (CFES), a test involving the Development Flight Instrumentation Pallet, communications tests with the Tracking and Data Relay Satellite (TDRS-1) that had been deployed during an earlier shuttle mission, and the flight of eight groups of U.S. postal commemoratives.

In addition to achieving the first night launch of the shuttle program, STS-8 also made the first night landing, following a mission of 6 days, 1 hour, 8 minutes and 43 seconds. At the end of the busy mission, which encompassed 98 orbits, Challenger returned to Earth at Edwards Air Force Base in California on September 5, 1983.

SPACE SHUTTLE MISSION 9: STS-9

(November 28–December 8, 1983)

U.S. Manned Spaceflight

The STS-9 mission featured the second space shuttle flight of Gemini and Apollo veteran John Young, the first flight of six crew members in a single spacecraft, the first U.S. flight of an astronaut representing the European Space Agency (ESA), and the first deployment of the Spacelab orbital laboratory.

In addition to John Young, who served as commander of STS-9, the crew aboard the shuttle Columbia included pilot Brewster Shaw, mission specialists Owen Garriott and Robert Parker, and Payload Specialists Byron Lichtenberg and ESA astronaut Ulf Merbold of Germany.

Young had spent over 20 hours on the surface of the Moon during Apollo 16, orbited the Moon in Apollo 10, and also flew in space during Gemini 3, Gemini 10 and STS-1—the first shuttle mission, in 1981.

Garriott had lived aboard the Skylab space station for nearly two months in 1973.

Ulf Merbold became the first representative of the ESA to fly in space.

A problem with the exhaust system of the shuttle's solid rocket booster (SRB) caused the originally scheduled launch date of September 30 to be delayed while the vehicle was removed from the launch pad, repaired, and then restacked. The launch took place on November 28, 1983.

Jointly developed by NASA and ESA, the Spacelab laboratory and observation platform enabled 73 the astronauts to perform scientific investigations across a variety of disciplines including astronomy, physics, atmospheric physics, Earth observation, life sciences, materials sciences, space plasma physics and technology.

A series of disturbing difficulties cropped up near the end of the otherwise successful mission when several instruments failed—including two of the spacecraft's on-board computers. Landing was delayed for about 8 hours, extending the total duration of the mission to 10 days, 7 hours, 47 minutes and 24 seconds. When the shuttle did land, on December 8, 1983 at Edwards Air Force Base in California, two of its three auxiliary power units caught fire.

Fortunately, no harm came to the crew, and no serious damage was done to the spacecraft.

SPACE SHUTTLE MISSION 10: STS-41B

(February 3–11, 1984)

U.S. Manned Spaceflight

Commanded by veteran astronaut Vance Brand, the STS-41B flight of the space shuttle Challenger saw the first untethered spacewalks of the space shuttle program, and the first use of the Manned Maneuvering Unit (MMU).

Lifting off on February 3, 1984, Brand—who was making his third flight into space, having previously served during the Apollo-Soyuz Test Flight (ASTP) mission and STS-5—was joined by pilot Robert Gibson and mission specialists Bruce McCandless, Ronald McNair, and Robert Stewart.

In response to the scary final moments of the previous shuttle mission (STS-9), when two of the shuttle Columbia's three auxiliary power units had caught fire as the shuttle landed, NASA officials delayed the originally-scheduled STS-41B launch on January 29 so all three of Challenger's auxiliary power units could be replaced. As a result, STS-41B launched on February 3, 1984.

Once in space, the mission took on aspects of the glory days of the U.S. space program's first era, as the American media and public took an avid interest in the adventures of McCandless and Stewart when the astronauts made the first use of the Manned Maneuvering Unit (MMU). Looking for all the world like a rocket backpack from a science fiction film, the MMU enabled McCandless and Stewart to direct themselves around the shuttle without having to remain tethered to the spacecraft. They spent 5 hours and 55 minutes in EVA during the first test of the MMU units, on February 7, 1984 and an additional 6 hours, 17 minutes on February 9.

Also tested for the first time were foot restraints attached to the Remote Manipulator System (RMS)—the idea being that an astronaut attached to the shuttle's robotic arm could be uniquely positioned and securely fastened while completing EVA tasks—in preparation for the retrieval and repair of the Solar Maximum satellite, which was planned for the next shuttle mission.

The thrilling exploits of McCandless and Stewart and their successful EVAs compensated for a series of frustrating failures. The mission's prime payloads, the WESTAR-VI and PALAPA-B2 satellites, were both placed in incorrect orbits due to faults in their Payload Assist Module-D (PAM-D) rocket motors. A fault in the remote manipulator system (RMS) prevented deployment of the German Shuttle Pallet Satellite (SPAS), which had originally flown during STS-7 and with this flight became the first satellite ever to be refurbished and flown again (even though it remained in the payload bay because of the RMS fault). Another failure foiled another experiment, the Integrated Rendezvous Target (IRT).

STS-41B also carried a Cinema-360 camera that the crew used to record their flight; the Acoustic Containerless Experiment System (ACES); a Monodisperse Latex Reactor (MLR); Radiation Monitoring Equipment (RME); and an Isoelectric Focusing (IEF) payload.

After a mission of 128 orbits, in 7 days, 23 hours, 15 minutes and 55 seconds, the crew made the first-ever shuttle landing at the Kennedy Space Center in Florida, when Challenger touched down on KSC Runway 15 on February 11, 1984.

SPACE SHUTTLE MISSION 11: STS-41C

(April 6–13, 1984)

U.S. Manned Spaceflight

With Robert Crippen as commander and Francis Scobee as pilot, space shuttle mission STS-41C launched on April 6, 1984. Also aboard the shuttle Challenger with Crippen and Scobee were mission specialists George Nelson, James van Hoften, and Terry Hart.

The two main objectives of STS-41C were the retrieval and repair of the Solar Maximum satellite (nicknamed "Solar Max" in the fairly large amount of media coverage accorded to the mission) and the deployment of the Long-duration Exposure Facility (LDEF), an orbiting cylinder containing 57 experiments designed to test the impact of long stays in space on a variety of items, including various materials and seeds. Intended to be left in orbit for retrieval during a later mission, the LDEF was successfully deployed.

In the mission's most dramatic task, George Nelson and James van Hoften used the nitrogen-propelled Manned Maneuvering Units (MMUs) to maneuver themselves alongside Solar Max and retrieve the satellite for repair in the shuttle's cargo bay. Wrestling the satellite into the cargo bay proved difficult; the planned duration of the mission was extended by one day to give the astronauts more time to complete their repair tasks.

Once they had managed to fasten Solar Max firmly to Challenger's cargo bay, Nelson and van Hoften replaced the satellite's malfunctioning altitude control system and its coronagraph/polarimeter, and then returned the craft to separate flight. A remarkable achievement, the repair mission was the first of its kind—the first-ever planned fix to be made on a spacecraft while it was in orbit.

Nelson and van Hoften carried out the retrieval on April 8, 1984 during an EVA of 2 hours and 38 minutes, and completed the repair during a spacewalk of 6 hours and 44 minutes on April 11.

STS-41C also featured the first direct ascent trajectory of the shuttle program—meaning that the shuttle Challenger achieved orbit by a longer-than-usual firing of its Space Shuttle Main Engines (SSMEs), and just one (rather than the usual two) firings of its Orbital Maneuvering System (OMS) engines. In the typical sequence, one OMS burn maneuvers the shuttle into

An historic first: STS-41C mission specialists George Nelson (right) and James van Hoften complete the first-ever planned repair of an orbiting spacecraft, April 11, 1984. [NASA/courtesy of nasaimages.org]

an elliptical orbit, and a second OMS firing places the vehicle in an almost-circular orbit of the Earth. In the case of STS 41-C, the direct ascent trajectory was utilized to propel Challenger directly into a higher-than-usual orbit (nearly 300 miles from Earth) so the shuttle could rendezvous with the Solar Max satellite.

The STS 41-C crew used IMAX and Cinema 360 cameras to document their flight, and also conducted an experiment that had been designed as part of the Shuttle Student Involvement Program (SSIP).

Originally scheduled to land at the Kennedy Space Center, STS 41-C was diverted to Edwards Air Force Base in California, where the crew touched down on April 13, 1984 after a mission of 6 days, 23 hours, 40 minutes and 7 seconds.

SPACE SHUTTLE MISSION 12: STS-41D

(August 30–September 5, 1984)

U.S. Manned Spaceflight

Shuttle Mission STS-41D featured the first flight of the space shuttle Discovery, and the first pad abort of the shuttle program. The flight was originally scheduled for launch on June 25, but had to be postponed when one of the shuttle's computers failed. Another near-launch the following day (at T-9 seconds, in the shuttle's first pad abort) had to be called off because of an apparent fault in one of the Space Shuttle Main Engines (SSMEs), which resulted in Discovery being removed from the launch pad entirely, so its engine could be replaced.

Changes in the flight schedule necessitated by the long delay caused NASA administrators to cancel the two missions scheduled to follow 41D, and to redistribute the payloads involved. An August 29 attempt to launch Discovery was delayed by a software problem, and the eventual launch on August 30, 1984 was briefly delayed when a plane strayed into the air space near the launch site, which is restricted during shuttle launches.

The crew for STS-41D featured commander Henry Hartsfield, Jr., pilot Michael Coats, mission specialists Judith Resnik, Steven Hawley, and Richard Mullane, and Payload Specialist Charles Walker.

With this flight, Judith Resnik became the second American woman to fly in space.

Discovery's maiden flight was a resounding success. The STS-41D crew launched three commercial satellites: Satellite Business System SBS-D, SYNCOM IV-2 (also known as LEASAT 2), and TELSTAR 3-C.

In another highlight, the crew tested a solar wing designed by the Office of Application and Space Technology. Equipped with solar arrays, the large (102 feet long by 13 feet wide) solar wing was designed to prove that large solar arrays could be used to provide power for future space facilities, including the International Space Station (ISS).

Other payloads included the Continuous Flow Electrophoresis System (CFES) III, which was operated by Charles Walker of McDonnell Douglas, the company that had manufactured the CFES; an Air Force experiment called Cloud Logic to Optimize Use of Defense Systems (CLOUDS); an IMAX camera; radiation monitoring equipment; and a Shuttle Student Involvement Program (SSIP) experiment.

Discovery's first flight came to an end on September 5, 1984 after 6 days, 56 minutes and 4 seconds. The shuttle landed at Edwards Air Force Base in California.

SPACE SHUTTLE MISSION 13: STS-41G

(October 5–13, 1984)

U.S. Manned Spaceflight

Launched aboard the shuttle Challenger on October 5, 1984, STS-41G featured commander Robert Crippen, who was flying in space for the fourth time, pilot Jon

McBride, mission specialists Kathryn Sullivan, Sally Ride, and David Leestma, and Payload Specialists Marc Garneau and Paul Scully-Power.

During the flight, Kathryn Sullivan became the first American woman to participate in a spacewalk.

Marc Garneau was the first Canadian citizen to fly in space, and during STS-41G, Paul-Scully Power became the first native of Australia to fly in space.

The crew successfully deployed the Earth Radiation Budget Satellite (ERBS), and conducted three experiments for NASA's Office of Space and Terrestrial Applications.

Garneau conducted a series of experiments collectively named CANEX, in honor of their having been designed by scientists in Canada, and as a representative of the U.S. Naval Research Laboratory, Scully-Power conducted oceanographic research.

On October 11, 1984, Kathryn Sullivan and David Leestma ventured outside the shuttle for a 3 hour, 29 minute spacewalk whose primary aim was to test the Orbital Refueling System (ORS). Their successful test demonstrated a means of transferring hydrazine fuel to refuel satellites in orbit. Sullivan became the first American woman (and second woman in history) to perform a spacewalk.

At the conclusion of STS-41G, which lasted 8 days, 5 hours, 23 minutes and 33 seconds, Challenger landed at the Kennedy Space Center in Florida on October 13, 1984.

SPACE SHUTTLE MISSION 14: STS-51A

(November 8–16, 1984)

U.S. Manned Spaceflight

The crew for STS-51A, which launched aboard the shuttle Discovery on November 8, 1984, featured commander Frederick Hauck, pilot David Walker, and Mission specialists Anna Fisher, Dale Gardner, and Joseph Allen.

STS-51A began with the successful deployment of the Canadian TELESAT-H (ANIK) communications satellite and the SYNCOM IV-1 (LEASAT-1) defense communications satellite.

Then, in two remarkable spacewalks, mission specialists Allen and Gardner donned Manned Maneuvering Unit (MMU) backpacks and floated free of the shuttle to capture two satellites that had each failed to reach their intended orbit when they were previously deployed, during STS-41B.

On November 12, 1984 the spacewalking astronauts captured the PALAPA-B2 satellite during an EVA of 6 hours, and 2 days later, they retrieved WESTAR-VI while accumulating an additional 5 hours and 42 minutes in EVA.

The unprecedented retrieval of the two satellites provided a dramatic demonstration of the space shuttle's far-reaching capabilities, and made clear the advantages of the shuttle as a platform for working in orbit.

The STS-51A crew landed at the Kennedy Space Center on November 16, 1984, after a flight of 7 days, 23 hours, 44 minutes and 56 seconds.

SPACE SHUTTLE MISSION 15: STS-51C

(January 24–27, 1985)

U.S. Manned Spaceflight

STS-51C was the first space shuttle mission devoted entirely to classified activities sponsored by the U.S. Department of Defense. Launched aboard the shuttle Discovery on January 24, 1985, the fifteenth mission of the space shuttle era was commanded by Apollo 16 and STS-4 veteran Thomas Mattingly, with pilot Loren Shriver, mission specialists Ellison Onizuka and James Buchli, and Payload Specialist Gary E. Payton.

Originally intended to fly aboard the shuttle Challenger, STS-51C was shifted to Discovery because of problems with Challenger's thermal tiles. On January 23, the launch was postponed for a day because of freezing weather.

Although missions devoted to the Department of Defense are considered classified and little information is known about their payloads, NASA has noted that the U.S. Air Force Inertial Upper Stage booster was successfully deployed during this mission.

After a flight of just 3 days, 1 hour, 33 minutes and 23 seconds, Discovery returned to Earth at the Kennedy Space Center in Florida on January 27, 1985.

SPACE SHUTTLE MISSION 16: STS-51D

(April 12–19, 1985)

U.S. Manned Spaceflight

The crew for STS-51D, which launched aboard the shuttle Discovery on April 12, 1985, included commander Karol Bobko, pilot Donald Williams, mission specialists Rhea Seddon, Jeffrey Hoffman, and David Griggs, Payload specialist Charles Walker, and Senator Jake Garn of Utah.

Garn was the first incumbent public official to fly in space.

STS-51D was delayed several times, first to accommodate the shift in payloads caused by the earlier cancellation of mission 51-E, and then again by an accident in which a suspended platform fell onto Discovery's payload bay door. The original schedule had called for a launch on March 19; a second try scheduled for March 28 was put off by the falling platform. The actual launch on April 12 was delayed for almost an hour when a ship strayed into the restricted waters where the shuttle's solid rocket boosters (SRBs) were expected to splash down prior to their recovery.

During the mission, the STS-51D crew successfully deployed the Canadian TELESAT-I (ANIK C-1) communications satellite.

Less successful was their attempt to deploy the SYNCOM IV-3 (LEASAT-3) satellite. A faulty lever—which was designed to start a sequence that would fire a motor designed to propel the satellite into orbit—had not engaged. Mission controllers decided to extend the flight for two days while they and the crew tried to devise a way to fix the problem.

Putting the improvised solution to work, mission specialists David Griggs and Jeffrey Hoffman made an EVA of three hours and six minutes on April 16 when they attached two flap-like devices (nicknamed "the flyswatters" within NASA) to the end of the shuttle's robotic arm, the Remote Manipulator System (RMS). Mission specialist Rhea Seddon then used the RMS to catch the stuck lever with the improvised flaps, in an effort to force the lever to engage and begin the planned sequence. The attempt demonstrated great skill and exceptional coordination between the crew and mission controllers, but it could not salvage the deployment. (The fixing of SYNCOM IV-3/LEASAT-3 would have to wait until the STS-51I mission of August 27–September 3, 1985, when the satellite was retrieved and repaired.)

STS-51D also featured the sixth flight of the Continuous Flow Electrophoresis System (CFES) III, which was operated by Payload Specialist Charles Walker of McDonnell Douglas (which manufactured the CFES). Other payloads included an American Flight Echocardiograph (AFE) for medical research, and a test designed to verify astronomical photography.

In a particularly engaging union of space science and popular culture, the flight also included a demonstration of the effects of microgravity on mechanical toys. The resulting lesson—"Toys in Space"—was destined for distribution to schools.

Following a mission of 6 days, 23 hours, 55 minutes and 23 seconds, Discovery returned to Earth on runway 33 at the Kennedy Space Center in Florida, in one of the most precarious landings of the shuttle program. The rough touchdown resulted in extensive damage to the shuttle's braking system and a blown tire, but the crew was not injured. As a result of the difficult landing, NASA officials decided that future flights would land at Edwards Air Force Base in California until the shuttle's nose wheel steering system—which would alleviate the need for excessive maneuvering of the spacecraft along the KSC runways—could be fully tested and put into common use.

SPACE SHUTTLE MISSION 17: STS-51B

(April 29–May 6, 1985)

U.S. Manned Spaceflight

The STS-51B crew of the space shuttle Challenger launched on April 29, 1985 with its primary mission being the operation of Spacelab-3, the orbital laboratory developed by the European Space Agency (ESA).

Robert Overmyer commanded STS-51B; Frederick Gregory served as pilot; and the crew included mission specialists Don Lind, Norman Thagard, and William Thornton and Payload Specialists Lodewijk van den Berg and Taylor Wang.

Spacelab-3 was originally intended to fly on mission 51-E, but that flight was canceled during the long delays prior to the shuttle Discovery's first flight (STS-41D). As a result, Spacelab-3 actually flew before Spacelab-2, which flew aboard STS-51F.

The Spacelab experiments covered materials sciences, life sciences, fluid mechanics, atmospheric physics and astronomy. This first operational flight for the orbital laboratory was designed to prove that the shuttle could serve as a stable microgravity environment for the sort of experiments in materials processing and fluid mechanics that ESA scientists had envisioned when they first designed Spacelab. The results were encouraging; 14 of the 15 primary Spacelab experiments were deemed successful.

STS-51B also marked the first time since the early 1960s test flights of the Mercury program that monkeys were carried into space on American spacecraft (the shuttle flight was also the first time that astronauts and monkeys flew in space together), as part of an investigation into the effects of weightlessness on animals.

STS-51B concluded with the shuttle landing at Edwards Air Force Base in California on May 6, 1985, after a flight of 7 days, 8 minutes and 46 seconds.

SPACE SHUTTLE MISSION 18: STS-51G

(June 17–24, 1985)

U.S. Manned Spaceflight

Three communications satellites were deployed during the STS-51G shuttle mission, which launched aboard the shuttle Discovery on June 17, 1985.

The commander of STS-51G was Daniel Brandenstein; he was accompanied by pilot John Creighton, mission specialists Shannon Lucid, Steven Nagel, and John Fabian, and Payload Specialists Patrick Baudry of France and Sultan Salman Al-Saud, the first citizen of Saudi Arabia to fly in space.

Baudry was the first representative of the French national space agency, Centre National d'Études Spatiales (CNES), to fly aboard the U.S. space shuttle.

Mexico's MORELOS-A communications satellite, the ARABSAT-A satellite of the Arab Satellite Communications Organization, and AT&T Corporation's TELSTAR-3D satellite were successfully deployed during the mission.

The crew also deployed and retrieved the Shuttle Pointed Autonomous Research Tool for Astronomy, SPARTAN-1, and operated the Automated Directional Solidification Furnace (ADSF), a materials processing furnace. Payload specialist Baudry carried out two French biomedical experiments.

Also included in the mission profile of STS-51G was the High Precision Tracking Experiment (HPTE), a test related to the proposed Strategic Defense Initiative (SDI).

Discovery landed at Edwards Air Force Base in California on June 24, 1985 after a mission of 7 days, 1 hour, 38 minutes and 52 seconds.

SPACE SHUTTLE MISSION 19: STS-51F

(July 29–August 6, 1985)

U.S. Manned Spaceflight

The STS-51F crew of the space shuttle Challenger experienced the first abort-to-orbit of the shuttle program.

Three Spacelab-2 pallets designed by the European Space Agency (ESA) were the primary payload of the flight. Because of the long delays prior to the shuttle Discovery's first flight (STS-41D), shuttle mission 51-E had been canceled, which resulted in a change in schedule for the Spacelab flights. As a result, the STS-51F flight of Spacelab-2 actually occurred after Spacelab-3 had already flown in space (during STS-51B).

The change in schedule also resulted in the STS-51F mission taking place after STS-51G, even though the "F" and "G" designations would appear to indicate otherwise.

STS-51F was commanded by veteran test pilot Gordon Fullerton, accompanied by pilot Roy Bridges and mission specialists Story Musgrave, Anthony England, and Karl Henize, and Payload Specialists Loren Acton and John-David Bartoe.

The mission actually suffered two separate abort procedures, each as a result of the shutdown of one or more Space Shuttle Main Engines (SSMEs). The originally scheduled launch on July 12 ended in a pad abort when a faulty coolant valve caused the spacecraft's three SSMEs to shut down just three seconds before ignition of the shuttle's solid rocket boosters (SRBs).

On launch day, July 29, 1985, there was an initial delay of a little more than an hour-and-a-half due to a software problem, and then, 5 minutes and 45 seconds into the launch, the number 1 SSME shut down prematurely, which meant that the shuttle would be unable to reach its planned orbit. As a result, NASA mission controllers called for an Abort-to-Orbit (ATO), which meant that the shuttle would enter a temporary orbit lower than the one called for in the original mission profile, by using the thrust from its two properly functioning SSMEs.

Once in the orbit dictated by the ATO trajectory, the crew used the shuttle's Orbital Maneuvering System (OMS) thrusters to move Challenger into its originally intended orbit. The combination of the ATO procedure and the on-orbit maneuvering saved the mission, but resulted in a need for extensive alteration of the original flight plan in order to accomplish the flight's objectives. The flight was extended by one day to accommodate the changes.

With the mission safely underway, the crew deployed the Spacelab-2 experiments in the shuttle's cargo bay. Spacelab-2 consisted of a wide array of experiments mounted on three pallets that were exposed to space, and a pressurized support module that became known as "the Igloo" (which it resembled).

Operated by the astronauts on Challenger or remotely from Earth, the instruments in the open cargo bay were used to perform experiments in life sciences, plasma physics, high-energy astrophysics, solar physics, atmospheric physics, astronomy, and technology research. The Spacelab systems worked well, and the shuttle was further proven to be a good platform for scientific research.

Despite their difficult start, the STS-51F crew was able to achieve most of their objectives in a flight of 7 days, 22 hours, 45 minutes and 26 seconds. They landed on August 6, 1985 at Edwards Air Force Base in California.

SPACE SHUTTLE MISSION 20: STS-51I

(August 27–September 3, 1985)

U.S. Manned Spaceflight

The twentieth space shuttle mission, STS-51I was launched aboard the shuttle Discovery on August 27, 1985.

Joe Engle commanded STS-51I; he was accompanied by pilot Richard Covey and mission specialists James van Hoften, John Lounge, and William Fisher.

A first attempt to launch Discovery, on August 24, had to be postponed because of bad weather, and a second try, on August 25, was delayed by a failure in one of the shuttle's on-board computers. The launch on August 27 was again delayed by bad weather, and by an unauthorized ship that strayed into the waters restricted for recovery of the shuttle's Solid Rocket Boosters (SRBs).

The STS-51I crew deployed three communications satellites: the American Satellite Company's ASC-1, the Synchronous Communications Satellite SYNCOM IV-4, and the Australian AUSSAT-1.

The sunshield for AUSSAT-1 became caught on the camera mounted on the shuttle's Remote Manipulator System (RMS), which forced the crew to deploy the satellite earlier than planned. That in turn led to the mission being shortened by one day.

Although it was properly deployed in the correct geosynchronous orbit, SYNCOM IV-4 failed to operate.

The failure was made particularly ironic by the fact that part of the mission profile for STS-51I was to attempt a repair on SYNCOM IV-3 (also known as LEASAT-3), which had been deployed in April of 1985 during STS-51D. The SYNCOM IV-3 satellite had failed because of a malfunctioning sequence lever; during its original deployment in April, the crew of STS-51D had tried to force the lever into the proper position, but they were not able to fix the problem at that time.

With that frustrating background in mind, mission specialists William Fisher and James van Hoften were given the task of trying to fix the recalcitrant SYNCOM IV-3 during STS-51I. They made two long spacewalks during the flight (7 hours and 20 minutes on August 31, 1985, and 4 hours and 26 minutes on September 1), and were able to successfully retrieve and repair SYNCOM IV-3 by installing a system that enabled ground controllers to move the satellite into the correct orbit after the end of the STS-51I mission.

The shortened mission lasted 7 days, 2 hours, 17 minutes and 42 seconds. Discovery returned to Earth on September 3, 1985 at Edwards Air Force Base in California.

SPACE SHUTTLE MISSION 21: STS-51J

(October 3–7, 1985)

U.S. Manned Spaceflight

The first flight of the space shuttle Atlantis and the second shuttle mission dedicated to the classified activities of the U.S. Department of Defense, STS-51J launched on October 3, 1985.

Making his third flight in space, Karol Bobko commanded STS-51J, accompanied by pilot Ronald Grabe, mission specialists David Hilmers and Robert Stewart, and Payload Specialist William Pailes.

The short mission ended after 4 days, 1 hour, 44 minutes and 38 seconds, when Atlantis touched down at Edwards Air Force Base in California on October 7, 1985.

SPACE SHUTTLE MISSION 22: STS-61A

(October 30–November 6, 1985)

U.S. Manned Spaceflight

Devoted to German scientific research using the Spacelab D-1 laboratory module, the STS-61A flight of the shuttle Challenger marked the first time eight astronauts traveled into space aboard a single spacecraft.

Henry Hartsfield commanded STS-61A, accompanied by pilot Steven Nagel, mission specialists James Buchli, Guion Bluford, and Bonnie Dunbar, and Payload Specialists Reinhard Furrer and Ernst Messerschmid of West Germany (Germany was at the time still divided into East and West, prior to the country's reunification after the fall of the Soviet Union) and European Space Agency (ESA) astronaut Wubbo Ockels—who with the STS-61A flight became the first citizen of The Netherlands to fly in space.

Launched on October 30, 1985, the mission featured a unique control arrangement, in which the shuttle's flight was controlled as usual from the Johnson Space Center in the United States, but the Spacelab operations were controlled by German space officials at the German Space Operations Center at Oberpfaffenhofen, in West Germany.

The Spacelab research included 75 experiments in basic and applied microgravity research covering materials science, life sciences and technology, communications and navigation. Also included was an innovative Vestibular Sled experiment that featured a sled on which an astronaut could ride along a set of rails while scientists studied the individual's reaction in the space environment.

STS-61A also featured the deployment of the Global Low Orbiting Relay (GLOMR) satellite, which was released from a Getaway Special canister.

After a flight of 7 days, 44 minutes and 51 seconds, STS-61A came to a close when Challenger landed on November 6, 1985 at Edwards Air Force Base in California.

SPACE SHUTTLE MISSION 23: STS-61B

(November 26–December 3, 1985)

U.S. Manned Spaceflight

Launched on November 26, 1985, the STS-61B mission featured the second flight of the space shuttle Atlantis and the second night launch of the shuttle program. It also marked the first flight in space by a citizen of Mexico, Dr. Rodolfo Neri Vela.

Brewster Shaw commanded STS-61B. He was joined by pilot Bryan O'Connor, mission specialists Mary Cleave, Sherwood Spring, and Jerry Ross, and Payload Specialists Rodolfo Neri Vela and Charles Walker.

The STS-61B crew deployed three communications satellites: the Mexican MORELOS-B and the Australian AUSSAT-2, which were pushed into orbit by Payload Assist Module-D (PAM-D) motors; and the RCA Americom SATCOM KU-2, which was deployed with a Payload Assist Module-D2 (PAM-D2) motor. The next-generation version of the payload assist motors, the PAM-D2 was designed to put heavier payloads into orbit.

mission specialists Jerry Ross and Sherwood Spring made two spacewalks while testing assembly procedures for building structures in space. They made a spacewalk of 5 hours and 32 minutes on November 29, 1985, and a second EVA of 6 hours and 41 minutes on December 1, while carrying out the Experimental Assembly of Structures in Extravehicular Activity (EASE) exercise, and the Assembly Concept for Construction of Erectable Space Structure (ACCESS) experiment.

Atlantis also carried several additional payloads during STS-61B, including the Morelos Payload Specialist Experiments (MPSE), which were conducted by payload specialist Rodolfo Neri Vela on behalf of the Mexican government, and the Continuous Flow Electrophoresis System (CFES), which payload specialist Charles Walker operated for his employer, McDonnell Douglas, which had manufactured the CFES.

The flight also tested the shuttle's digital autopilot, and it featured an experiment in the Diffusive Mixing of Organic Solutions (DMOS), a Getaway Special canister for Telesat of Canada, and the inclusion of an IMAX camera mounted in the shuttle's cargo bay, which recorded the spacewalking construction activities of Ross and Spring.

Following a flight of 6 days, 21 hours, 4 minutes and 49 seconds, the crew returned to Earth at Edwards Air Force Base in California on December 3, 1985. Weather conditions at Edwards lent some drama to the landing; the mission was shortened by one orbit (to 109) because of lightning, and mission controllers had the shuttle land on the base's concrete Runway 22 after heavy rains drenched the dry lake bed site that had originally been scheduled for the landing.

SPACE SHUTTLE MISSION 24: STS-61C

(January 12–18, 1986)

U.S. Manned Spaceflight

The STS-61C flight of the space shuttle Columbia launched on January 12, 1986, after a series of frustrating delays. The original launch, scheduled for December 18, 1985, was first delayed to allow additional time to prepare the shuttle, and then delayed again the following day because of what was thought at the time to be a problem with one of the Solid Rocket Boosters (SRBs)—which was later determined to be due to a faulty instrument. Another launch attempt, on January 6, 1986, was halted just prior to lift-off because of a leak of liquid oxygen.

The next day, January 7, another attempted launch had to be postponed when bad weather set in at the two overseas sites reserved for emergency landings in the event that the shuttle might need to make a Trans-Atlantic Abort Landing. At least one of the two sites, which are located in Moron, Spain and Dakar, Senegal, would have to be available for the shuttle to be cleared for launch; in this case, the bad weather caused both sites to be inaccessible at the same time. On January 9, the next scheduled launch was called off when a liquid oxygen sensor broke off of the launch pad and became wedged in one of the valves leading to the shuttle's number two Space Shuttle Main Engine (SSME). Bad weather further delayed the next launch attempt, on January 10, for two more days.

Launch attempt number seven proved to be the lucky number; when the actual lift-off occurred on January 12, the countdown proceeded smoothly and there were no further delays.

Robert Gibson commanded STS-61C; he was accompanied by pilot Charles Bolden, mission specialists Franklin Chang-Diaz, Steven Hawley, and George Nelson, and Payload Specialists Robert Cenker and Florida Congressman Bill Nelson.

Serving on the first of his many space missions, during STS-61C Franklin Chang-Diaz became the first native of Costa Rica to fly in space.

Bill Nelson was the second political figure to fly aboard the shuttle; Utah Senator Jake Garn had been the first, during STS-51D in April 1985.

The STS-61C crew deployed the RCA Americom satellite SATCOM KU-1, which was propelled into orbit by a Payload Assist Module-D2 (PAM-D2) motor—one of the second-generation payload motors used to place heavier payloads into orbit.

Faulty batteries scuttled a 35 mm camera that had been designed to photograph the comet Halley—an experiment known as the Comet Halley Active Monitoring Program (CHAMP). Other experiments included the Materials Science Laboratory-2 (MSL-2), Hitchhiker G-1, the Initial Blood Storage experiment, a hand-held protein crystal growth experiment, the Infrared Imaging Experiment (IR-IE), three experiments designed by students as part of the Shuttle Student Involvement Program (SSIP), and 13 Getaway Special experiments.

As it neared its conclusion, STS-61C ran into delays similar to the ones that had troubled the flight at its start. Landing was originally scheduled for the Kennedy Space Center in Florida on January 17, but it was rescheduled for January 16 to save time in preparing the shuttle for its next flight. That landing attempt was then postponed by bad weather, as was another attempt on the originally scheduled target of January 17. The continuing foul weather in Florida eventually led mission controllers to divert the landing to Edwards Air Force Base in California, where the shuttle finally returned to Earth during a night landing on January 18, 1986, after 98 orbits in 6 days, 2 hours, 3 minutes and 51 seconds. The shuttle was returned to the Kennedy Space Center on January 23, just five days before the start of the next flight, which was scheduled to be made by the shuttle Challenger on January 28.

SPACE SHUTTLE MISSION 25: STS-51L

(January 28, 1986)

U.S. Manned Spaceflight

The twenty-fifth mission of the space shuttle program, the STS-51L flight of the space shuttle Challenger was scheduled to be the first to include a professional teacher, as part of NASA's Teacher in Space project.

The mission profile also called for deployment of the Tracking and Data Relay Satellite-2 (TDRS-2), and a free-flying module equipped with ultraviolet spectrometers and cameras to make a close study of the tail and coma of the comet Halley. Also aboard were an experiment in fluid dynamics, the Comet Halley Active Monitoring Program (CHAMP), a Phase Partitioning Experiment (PPE), and three experiments designed by students as part of the Shuttle Student Involvement Program (SSIP).

The commander of STS-51L was Francis Scobee, who was making his second flight into space. Pilot Michael Smith was on his first space mission; mission specialists Judith Resnik, Ellison Onizuka, and Ronald McNair had each flown in space once before. Payload Specialists Gregory Jarvis and Christa McAuliffe were each making their first spaceflight.

The STS-51L crew of the space shuttle Challenger. (Front, left to right): Michael Smith, Francis Scobee, Ronald McNair, (back, left to right): Ellison Onizuka, Christa McAuliffe, Gregory Jarvis, Judith Resnik. [NASA/courtesy of nasaimages.org]

During her first flight, aboard the shuttle Discovery during STS-41D in 1984, Judith Resnik had become the second American woman to fly in space (Sally Ride had been the first, during STS-7 in June, 1983).

An employee of Hughes Aircraft Co., aerospace engineer Gregory Jarvis had originally been scheduled to fly aboard Columbia during the mission just prior to this one (STS-61C, earlier in January).

Christa McAuliffe had been selected for the honor of being America's first teacher in space from an applicant pool of 11,000. She was to have conducted televised lessons aboard Challenger that would be watched by children in schools across the country.

For its tenth mission, Challenger was scheduled to lift off from Pad 39B at the Kennedy Space Center in Florida. It was to be the first launch from that location since the last Apollo spacecraft had lifted off from Pad 39B during the Apollo-Soyuz Test Project in 1975.

The landing was also scheduled to take place at the Kennedy Space Center, after an envisioned flight of 6 days and 34 minutes.

Originally scheduled for January 22, the launch of STS-51L was delayed by the ongoing scheduling problems that postponed the previous mission (STS-61C). A launch attempt scheduled for January 25 had to be put off because of bad weather at the overseas site in Dakar, Senegal where the shuttle would have landed in the event that the vehicle might have needed to make a Trans-Atlantic Abort Landing. The timing of the launch was adjusted so that the Trans-Atlantic Abort Landing site in Casablanca—which was not equipped for a night landing—could be available instead of the Senegal site; but launch preparations couldn't be made quickly enough to meet the new time, so the launch was delayed another day. An anticipation of bad weather at the Kennedy Space Center in Florida caused the launch to be rescheduled for January 27.

The January 27 launch was waylaid by one of the strangest delays of the entire American space program. The equipment the ground crew used to close the shuttle's hatch became stuck on the hatch, and couldn't be disengaged. After holding the shuttle in place on the ground for 24 hours, the faulty equipment was finally sawed off and the bolt that had connected it to the spacecraft was drilled out. The January 28 launch was delayed for two hours by a fault in the launch processing system.

When the launch finally came, televised to an audience of millions and closely watched by whole classes of schoolchildren as part of the Teacher in Space project, the festive pre-launch anticipation was quickly transformed into a searing horror, as Challenger caught fire and exploded 73 seconds after liftoff.

The shuttle disintegrated. The crew compartment was jettisoned from the wrecked spacecraft and plummeted into the Atlantic Ocean. It was unclear at the time if any of the crew had survived the initial explosion; in any case, the velocity with which the pieces of the shuttle hit the water made it clear that no one would emerge from the accident alive.

Although later inquiry would attribute the accident to a variety of factors, including unrealistic scheduling expectations and faulty communications within NASA and among the space agency and its contractors, the proximate cause of the accident was fixed as a faulty "O ring"—the rubber gasket that normally sealed the joints of the shuttle's huge Solid Rocket Boosters (SRBs). During past missions, there had been concern on the part of some NASA officials and engineers at the Morton Thiokol Corporation, which manufactured the SRBs, about the possible effect that extremely cold weather might have on the O rings' ability to maintain a tight seal. There had previously been evidence of deterioration of the O rings on SRBs that had been recovered after prior missions, but the damage had not been thought to pose an immediate danger during launches prior to STS-51L. In any case, those launches that took place in warmer weather conditions did not appear to be at risk for the catastrophic failure that the icy weather precipitated in the case of the Challenger disaster.

In a sad "if only" scenario, it was revealed after the accident that some of the engineers at Morton Thiokol had actually advised against a launch attempt on January 28 because of the potential impact of the cold weather; but they were overruled, both by NASA officials and by the senior management of their own company.

The Challenger disaster struck an enormous chord of deeply felt emotion in the American public. The frightening images of the explosion instantly became an indelible symbol of technology gone wrong, and of the fragile beauty of human courage in the pursuit of lofty goals.

With the loss of the Challenger crew, the U.S. space program came to an abrupt halt. An investigatory commission led by former U.S. Secretary of State William Rogers was convened, its members including Neil Armstrong, veteran test pilot Chuck Yeager and Sally Ride. The commission presented its report to President Ronald Reagan on June 9, 1986, citing the administrative and scheduling problems as well as the faulty O-rings as contributing causes of the accident. The report set out in grim detail the mechanism by which the icy weather had caused the O-ring to become brittle and fail, which allowed the SRB to leak fuel onto the shuttle's huge external fuel tank, which in turn had ignited in a massive explosion that destroyed the shuttle and killed the crew.

But the report also served as an indictment of the culture that had rushed the shuttle program into greater productivity at the expense of NASA's long-held insistence on the highest possible standard of safety.

The Rogers Commission report described a blinkered corporate culture and shortsighted management process at NASA, and it mattered little that many other public and private institutions throughout the country could increasingly be characterized in similar fashion. As the embodiment of the nation's high ideals for scientific and technological progress, the space agency was expected to serve as an example of excellence in all its endeavors, and particularly in any activity involving the safety of its personnel.

The Challenger disaster and the crisis of confidence that it engendered both within and outside of NASA led even ordinary Americans to question the wisdom of even the basic premise of the space shuttle program. Thoughtful commentators and opinion leaders openly wondered if the nation had been well served by the space agency having spent years developing a "space transportation system" that placed such great priority on the delivery of satellites into space for paying customers, at the expense of exploration that might have built on the successes of the Apollo and Skylab programs.

Fears that preoccupation with bureaucracy and commercial efficacy had stifled the sort of thinking that had given birth to the agency's most cherished ideals gradually receded as NASA officials demonstrated their willingness to work through the aftermath of the Challenger disaster in a forthright and transparent manner.

As a result, the Challenger legacy is one of reestablishing the primacy of the individual human being in the overall scheme of the U.S. space program. The accident reacquainted the American public with the dangers inherent in spaceflight—even in an age where regularly scheduled missions seemed to promise a welcome absence of dramatic moments—and by extension, reawakened a shared appreciation of the courage that spacefarers display even on missions that are deemed routine.

The loss of the STS-51L crew also renewed the ideal that from the earliest days of the U.S. space program recognized that each individual who participates in the long reach toward the stars does so as the representative of an entire nation of individuals, who in turn share some moral responsibility for ensuring the best possible chance of his or her safe journey.

SPACE SHUTTLE MISSION 26: STS-26

(September 29–October 3, 1988)

U.S. Manned Spaceflight

Two years and eight months after the Challenger disaster, America returned to manned spaceflight with the STS-26 mission of the space shuttle Discovery, which launched on September 29, 1988.

Frederick Hauck was the commander of STS-26; the flight was his third space mission. Richard Covey served as pilot, and John Lounge, George Nelson, and David Hilmers were mission specialists for the flight. Each crew member had flown in space at least once before; and Nelson, like Hauck, was making his third spaceflight.

Reflecting just one of the many modifications that had followed in the wake of the Challenger accident, STS-26 crew members were outfitted with new spacesuits that were partially pressurized, for use during launch and re-entry. The new spacesuits inadvertently led to a delay of about 98 minutes at launch, when technicians had to replace fuses in the suits' cooling systems.

During the flight, the crew successfully deployed the Tracking and Data Relay Satellite-3 (TDRS-3)—which despite its numerical designation was the second TDRS to be placed in orbit. The satellite was propelled into geosynchronous orbit by its attached Inertial Upper Stage (IUS).

STS-26 also featured a number of secondary payloads, including the Orbiter Experiments Autonomous Supporting Instrumentation System-1 (OASIS-1), the Physical Vapor Transport of Organic Solids (PVTOS), the Earth-Limb Radiance Experiment (ELRAD) and an Automated Directional Solidification Furnace (ADSF). Experiments included a study of protein crystal growth, an experiment in infrared communications, a test involving the aggregation of red blood cells, an Isoelectric Focusing Experiment (IFE), a lightning experiment, a Phase Partitioning Experiment (PPE), and two experiments that were designed as part of the Shuttle Student Involvement Program (SSIP).

The crew also deployed a Ku-band antenna, but the command given to aim the antenna did not correspond with the actual telemetry, leaving a mystery for technicians to sort out at mission's end.

A potentially more serious fault occurred in the shuttle's environmental control system, when ice formed in the equipment designed to keep the crew cabin cool, resulting in temperatures rising to about 85 degrees (Fahrenheit). Fortunately, the temperature rose no higher, and the crew was in no serious danger.

Discovery returned to Earth after 64 orbits, in 4 days, 1 hour and 11 seconds, landing at Edwards Air Force Base in California on October 3, 1988.

SPACE SHUTTLE MISSION 27: STS-27

(December 2–6, 1988)

U.S. Manned Spaceflight

The second space shuttle flight of the post-Challenger era, STS-27, was a mission dedicated to the U.S. Department of Defense. It launched on December 2, 1988 aboard the shuttle Atlantis.

Robert Gibson commanded STS-27; Guy Gardner piloted the shuttle, and Richard Mullane, Jerry Ross, and William Shepherd served as mission specialists.

Originally scheduled for December 1, the first attempt at launch was postponed by unsatisfactory weather conditions.

During the launch on December 2, insulating material from one of the shuttle's Solid Rocket Boosters (SRBs) came loose and hit Atlantis about a minute-and-a-half after the shuttle left the launch pad. It was one of the earliest occasions of appreciable tile damage being caused to the shuttle by detached SRB insulation; many years later, similar damage would lead to the loss of the shuttle Columbia and its crew, during the STS-107 flight in February 2003.

Despite the damage caused by the errant insulation, the Atlantis crew returned safely after a little more than four days in space.

Because STS-27 was dedicated to the classified activities of the U.S. Department of Defense, few details of the flight have been released to the public. This was the third space shuttle mission devoted to the Department of Defense.

After a mission of 4 days, 9 hours, 5 minutes and 37 seconds, Atlantis landed at Edwards Air Force Base in California on December 6, 1988.

SPACE SHUTTLE MISSION 28: STS-29

(March 13–18, 1989)

U.S. Manned Spaceflight

Launched on March 13, 1989, the STS-29 crew of the space shuttle Discovery successfully deployed the third Tracking and Data Relay Satellite, TDRS-4 (the satellites were built and prepared for launch in numerical order, but were deployed in space in the order in which particular shuttle flights became available and were flown).

Michael Coats was commander of STS-29. John Blaha served as pilot of Discovery, and James Bagian, James Buchli, and Robert Springer were mission specialists for the flight.

Launch was originally scheduled for February 18, but problems with the shuttle's master events controller and the liquid oxygen turbopumps on the shuttle's three Space Shuttle Main Engines (SSMEs) led to a delay while repairs were made. The faulty components were all replaced prior to the launch on March 13, which was delayed for about two hours by weather concerns.

The TDRS-4 satellite, which was part of a network of six satellites launched over the course of more than a decade, was pushed into geosynchronous orbit by its attached Inertial Upper Stage (IUS). The first of the six satellites in the TDRS network was deployed during STS-6 in April 1983, and the sixth and final satellite in the series was deployed during STS-70 in July 1995.

Also carried aboard Discovery during STS-29 were the Orbiter Experiments Autonomous Supporting Instrumentation System-1 (OASIS-1), the Space Station Heat Pipe Advanced Radiator Experiment (SHARE), and a hand-held IMAX camera that the crew used to photograph the Earth.

Other payloads included experiments in protein crystal growth and chromosome and plant cell division, the Air Force Maui Optical Site (AMOS) experiment, and two experiments that were designed as part of the Space Shuttle Student Involvement Program (SSIP).

After 80 orbits, in 4 days, 23 hours, 38 minutes and 50 seconds, STS-29 came to a close when Discovery landed at Edwards Air Force Base in California on March 18, 1989.

SPACE SHUTTLE MISSION 29: STS-30

(May 4–8, 1989)

U.S. Manned Spaceflight

In NASA's first planetary exploration mission in more than a decade, the Magellan Venus radar mapping probe was deployed during the STS-30 flight of the space shuttle Atlantis. It was the first time a planetary probe had ever been deployed from a manned spacecraft.

Launched on May 4, 1989, STS-30 was commanded by David Walker, who was accompanied by Ronald Grabe, who served as pilot for the flight, and by Norman Thagard, Mary Cleave, and Mark Lee, who served as mission specialists.

The original launch, scheduled for April 28, had to be postponed because of a faulty liquid hydrogen pump in one of the Space Shuttle Main Engines (SSMEs) and a leak in the liquid hydrogen link between the shuttle and its external tank. The May 4 launch was briefly delayed by poor weather conditions.

Once underway, the STS-30 crew quickly accomplished its primary objective. The Magellan probe was successfully deployed a little more than six hours into the flight, and its attached Inertial Upper Stage (IUS) and second stage propelled it into the proper trajectory for its 15 month trip to Venus.

Following the successful Pioneer Venus Orbiter mission (which launched in 1978), Magellan was designed to further scientists' knowledge of Venus through the use of next-generation radar mapping equipment. Building on the work of the less sophisticated radar mapper of the Pioneer Venus Orbiter, Magellan made a more detailed study of Venus at greater resolution. Magellan also carried instruments to measure surface temperatures and geological formations, and by the mid-1990s the probe had successfully mapped nearly the entire surface of Venus and returned a diverse treasure trove of scientific data about the planet.

During STS-30, Atlantis also carried the Mesoscale Lightning Experiment, an experiment that observed the effects of the microgravity environment on fluids, and the Air Force Maui Optical Site (AMOS) experiment.

One of the shuttle's five onboard general-purpose computers failed during the flight and was replaced—the first time an onboard computer switch was made during a shuttle mission.

STS-30 came to a close after 4 days, 56 minutes and 28 seconds, landing at Edwards Air Force Base in California on May 8, 1989.

SPACE SHUTTLE MISSION 30: STS-28

(August 8–13, 1989)

U.S. Manned Spaceflight

The space shuttle Columbia launched on its first flight of the post-Challenger era, STS-28, on August 8, 1989. STS-28 was dedicated to the classified activities of the U.S. Department of Defense.

Making his third spaceflight, Brewster Shaw commanded STS-28. He was joined by pilot Richard Richards, and mission specialists James Adamson, David Leestma, and Mark Brown.

In a flight of 81 orbits, the crew performed activities for the Department of Defense which remain classified and are therefore unavailable to the public. STS-28 was the fourth shuttle mission devoted to the Department of Defense.

The flight ended after 5 days, 1 hour and 8 seconds, with Columbia returning to Earth at Edwards Air Force Base in California on August 13, 1989.

SPACE SHUTTLE MISSION 31: STS-34

(October 18–23, 1989)

U.S. Manned Spaceflight

The shuttle Atlantis lifted off on its fifth flight, STS-34, on October 18, 1989. The main payload for the short mission was the Galileo Jupiter orbiter and probe—which would ultimately become the first spacecraft to orbit Jupiter and the first probe to enter the planet's atmosphere.

Donald Williams commanded STS-34; Michael McCulley served as pilot, and Franklin Chang-Diaz, Shannon Lucid, and Ellen Baker were mission specialists for the flight.

Launch was originally scheduled for October 12, but was delayed by a mechanical fault in a Space Shuttle Main Engine (SSME) controller. A second try at launch, on October 17, was postponed by weather considerations at the Kennedy Space Center in Florida that mitigated against a return-to-launch-site landing in the event of an emergency.

The Galileo spacecraft was deployed six-and-a-half hours into the flight. The probe's Inertial Upper Stage (IUS) and second stage rockets fired as planned, propelling Galileo into the proper trajectory for its six-year journey to Jupiter.

At the end of that journey, Galileo successfully entered orbit around Jupiter and released its probe into the planet's atmosphere on July 13, 1995. Both the orbiter and the probe returned a great deal of scientific data, including evidence of a previously unknown radiation belt around the planet, the possibility of water beneath the surface of Jupiter's moon Europa, and ongoing volcanic activity on the moon Io. Galileo has also discovered magnetic fields on Io and Ganymede, and measured winds of more than 600 kilometers on Jupiter.

In addition to deploying Galileo, the STS-34 crew also carried a diverse array of payloads, including the Shuttle Solar Backscatter Ultraviolet (SSBUV) experiment, the Growth Hormone Crystal Distribution and Polymer Morphology experiments, the Mesoscale Lightning Experiment (MLE), the Air Force Maui Optical Site (AMOS) experiment, a study of sensor technology, and an investigation of the impact of the space environment on the formation of ice crystals that was designed as part of the Shuttle Student Involvement Program (SSIP). The crew also used an IMAX camera to document their activities during the mission.

After a flight of 4 days, 23 hours, 39 minutes and 20 seconds, Atlantis returned to Earth on October 23, 1989, at Edwards Air Force Base in California.

SPACE SHUTTLE MISSION 32: STS-33

(November 22–27, 1989)

U.S. Manned Spaceflight

Dedicated to classified activities on behalf of the U.S. Department of Defense, STS-33 featured the first night launch of the post-Challenger era. The STS-33 crew launched aboard the space shuttle Discovery on the evening of November 22, 1989.

Frederick Gregory served as commander of STS-33. He was accompanied by pilot John Blaha and mission specialists Story Musgrave, Manley Carter, and Kathryn Thornton.

The launch had originally been planned for November 20, but concerns about the integrity of the electronic assemblies on the shuttle's Solid Rocket Boosters (SRBs) forced a postponement while the assemblies were replaced.

Details of the STS-33 flight remain classified; the mission was the fifth shuttle flight devoted to the activities of the U.S. Department of Defense.

Discovery returned to Earth after a mission of 79 orbits, in 5 days, 6 minutes and 49 seconds, landing at Edwards Air Force Base in California on November 27, 1989.

SPACE SHUTTLE MISSION 33: STS-32

(January 9–20, 1990)

U.S. Manned Spaceflight

Launched aboard the shuttle Columbia on January 9, 1990, the 11 day STS-32 flight set a new record for the longest shuttle mission up to that time (previously the longest flight had been that of the STS-9 crew, from November 28 to December 8, 1983).

The STS-32 crew: (front, left to right) Daniel Brandenstein, James Wetherbee, (back, left to right) Marsha Ivins, David Low, and Bonnie Dunbar. [NASA/courtesy of nasaimages.org]

In many ways, STS-32 represented an important milestone in the healing process for NASA as the agency moved forward from the Challenger accident that had killed the crew of STS-51L in January 1986. STS-32 was the first launch from launch pad 39A in four years (pad 39B had been used for all launches since the resumption of flight following the Challenger accident); it was the first long mission of the post-Challenger era; and as one its prime objectives, it had the task of retrieving the Long Duration Exposure Facility (LDEF), which had been deployed during STS-41C in April 1984 and had then been left in space for far longer than originally planned because of the long suspension of shuttle flights that followed the Challenger disaster.

Daniel Brandenstein commanded STS-32; he was accompanied aboard Columbia by pilot James Wetherbee and mission specialists Bonnie Dunbar, David Low, and Marsha Ivins.

Originally set for December 18, 1989, the launch was delayed by the ongoing refurbishing of launch pad 39A. A second launch, scheduled for January 8, was put off by bad weather.

The crew successfully deployed the SYNCOM IV-F5 (also known as LEASAT 5) defense communications satellite into a geosynchronous orbit.

Then, on the fourth day of the flight, they successfully retrieved the LDEF. An orbiting cylinder weighing 21,400 pounds, the LDEF contained 57 experiments designed to test the effects of long exposure to the space environment on a variety of materials, including seeds. It had been in space for nearly six years.

STS-32 also carried several scientific experiments and instruments, including the Characterization of Neurospora Circadian Rhythms experiment, an American Flight Echocardiograph (AFE), a Mesoscale Lightning Experiment, the Air Force Maui Optical Site (AMOS) experiment, a latitude and longitude locator, an IMAX camera, a protein crystal growth exercise, and an experiment measuring the behavior of fluids in orbit.

After 172 orbits, in 10 days, 21 hours and 36 seconds, the flight featured a night landing at Edwards Air Force Base in California, touching down a little after 1:30 A.M. on January 20, 1990.

SPACE SHUTTLE MISSION 34: STS-36

(February 28–March 4, 1990)

U.S. Manned Spaceflight

The sixth shuttle mission dedicated to the classified activities of the U.S. Department of Defense, STS-36 launched on February 28, 1990 on the space shuttle Atlantis.

The commander of STS-36 was John Creighton; he was joined on the flight by pilot John Casper and mission specialists Richard Mullane, David Hilmers, and Pierre Thuot.

Originally scheduled for February 22, the launch was postponed when Creighton became ill. It was the first time that an astronaut's health had been a factor in a space mission since the Apollo 13 flight in April 1970, when NASA medical personnel concerned about Thomas Mattingly's possible exposure to German Measles had removed Mattingly from the Apollo 13 mission.

Subsequent launch attempts on the following three days were stymied by Creighton's illness, poor weather, and by the malfunction of a range safety computer. A launch attempt on February 26 was put off by bad weather.

The launch on February 28 was a night launch. Details of the mission are classified, as all activities were in support of the U.S. Department of Defense.

Atlantis landed on March 4, 1990, after 72 orbits and a flight of 4 days, 10 hours, 18 minutes and 22 seconds, at Edwards Air Force Base in California.

SPACE SHUTTLE MISSION 35: STS-31

(April 24–29, 1990)

U.S. Manned Spaceflight

The thirty-fifth space shuttle mission, STS-31 launched aboard the shuttle Discovery on April 24, 1990. The STS-31 crew deployed the Hubble Space Telescope (HST).

Loren Shriver commanded STS-31. Charles Bolden served as pilot for the flight—Discovery's tenth mission—and STS-31 mission specialists were Steven Hawley, Bruce McCandless, and Kathryn Sullivan.

Launch scheduling for STS-31 originally envisioned lift-off on April 18, then on April 12, and then on April 10. The April 10 attempt was postponed just prior to launch by a fault in an auxiliary power unit (APU).

To deploy the Hubble Space Telescope (HST) properly, Discovery traveled to a much-higher-than-usual orbit for a shuttle, achieving an altitude of some 380 miles during the flight.

Part of NASA's Great Observatories program, the Hubble Space Telescope has vastly expanded scientific study of the universe for nearly two decades by returning photos of celestial objects and phenomena at far greater distances in deep space than was previously possible.

The HST was initially equipped with a faint object camera, a faint object spectrograph, the Goddard High Resolution Spectrograph, a high-speed photometer, and the Wide Field/Planetary Camera.

A slight manufacturing defect in the telescope's main mirror initially caused it to return blurry photos; the defect was fixed during the first HST servicing and repair mission, STS-61, in 1993. Subsequent shuttle crews mounted additional servicing missions at regular intervals; the second servicing flight was STS-82, in 1997, followed by STS-103 in December 1999 and STS-109 in March 2002.

Other STS-31 payloads included instruments designed to detect particulate matter in the shuttle's payload bay, and radiation monitoring equipment to measure gamma ray levels in the crew cabin. Experiments included a polymer membrane processing test to examine methods of controlling porosity in the microgravity environment, a Shuttle Student Involvement Program (SSIP) experiment studying the behavior of electrical arcs in the space environment, and the Air Force Maui Optical Site (AMOS) experiment.

The shuttle was also equipped with IMAX cameras—one mounted in the cargo bay, and a second hand-held unit for use by the crew—to record highlights of the mission for later distribution.

After a mission of 5 days, 1 hour, 16 minutes and 6 seconds, Discovery touched down at Edwards Air Force Base in California on April 29, 1990. The landing featured the first use of the new carbon brakes that had been designed for the shuttle in the wake of difficult landings at the Kennedy Space Center in Florida; the carbon brakes replaced an earlier version that had used beryllium.

SPACE SHUTTLE MISSION 36: STS-41

(October 6–10, 1990)

U.S. Manned Spaceflight

The STS-41 crew of the shuttle Discovery launched on October 6, 1990, with a primary objective of deploying the Ulysses solar polar orbiter. The Ulysses spacecraft was a project of the European Space Agency (ESA).

Richard Richards commanded STS-41. He was accompanied by pilot Robert Cabana and mission specialists William Shepherd, Bruce Melnick, and Thomas Akers.

The deployment of the Ulysses spacecraft featured the first combination of an Inertial Upper Stage (IUS) and a Payload Assist Module-S (PAM-S) motor, which together placed the probe on the appropriate trajectory for its journey.

The Ulysses project had evolved out of a plan that originally envisioned NASA and ESA collaborating on two spacecraft designed to study the Sun. NASA had withdrawn from the venture in 1981, but ESA continued development of its portion of the project, which ultimately became Ulysses. The mission profile for Ulysses called for intensive study of the Sun's polar areas.

Also carried aboard Discovery during STS-41 were the INTELSAT Solar Array Coupon, a Shuttle Solar Backscatter Ultraviolet experiment, and the first flight of the Voice Command System, which allowed the crew to operate on-board cameras by using voice commands.

Other activities included an experiment in chromosome and plant cell division, a Solid Surface Combustion Experiment, investigations of polymer membrane processing, an experiment in physiological systems, an experiment for the Shuttle Student Involvement Program (SSIP), and the Air Force Maui Optical Site (AMOS) experiment.

After 66 orbits and a flight of 4 days, 2 hours, 10 minutes and 4 seconds, Discovery landed at Edwards Air Force Base in California on October 10, 1990.

SPACE SHUTTLE MISSION 37: STS-38

(November 15–20, 1990)

U.S. Manned Spaceflight

Dedicated to the classified activities of the U.S. Department of Defense, the STS-38 flight of the shuttle Atlantis featured the first landing at the Kennedy Space Center (KSC) in Florida in more than five years.

As commander of STS-38, Richard Covey made his third flight in space; he was accompanied by pilot Frank Culbertson, and mission specialists Robert Springer, Carl Meade, and Charles Gemar.

A liquid hydrogen leak that had been discovered during an early attempt to launch the shuttle Columbia during the STS-35 mission caused NASA controllers to delay the first scheduled launch of STS-38, which had been planned for July 1990. Atlantis was tested as a precaution after the leak had been found on Columbia, and it too was found to be leaking hydrogen fuel, from a tube that connected the shuttle to its huge external tank. As a result, Atlantis was removed from the launch pad for repair.

En route to being fixed, the shuttle suffered damage from foul weather, with a hailstorm damaging its protective heat tiles, only to then receive another minor blow from a falling beam.

With all damage repaired, Atlantis was set for a launch on November 9, which was then rescheduled for November 15, when the mission at last began, with a night launch.

Because STS-38 was dedicated to the classified activities of the U.S. Department of Defense, few details about the flight have been made public. STS-38 was the seventh space shuttle mission devoted to the Department of Defense.

The end of the mission brought a new set of challenges for the crew and for NASA officials at Mission Control. The originally scheduled landing at Edwards Air Force Base in California had to be put off for a day because of crosswinds at the Base, and the mission was extended for twenty-four hours. When the winds did not let up the following day, mission controllers decided that the shuttle should instead land at the Kennedy Space Center in Florida.

No shuttle had landed at KSC in more than five years, since the STS-51D flight of the shuttle Discovery in April 1985. At the end of that mission, Discovery had made a harsh landing that resulted in a blown tire and extensive brake damage, and as a result, NASA officials had decided that all future flights would land at Edwards until the shuttle's steering system could be overhauled. That problem had since been solved, and new brakes made of carbon (the earlier versions used beryllium) had been inaugurated into the shuttle system with the STS-31 flight of Discovery in April 1990.

After 79 orbits, in 4 days, 21 hours, 54 minutes and 31 seconds, STS-38 Atlantis landed safely at KSC on November 20, 1990.

SPACE SHUTTLE MISSION 38: STS-35

(December 2–10, 1990)

U.S. Manned Spaceflight

The tenth flight of the space shuttle Columbia was the STS-35 ASTRO-1 mission, launched on December 2, 1990.

As commander of STS-35, veteran astronaut Vance Brand made his fourth flight in space. He was accompanied by pilot Guy Gardner, mission specialists

Jeffrey Hoffman, John Lounge, and Robert Parker, and payload specialists Samuel Durrance and Ronald Parise.

Launch had originally been scheduled for more than six months earlier, on May 16, 1990, but a problem with the shuttle's coolant system caused a delay until May 30, when a second, more serious difficulty arose: ground crews discovered hydrogen in several areas on the shuttle, including Columbia's aft compartment, and in the surrounding launch facilities. The technical personnel traced the anomaly to two breaches in the systems that support the shuttle's launch. The first was a minor hydrogen leak on the mobile launcher platform, and the second was a major leak in the 17-inch tube that connected the shuttle to its External Tank.

The liquid hydrogen leaks forced the postponement of the mission, and Columbia had to be removed from the launch pad for repairs. The initial problems were quickly fixed (the leak-free 17-inch assembly of the shuttle Endeavour was borrowed for use on Columbia, and a new assembly was installed on the External Tank), and a new launch attempt was set for September 1.

The ASTRO-1 astronomical observatory had remained in Columbia's cargo bay throughout the repairs to the shuttle, and had been regularly checked; now, just two days before the scheduled September 1 launch, the observatory suffered a malfunction, delaying the launch for another five days while engineers replaced electronic parts of the Broad Band X-ray Telescope (BBXRT). Then, as preparations were made for a launch on September 6, the hydrogen leak problem cropped up again, with hydrogen once again filling the aft compartment of the shuttle. Stymied mission controllers were forced to delay the flight again.

It was subsequently found that a hydrogen leak separate from those discovered during the first launch attempt was at the heart of the new problem. Hydrogen recirculation pumps in the aft compartment were replaced, as was a damaged Teflon seal in the hydrogen prevalve of the number three Space Shuttle Main Engine (SSME), and launch was set for September 18. That attempt was foiled during tanking operations, when hydrogen was again found in the aft compartment.

At that point, the persistent hydrogen problem was assigned to a special team of engineers, and the STS-35 flight was delayed indefinitely until the source of the problem could be determined and the leaks could be eliminated.

In the meantime, the scheduled July launch of STS-38, dedicated to the Department of Defense, had also been delayed when NASA officials struggling with Columbia's hydrogen leaks also decided to test the shuttle Atlantis as a precautionary measure. Atlantis was also found to have a similar problem, but it proved easier to fix, and as things worked out, STS-38 jumped ahead of STS-35 and launched on November 15.

In early October, Columbia was moved to launch pad 39-B to make room for Atlantis, and Columbia subsequently had to be removed from the pad to avoid being damaged by tropical storm Klaus.

By the end of October the persistent hydrogen leak had been attributed to a seal in the prevalve of the shuttle's main propulsion system; the seal had been damaged when it was installed, just after Columbia's previous flight. A test on October 30 tracked the amount of hydrogen in the aft compartment with special sensors and also used video cameras to observe the compartment, whose door had been replaced by a see-through Plexiglas sheet for the occasion. No excessive amounts of hydrogen were detected, and the mission was, finally, back on track.

The crew of STS-35 still had to wait until December 2 to launch, as the Atlantis STS-38 flight had in the meantime been inserted into the launch rotation; but after the long, half-year postponement of the originally scheduled launch, and a brief 21 minute weather-related delay, STS-25 launched without incident on the night of December 2.

Equipped with four telescopes—the Hopkins Ultraviolet Telescope, the Wisconsin Ultraviolet Photo-Polarimeter Experiment, the Ultraviolet Imaging Telescope and the Broad Band X-ray Telescope—the ASTRO-1 observatory was designed for around the clock ultraviolet and X-ray observations.

The crew's use of the ultraviolet telescopes was impaired by the failure of both of the data display units that controlled the instruments' pointing mechanism. Mission support teams at the Marshall Space Flight Center had to take over the job of aiming the ultraviolet telescopes, with crew members manually fine-tuning the positioning during flight. The Broad Band X-ray Telescope was not affected by the loss of the data display units because it was designed to be controlled by ground-based teams at the Goddard Space Flight Center.

A far less serious but no less frustrating fault occurred in the shuttle's plumbing system, when a clogged drain made it difficult for the crew to dump wastewater. They implemented a temporary solution by using spare storage containers.

During STS-35 Columbia also carried the Shuttle Amateur Radio Experiment-2 (SAREX-2) and the Air Force Maui Optical Site (AMOS) experiment; and as part of the Space Classroom Program, crew members conducted a lesson entitled "Assignment: The Stars" that was designed to encourage students' interest in science, math and technology.

Despite the enormously frustrating delays and the in-flight difficulties, scientists evaluating the mission estimated that it achieved about 70 percent of its planned science objectives.

When the time came for landing, a forecast of poor weather forced mission controllers to shorten the flight by one day. Columbia landed at night on December 10, 1990, at Edwards Air Force Base in California, after 144 orbits in 8 days, 23 hours, 5 minutes and 8 seconds.

SPACE SHUTTLE MISSION 39: STS-37

(April 5–11, 1991)

U.S. Manned Spaceflight

Launched aboard the shuttle Atlantis on April 5, 1991, the STS-37 crew deployed the Compton Gamma Ray Observatory (GRO).

Steven Nagel commanded STS-37. He was accompanied by pilot Kenneth Cameron and mission specialists Jerry Ross, Jerome "Jay" Apt, and Linda Godwin.

In sharp contrast to the just-previous shuttle mission, STS-35—which suffered a long series of postponements and scrubbed launches—STS-37 launched on April 5 with only a brief weather-related delay.

The STS-37 crew deployed the Compton GRO on the third day of the flight. A problem quickly presented itself: the observatory's high-gain antenna was stuck and would not extend. Mission controllers pondered a variety of potential solutions, and finally settled on an innovation that had last been used (and then, unsuccessfully) during STS-51D in April 1985. It was decided that Jerry Ross and Jay Apt would make an unscheduled spacewalk to try to fix the balky antenna.

During an EVA of 4 hours and 26 minutes on April 7, 1991, Ross and Apt were able to force the antenna free, and then oriented it manually—salvaging the work of the observatory.

Named for Arthur Holly Compton, the scientist who discovered that X-rays act like particles as well as waves (the scattering action that X-rays produce is known as the Compton Effect), the Compton GRO was part of NASA's Great Observatories program. Compton also played an instrumental role in the research that led the United States to develop the atomic bomb during World War II.

Equipped with four instruments—the Burst and Transient Source Experiment (BATSE), the Oriented Scintillation Spectrometer Experiment (OSSE), the Imaging Compton Telescope (Comptel) and the Energetic Gamma Ray Experiment Telescope (EGRET)—that made observations over an energy range from 30 keV to 30 GeV, the Compton GRO operated in space for over nine years. Following the end of its mission on June 4, 2000, NASA controllers maneuvered the observatory into Earth atmosphere where it gradually disintegrated, as it was too big to burn up completely during a normal re-entry.

The day after their successful spacewalk to fix the observatory's antenna, Apt and Ross made the first scheduled EVA since the STS-61B flight of Atlantis in November 1985—when Ross and Sherwood Spring had made two spacewalks to test procedures for assembling structures in space.

During their second spacewalk, on April 8, Ross and Apt spent 5 hours and 47 minutes outside the shuttle while they tried out innovative equipment and techniques for future EVAs.

In their two STS-37 spacewalks, Ross and Apt accumulated a total of 10 hours and 13 minutes in EVA.

Other payloads aboard Atlantis during STS-37 included the Ascent Particle Monitor (APM), the Shuttle Amateur Radio Experiment II (SAREX II), the Air Force Maui Optical Site (AMOS) experiment, the Bioserve/Instrumentation Technology Associates Materials Dispersion Apparatus (BIMDA), a protein crystal growth experiment, and radiation monitoring equipment (RME III).

The shuttle was originally scheduled to land on April 10, but poor weather at both Edwards Air Force Base in California and at the Kennedy Space Center in Florida led to the mission being extended by one day. After 93 orbits, in 5 days, 23 hours, 32 minutes and 44 seconds, Atlantis returned to Earth at Edwards Air Force Base on April 11, 1991.

SPACE SHUTTLE MISSION 40: STS-39

(April 28–May 6, 1991)

U.S. Manned Spaceflight

The fortieth flight of the space shuttle program, STS-39 was launched aboard the shuttle Discovery on April 28, 1991.

STS-39 was the first shuttle mission to carry both classified and unclassified payloads.

As commander of STS-39, Michael Coats made his third flight into space. He was accompanied by pilot Blaine Hammond and mission specialists Guion Bluford (who was also making his third spaceflight), Gregory Harbaugh, Richard Hieb, Donald McMonagle, and Charles Veach.

The originally scheduled launch date for STS-39 had been March 9, but a set of cracked hinges on a mechanical assembly on the shuttle's external fuel tank resulted in a delay while the faulty hinges were replaced with an intact set borrowed from the shuttle Columbia. A second try at launch, on April 23, fell to a further delay that was caused by an unexpected reading from a transducer on a pump for one of the Space Shuttle Main Engines (SSMEs). The transducer was replaced prior to the launch on April 28.

Although the flight was dedicated in part to the classified activities of the U.S. Department of Defense (DOD) and the details of that part of the mission were as a result not made available to the general public, STS-39 differed significantly from the seven previous DOD-dedicated shuttle flights because it also carried unclassified payloads.

The unclassified portion of the flight included the Infrared Background Signature Survey (IBSS), and the Space Test Payload-1 (STP-1). Experiments included the Chemical Release Observation (CRO) and the Shuttle Pallet Satellite-II (SPAS-II).

High winds at the originally scheduled landing site, Edwards Air Force Base in California, caused the landing to be diverted to the Kennedy Space Center (KSC) in Florida. Discovery touched down at KSC on May 6, 1991, after a flight of 8 days, 7 hours, 22 minutes and 23 seconds.

SPACE SHUTTLE MISSION 41: STS-40

(June 5–14, 1991)

U.S. Manned Spaceflight

A milestone flight in the study of aerospace medicine and biomedical research, the STS-40 Spacelab Life Sciences mission launched aboard the shuttle Columbia on June 5, 1991.

Bryan O'Connor commanded STS-40. He was accompanied by pilot Sidney Gutierrez, mission specialists James Bagian, Tamara Jernigan, and Rhea Seddon, and payload specialists Drew Gaffney and Millie-Hughes Fulford.

Two days before the originally scheduled launch date of May 22, a transducer that had been replaced during a test for a liquid hydrogen leak in 1990 failed a test by its manufacturer. The faulty component led caution-minded NASA officials to delay the launch as they studied the possibility that a similar failure in one or more of the nine liquid hydrogen and liquid oxygen transducers in the shuttle's fuel and oxidizer lines could cause those devices to break loose and become caught in the turbopumps of one of the Space Shuttle Main Engines (SSMEs), causing the SSME to fail.

In addition to the transducer problems, one of the shuttle's five on-board computers failed, and another faulty component was discovered in the shuttle's maneuvering system.

The computer and components were replaced, as were five of the suspect liquid oxygen and three liquid hydrogen transducers. An additional three liquid hydrogen transducers were removed entirely.

A launch attempt on June 1 was put off by a fault in an inertial measurement unit, which was also replaced prior to the launch on June 5.

Although it was the fifth Spacelab mission, STS-40 was the first to concentrate entirely on life sciences. The crew conducted a wide range of tests designed to study the effects of the space environment on six body systems, including the cardiovascular and cardiopulmonary system (the heart, lungs and blood vessels), the renal and endocrine system (the kidneys and hormone-secreting organs and glands), the blood system (blood plasma), the immune system (the white blood cells), the musculoskeletal system (muscles and bones), and the neurovestibular system (the brain and nervous system, and the eyes and inner ear).

The STS-40 crew included three MDs (Bagian, Seddon and Gaffney) who had been specially chosen for their areas of expertise. Over the course of the mission, crew members conducted 18 medical experiments using human, rodent and jellyfish test subjects, and returned more medical data than any previous NASA mission.

Columbia also carried 12 Getaway Special canisters that included experiments in materials science, plant biology and cosmic radiation; and the flight also featured the Middeck Zero-Gravity Dynamics Experiment (MODE), and seven shuttle Orbiter Experiments (OEX).

Following a mission of 9 days, 2 hours, 14 minutes and 20 seconds, Columbia returned to Earth at Edwards Air Force Base in California on June 14, 1991.

SPACE SHUTTLE MISSION 42: STS-43

(August 2–11, 1991)

U.S. Manned Spaceflight

When the space shuttle Atlantis launched on August 2, 1991 at the start of STS-43, its payloads—which included the Tracking and Data Relay Satellite-5 (TDRS-E)—made it more heavily laden than any shuttle had been on any previous flight. The total weight at launch (including the shuttle itself) was 235,000 pounds.

As commander of STS-43, John Blaha made his third flight into space. He was accompanied by pilot Michael Baker and mission specialists Shannon Lucid (who was also on her third space mission), James Adamson, and David Low.

The mission had originally been scheduled to begin on July 23 but was delayed for a day by a malfunctioning electronics assembly. Launch the following day was postponed by a malfunctioning controller on one of the Space Shuttle Main Engines (SSMEs); and another try at launch on August 1 was put off by an erroneous gauge on a cabin pressure vent valve and by weather considerations.

The TDRS-E satellite was deployed early in the flight, about six hours after launch. Propelled into geosynchronous orbit by its Inertial Upper Stage (IUS), the satellite was the fourth in the TDRS series. The first of the six satellites in the TDRS network had been deployed during STS-6 in April 1983, and the sixth and final satellite in the series would be deployed during STS-70 in July 1995.

Other payloads aboard Atlantis during STS-43 included the Space Station Heat Pipe Advanced Radiator Element II (SHARE II) and the Shuttle Solar Backscatter Ultra-Violet instrument.

Experiments included an exercise in auroral photography, a study of protein crystal growth, investigations into polymer membrane processing, the Solid Surface Combustion Experiment (SSCE), and the Air Force Maui Optical Site (AMOS) experiment.

After a successful flight of 8 days, 21 hours, 21 minutes and 25 seconds and 142 orbits, Atlantis landed at the Kennedy Space Center in Florida on the morning of August 11, 1991.

SPACE SHUTTLE MISSION 43: STS-48

(September 12–18, 1991)

U.S. Manned Spaceflight

The STS-48 crew of the space shuttle Discovery inaugurated NASA's Mission to Planet Earth initiative—a series of missions designed to study the Earth's environment.

John Creighton commanded STS-48. He was accompanied by pilot Kenneth Reightler, Jr. and mission specialists James Buchli, Charles Gemar, and Mark Brown.

The STS-48 crew launched on September 12, 1991. The launch was briefly delayed by problems with a communications link between the launch site, at the Kennedy Space Center (KSC) in Florida, and mission controllers at the Johnson Space Center (JSC) in Houston, but the difficulties were quickly resolved.

The crew successfully deployed the Upper Atmosphere Research Satellite (UARS), the first spacecraft in NASA's Mission to Planet Earth series, which consisted of a series of detailed studies of the Earth's environment. The UARS satellite was a 14,500-pound orbital observatory designed to examine the Earth's troposphere, including the ozone layer, to a degree never before attempted. Equipped with 10 instruments to observe and gather data about the life-sustaining gases of the troposphere, the UARS observatory was designed to make an 18 month study, but ultimately returned data until December 14, 2005 when it stopped functioning after 14 years and 78,000 orbits of the Earth.

The UARS sensing and measuring devices included the Cryogenic Limb Array Etalon Spectrometer (CLAES), the Improved Stratospheric and Mesospheric Sounder (ISAMS), a Microwave Limb Sounder (MLS), the Halogen Occultation Experiment (HALOE), a high resolution Doppler imager, a wind imaging interferometer, the Solar Ultraviolet Spectral Irradiance Monitor (SUSIM), the Solar/Stellar Irradiance Comparison Experiment (SOLSTICE), a particle environment monitor, and the Active Cavity Radiometer Irradiance Monitor (ACRIM II).

The Mission to Planet Earth studies later became part of NASA's Earth Science Enterprise, which promotes the study of the total Earth system and draws upon data from a wide array of past and present Earth observation studies.

Also carried aboard Discovery during STS-48 were an Ascent Particle Monitor; the Middeck 0-Gravity Dynamics Experiment (MODE), the Shuttle Activation Monitor, the Cosmic Ray Effects and Activation Monitor (CREAM), experiments involving the biological study of rodents, a protein crystal growth exercise, polymer membrane processing investigations, and the Air Force Maui Optical Site (AMOS) experiment.

An unexpected bit of drama cropped up during the flight when the crew had to move the shuttle out of the path of a fairly large piece of "space junk"—a discarded Russian rocket floating in orbit nearby. The move was more prudent precaution than an absolute necessity, and the shuttle was in no grave danger, but the specter of an orbital collision was a disturbing reminder of the unexpected dangers of spaceflight.

STS-48 came to a close after 81 orbits, in 5 days, 8 hours, 27 minutes and 38 seconds. Discovery had been scheduled to land at KSC, but poor weather there forced mission controllers to divert the landing to Edwards Air Force Base in California, where the crew set down safely in a night landing in the early hours of September 18, 1991.

SPACE SHUTTLE MISSION 44: STS-44

(November 24–December 1, 1991)

U.S. Manned Spaceflight

Dedicated to the classified activities of the U.S. Department of Defense but also carrying unclassified payloads, STS-44 featured some tense moments.

Frederick Gregory was commander of STS-44; he was joined by pilot Terence Henricks, mission specialists Story Musgrave, Mario Runco, and James Voss, and payload specialist Thomas Hennen.

The tenth flight of the space shuttle Atlantis, STS-44 was originally scheduled to launch on November 19, but a problem with the Inertial Upper Stage (IUS) designed to place the Defense Support Program satellite into orbit caused a delay of nearly a week. The night launch on November 24 was delayed for a variety of minor repairs, and for a novel traffic problem: the shuttle had to endure a brief wait while another spacecraft, orbiting high overhead, passed by.

Although it was devoted in part to the classified activities of the U.S. Department of Defense and therefore no details of that aspect of the mission have been made public, STS-44 also featured several unclassified payloads, including the Interim Operational Contamination Monitor (IOCM), the Terra Scout, the Air Force Maui Optical Site (AMOS) experiments, the Cosmic Radiation and Activation Monitor (CREAM), the shuttle activation monitor (SAM), visual function tester-1, an ultraviolet plume instrument, and radiation monitoring equipment. Other experiments included the Bioreactor Flow and Particle Trajectory experiment, and the Extended Duration Orbiter Medical Project.

Just as Discovery had had to be maneuvered to avoid a discarded rocket during the just-previous STS-48 flight in September, the crew of Atlantis was forced to alter their orbit when they encountered a discarded rocket looming nearby their orbit.

A more serious difficulty arose on November 30, when one of the shuttle's three orbiter inertial measurement units failed. Mission controllers were forced to shorten the planned 10-day mission by 2 days, and the landing was diverted to Edwards Air Force Base in California, where Atlantis returned to Earth on December 1, 1991.

The revised STS-44 mission lasted 6 days, 22 hours, 50 minutes and 44 seconds.

SPACE SHUTTLE MISSION 45: STS-42

(January 22–30, 1992)

U.S. Manned Spaceflight

The forty-fifth space shuttle mission, launched aboard the shuttle Discovery on January 22, 1992, STS-42 was also a milestone in international scientific cooperation in space. The crew conducted a wide array of experiments designed by scientists from six international organizations as part of the first flight of the International Microgravity Laboratory (IML-1).

STS-42 was commanded by Ronald Grabe. He was accompanied aboard Discovery by pilot Stephen Oswald, mission specialists Norman Thagard, David Hilmers, and William Readdy, and payload specialists Roberta Bondar of the Canadian Space Agency and Ulf Merbold of the European Space Agency (ESA).

Crew members alternated in two shifts to observe the effects of the microgravity environment on the human nervous system, and on other forms of life, including shrimp eggs, lentil seedlings, fruit fly eggs and bacteria. They also conducted experiments in materials processing, growing crystals from enzymes, mercury iodide, and a virus.

Other payloads carried aboard Discovery during STS-42 included 12 Getaway Special (GAS) canisters mounted on a GAS Bridge Assembly in the shuttle's cargo bay, experiments in the gelation of solids and in polymer membrane processing, and two experiments designed as part of the Shuttle Student Involvement Program (SSIP). The shuttle was also equipped with the Radiation Monitoring Experiment III, and an IMAX camera to record the activities of the crew.

Mid-way through the flight, mission controllers decided to extend the mission by one day to allow time for the crew to continue their experiments.

After a flight of 129 orbits, in 8 days, 1 hour, 14 minutes and 44 seconds, the STS-42 crew returned to Earth at Edwards Air Force Base in California on January 30, 1992.

SPACE SHUTTLE MISSION 46: STS-45

(March 24–April 2, 1992)

U.S. Manned Spaceflight

Launched aboard the shuttle Atlantis on March 24, 1992, the crew of STS-45 successfully deployed the Atmospheric Laboratory for Applications and Science (ATLAS-1).

Charles Bolden commanded STS-45. He was accompanied by pilot Brian Duffy, payload commander Kathryn Sullivan, mission specialists David Leestma and Michael Foale, and payload specialists Byron Lichtenberg and Dirk Frimout of Belgium, who with this mission became the first Belgian citizen to fly in space.

The first attempt at launch, on March 23, had to be postponed because of fears of a liquid hydrogen and liquid oxygen leak in the shuttle's aft compartment. The higher-than-acceptable concentrations of hydrogen and oxygen were found to be the result of a problem with the plumbing of the shuttle's main propulsion system rather than resulting from leaks, and the problem was quickly fixed.

Equipped with 12 instruments to conduct investigations designed by scientists from the United States, the United Kingdom, Belgium, France, Germany, Japan, Switzerland and The Netherlands, the Atlas-1 laboratory was mounted on Spacelab pallets in the shuttle's cargo bay. The crew used the battery of equipment to conduct observations in four areas: atmospheric science, solar science, space plasma physics, and ultraviolet astronomy.

Atmospheric science investigations included a study of atmospheric Lyman-Alpha emissions (ALAE), atmospheric trace molecule spectroscopy, observations with a Grille spectrometer and the Imaging Spectrometric Observatory (ISO), and the use of the Millimeter-Wave Atmospheric Sounder (MAS).

Instruments for solar science included the Active Cavity Radiometer Irradiance Monitor (ACRIM) and the Solar Ultraviolet Spectral Irradiance Monitor (SUSIM); investigations included measurement of the Solar Constant, and observation of the solar spectrum from 180 to 3,200 Nanometers.

For space plasma physics, ATLAS-1 was equipped to capture data with the Atmospheric Emissions Photometric Imaging (AEPI) payload and to conduct investigations with the Space Experiments with Particle Accelerators (SEPAC) payload.

The Far Ultraviolet Space Telescope (FAUST) was also included among the ATLAS-1 instruments, to make observations in ultraviolet astronomy.

In addition to the ATLAS-1 scientific instruments, Atlantis also carried the Shuttle Solar Backscatter Ultraviolet/A payload; experiments in crystal growth, polymer membrane processing, and space tissue loss; the visual function tester-2; Cloud Logic to Optimize Use of Defense System (CLOUDS); radiation monitoring equipment; and the Shuttle Amateur Radio Experiment II (SAREX II).

STS-45 was extended for one day mid-way through the mission to allow the crew more time to continue their scientific experiments.

After 143 orbits, in 8 days, 22 hours, 9 minutes and 28 seconds, Atlantis returned to Earth at the Kennedy Space Center in Florida on April 2, 1992.

SPACE SHUTTLE MISSION 47: STS-49

(May 7–16, 1992)

U.S. Manned Spaceflight

The first flight of the new space shuttle Endeavour, STS-49 launched on May 7, 1992 and achieved a milestone satellite retrieval that featured the first-ever three-person EVA and two spacewalks longer than any previously attempted. The flight was also the first U.S. mission to include four EVAs, and it set a new standard for the longest spacewalk by a female astronaut.

As commander of STS-49, Daniel Brandenstein made his fourth flight in space. He was accompanied by pilot Kevin Chilton and mission specialists Pierre Thuot, Kathryn Thornton, Richard Hieb, Thomas Akers, and Bruce Melnick.

The mission got off to an inauspicious start when problems with two of the high-pressure oxidizer turbopumps in Endeavour's Space Shuttle Main Engines (SSMEs) led to all three engines being replaced prior to launch. The original launch date of May 4 was switched to May 7 to facilitate better conditions for photographic coverage of the shuttle during launch.

Once in orbit, the new shuttle—and its crew—performed exceptionally well. The crew achieved its prime objective of rescuing the Intelsat VI satellite, which had been stranded in a useless orbit since its launch in 1990, through an unprecedented four spacewalks. The task of capturing the satellite, attaching a motor to propel it into the proper orbit, and then redeploying it, proved to be an arduous one.

Tracking down the Intelsat spacecraft and making a proper rendezvous with it were in themselves no easy task, and wrestling the errant satellite into the shuttle's cargo bay required a Herculean effort from mission specialists Thomas Akers, Richard Hieb, and Pierre Thuot. Their record-setting spacewalk of 8 hours and 29 minutes—the third EVA of the mission—set a new single EVA duration record that replaced the Apollo 17 EVA mark of Moon walkers Eugene Cernan and Harrison Schmitt.

Thuot and Hieb were unable to capture the 17-foot by 12-foot satellite during the first STS-49 spacewalk, on May 10. Perched at the end of the shuttle's robotic arm, Thuot tried to latch onto the Intelsat spacecraft but had to back off the attempt when the satellite began to sway back and forth. With Hieb monitoring his effort from Endeavour's open payload bay, ready to help his EVA partner maneuver the satellite into the shuttle, Thuot made three attempts to capture

The first-ever three person spacewalk. STS-49 crew mates (left to right) Richard Hieb, Thomas Akers and Pierre Thuot work on the Intelsat VI, May 13, 1992. [NASA/courtesy of nasaimages.org]

the Intelsat craft during the first spacewalk, which lasted 3 hours and 43 minutes, but he was unable to secure the satellite.

On May 11, during the second spacewalk of the flight, Thuot and Hieb spent five-and-a-half hours in EVA while Thuot made five more attempts to capture the Intelsat VI satellite. Again, the effort proved far more difficult than expected, and again the satellite remained beyond the reach of the STS-49 spacewalkers.

The difficulties in retrieving the satellite led to a remarkable, innovative solution that made the third EVA of STS-49—which incidentally was the one hundredth spacewalk in history—a milestone of productive work in space.

On May 13, 1992, Pierre Thuot, Richard Hieb and Thomas Akers became the first individuals ever to perform a three-person spacewalk when they latched onto the recalcitrant Intelsat VI long enough for Thuot to attach a line to it.

Once the Intelsat VI craft was in the shuttle's payload bay, the astronauts attached a perigee kick motor (PKM) to it, and on the eighth day of the mission, the satellite was successfully propelled by the PKM into a functional orbit.

Thuot, Hieb and Akers accumulated 8 hours and 29 minutes in EVA during the "hand capture" of the satellite—setting a new record for the longest spacewalk in history. In addition to the extraordinary effort it required, the capture also involved great skill and great risk. NASA mission managers had to be convinced first that the effort was worth the dangers inherent in having three crew members in EVA simultaneously for the first time, and they conferred with engineers from Hughes Aerospace Corp. to ensure that the astronauts could handle the satellite without incurring injury.

In addition to the skills of the EVA crew, the successful capture also demanded precise maneuvering of the shuttle by commander Daniel Brandenstein, who was consistently able to rendezvous with the satellite at extremely close distances, and careful use of the RMS by Bruce Melnick to position Thuot and to maneuver the captured Intelsat VI into the shuttle's payload bay.

The fourth EVA of the flight, an exercise designed to test procedures for assembling portions of the planned space station that was at the time still known as space station Freedom (it would later become the International Space Station), took place on May 14, 1992. Conducted by Thomas Akers and Kathryn Thornton, the Assembly of Station by EVA Methods (ASEM) exercise took 7 hours and 44 minutes—resulting in another new record for the milestone flight, as Kathryn Thornton set a new mark for the longest spacewalk by a female astronaut.

The STS-49 crew landed at Edwards Air Force Base in California on May 16, 1992, after 8 days, 21 hours, 17 minutes and 38 seconds. The touch down featured another first, as a drag chute was tested for the first time during a space shuttle landing.

SPACE SHUTTLE MISSION 48: STS-50

(June 25–July 9, 1992)

U.S. Manned Spaceflight

The crew of STS-50, which launched aboard the shuttle Columbia on June 25, 1992, set a new record for the longest shuttle mission up to that time, and advanced the study of microgravity through the first flight of the U.S. Microgravity Laboratory (USML-1).

Richard Richards commanded STS-50; he was accompanied by pilot Kenneth Bowersox, payload commander Bonnie Dunbar, mission specialists Ellen Baker and Carl Meade, and payload specialists Lawrence DeLucas and Eugene Trinh.

Columbia had undergone an extensive series of modifications prior to STS-50. Engineers at Rockwell International Corp. outfitted the shuttle with more than 50 new or altered pieces of equipment, many of which were designed to make Columbia more suitable for longer-duration flights. With the installation of the Extended Duration Orbiter (EDO) hardware, Columbia became the first shuttle equipped for longer missions.

Utilizing the USML-1 and its pressurized Spacelab module, the crew conducted experiments in a variety of areas. USML-1 was equipped with a crystal growth furnace, a drop physics module, the Surface Tension Driven Convection Experiments, experiments in Zeolite crystal growth and protein crystal growth, a Glovebox facility, the Space Acceleration Measurement System (SAMS), a Generic Bioprocessing Apparatus (GBA), Astroculture-1, the Extended Duration Orbiter Medical Project, and a solid surface combustion experiment.

During STS-50 Columbia also carried investigations into polymer membrane processing (IPMP), an ultraviolet plume instrument, and the Shuttle Amateur Radio Experiment II (SAREX II).

The mission was extended by one day because of poor weather at the originally scheduled landing site at Edwards Air Force Base in California, and the landing was subsequently diverted to the Kennedy Space Center (KSC) in Florida. After completing the longest shuttle mission up to that time—221 orbits, in 13 days, 19 hours, 30 minutes and 4 seconds—Columbia made the tenth KSC landing of the shuttle program, on July 9, 1992. It was the first time that Columbia had landed at KSC at the end of a mission, and the landing included both a test of the shuttle drag chute and the first use of the redesigned synthetic tread shuttle tires.

SPACE SHUTTLE MISSION 49: STS-46

(July 31–August 8, 1992)

U.S. Manned Spaceflight

Dedicated to international flight experiments and marking the first spaceflight of an Italian citizen, STS-46 was launched aboard the shuttle Atlantis on July 31, 1992.

Loren Shriver served as commander of STS-46; he was joined by pilot Andrew Allen, mission specialists Jeffrey Hoffman, Franklin Chang-Diaz, and Marsha Ivins, mission specialist Claude Nicollier—the first citizen of Switzerland to fly in space, and payload specialist Franco Malerba of Italy—the first Italian citizen to fly in space.

A project of the European Space Agency (ESA), the European Retrievable Carrier (EURECA) was deployed by Nicollier, who maneuvered the spacecraft out of Atlantis' cargo bay with the shuttle's Remote Manipulator System (RMS) robotic arm. The EURECA deployment was hampered by several malfunctions, including a fault in its data processing apparatus and an incorrect attitude reading from the spacecraft after its release. The problems were quickly solved, however, and EURECA was successfully placed in the proper orbit on the sixth day of the flight, positioned to carry out its experiments until its retrieval by the shuttle Endeavour during STS-57 in 1993.

The flight also featured the first major test of tethered spaceflight, albeit with disappointing results. The concept of joining two spacecraft with a tether for linked flight had been tested in space as early as Gemini XI and Gemini XII in September and November 1966, with a 100-foot dacron tether connecting each Gemini capsule to an unmanned Agena target vehicle for a brief period of tethered flight.

Prior to STS-46, teams of engineers from NASA and the Italian space agency Agenzia Spaziale Italiana (ASI) had collaborated for more than a decade to design the equipment, the plan, and the procedures that were to have been used during the Tethered Satellite System (TSS-1) test aboard Atlantis. The exercise was designed to establish a link between the shuttle and a satellite that would be deployed at a different orbital altitude from Atlantis, for tethered flight at a distance of as much as 12.5 miles. Despite all the careful preparations, however, a mechanical hitch foiled the test, when the tether jammed after it had been unwound to a distance of just 840 feet. The crew attempted to release the stuck line for several days, but the experiment finally had to be written off as a failure. The TSS satellite was retrieved and returned to Earth.

Other payloads carried aboard Atlantis during STS-46 included a study of the integration of oxygen in materials and thermal management processing, the Consortium for Materials Development in Space Complex Autonomous Payload (CONCAP II and III), an IMAX camera mounted in the shuttle's cargo bay, the Limited Duration Space Environment Candidate Materials Exposure experiment, a study of pituitary growth hormone cell function, an ultraviolet plume instrument, and the Air Force Maui Optical Site (AMOS) experiment.

The mission ended after 7 days, 23 hours, 15 minutes and 3 seconds, with Atlantis landing at the Kennedy Space Center in Florida on the morning of August 8, 1992.

After STS-46, Atlantis was moved to the Rockwell International Corp. site in California for scheduled check-out and maintenance. While it was at the Rockwell facility, the shuttle underwent a series of modifications that would enable it to dock with the Russian space station Mir.

SPACE SHUTTLE MISSION 50: STS-47

(September 12–20, 1992)

U.S. Manned Spaceflight

The fiftieth mission of the space shuttle program, STS-47 began on September 12, 1992 when the shuttle Endeavour lifted off from launch pad B at the Kennedy Space Center (KSC) in Florida.

As commander of the milestone flight, Robert Gibson made his fourth trip into space. He was joined on the mission by pilot Curtis Brown, payload commander Mark Lee, mission specialists Jan Davis, Jay Apt and Mae Jemison, and payload specialist Mamoru Mohri of Japan.

During STS-47, Mae Jemison became the first African American woman to fly in space. A medical doctor, Jemison served as the science mission specialist for the flight.

The flight also featured several other milestones, as Mark Lee and Jan Davis became the first married couple to fly in space, and Mamorou Mohri became the first Japanese citizen to fly on the space shuttle (the journalist Toyohiro Akiyama had been the first Japanese citizen to fly in space, during a visit to the space station Mir in 1990, aboard Soyuz TM-11).

There was another impressive fact about the fiftieth shuttle flight: STS-47 marked the first time since the November 1985 launch of STS-61B that a shuttle mission launched with no delays or postponements.

The primary payload for STS-47 was Spacelab-J, which was co-sponsored by NASA and the National Space Development Agency (NASDA) of Japan.

Spacelab-J lived up to the achievements of its predecessors in the Spacelab series, in both the quantity and quality of scientific data it gathered and in the goodwill it demonstrated through international scientific cooperation. The Spacelab-J program featured 43 experiments in two main areas of investigation: life sciences and materials science.

Using a variety of test subjects that included the crew, Japanese koi, animal and plant cells, chicken embryos, fruit flies, seeds, and frogs and their eggs, the crew conducted life sciences experiments in human health, cell and developmental biology, animal and human physiology and behavior, space radiation, and biological rhythms.

The 24 materials science investigations included experiments in biotechnology, electronics, fluid dynamics and transport phenomena, glass and ceramics, metal and alloys, and acceleration measurements.

Endeavour also carried a variety of additional payloads during STS-47, including 12 Getaway Special canisters in its payload bay, the Israeli Space Agency Investigation about Hornets (ISAIAH), an ultraviolet plume instrument, the Solid Surface Combustion Experiment (SSCE), the Shuttle Amateur Radio Experiment (SAREX II), and the Air Force Maui Optical Site (AMOS) experiment.

During the flight STS-47 was extended by one day to allow the crew more time to complete the large number of experiments.

With their milestone mission complete after 7 days, 22 hours, 30 minutes and 23 seconds (and 126 orbits), the crew of Endeavour returned to Earth at KSC on September 20, 1992. It was the twelfth shuttle landing at KSC, and the first operational use of a drag chute during landing. The drag chute proved less than optimal; post-flight analysis revealed that the shuttle was pulled to one side of the landing strip, most likely due to the drag chute.

SPACE SHUTTLE MISSION 51: STS-52

(October 22–November 1, 1992)

U.S. Manned Spaceflight

Launched on October 22, 1992 aboard the shuttle Columbia, STS-52 featured a wide array of international scientific payloads for sponsors that included the

Italian space agency Agenzia Spaziale Italiana (ASI), the Canadian Space Agency (CSA), and the European Space Agency (ESA).

The commander of STS-52 was James Wetherbee; he was accompanied on the flight by pilot Michael Baker, mission specialists Charles Veach, William Shepherd, and Tamara Jernigan, and CSA payload specialist Steven MacLean.

The mission's original launch was delayed by about a week by the replacement of the number three Space Shuttle Main Engine (SSME), which mission controllers deemed a necessary precaution when cracks were found in the engine's liquid hydrogen coolant system.

Jointly sponsored by NASA and the Italian space agency, the LAGEOS II satellite was propelled into orbit by the Italian Research Interim Stage (IRIS), which was used for the first time during STS-52.

Over the course of 11 years, researchers correlated data they gained from observing the LAGEOS spacecraft with measurements of the speed at which California's tectonic plates are known to have moved, in an effort to confirm the "Frame Dragging" effect described in Albert Einstein's theory of relativity.

The STS-52 crew also conducted a series of experiments that were bundled together as the U.S. Microgravity Payload-1 (USMP-1).

The international flavor of the mission was reflected in the Canadian Experiment-2 (CANEX-2), whose experiments included the Space Vision System—which entailed the deployment, on the ninth day of the mission, of the Canadian Target Assembly (CTA) satellite; a study of the impact of low-Earth orbit exposure on various materials; the Queen's University Experiment in Liquid-Metal Diffusion (QUELD); a study of phase partitioning in liquids; the Sun Orbiter Glow-2 (OGLOW-2) experiment; and a series of Space Adaptation Tests and Observations (SATO).

The ESA was represented by the Attitude Sensor Package. Mounted on a "Hitchhiker" plate, the sensor package included a modular star sensor, a yaw Earth sensor and a low altitude conical Earth sensor.

During the mission the shuttle was also used by an orbiting satellite of the Strategic Defense Initiative Organization as a reference point to calibrate an ultraviolet plume instrument on the satellite.

After a flight of 9 days, 20 hours, 56 minutes and 13 seconds, Columbia and her crew landed at the Kennedy Space Center in Florida on November 1, 1992, successfully deploying a drag chute during the landing.

SPACE SHUTTLE MISSION 52: STS-53

(December 2–9, 1992)

U.S. Manned Spaceflight

The last space shuttle mission scheduled to include classified activities undertaken on behalf of the U.S. Department of Defense, the STS-53 mission began with the launch of the shuttle Discovery on December 2, 1992. It was Discovery's fifteenth flight.

David Walker was commander of STS-53; he was joined on the flight by pilot Robert Cabana and mission specialists Guion Bluford, James Voss, and Michael Clifford.

Prior to STS-53, Discovery had undergone extensive modifications throughout much of 1992. Approximately 78 changes were made to the shuttle as a result of its scheduled maintenance checkup.

The launch on December 2 was delayed for several hours to allow for the melting of ice that had formed on the external fuel tank overnight.

The first day of the mission was devoted to the classified activities of the Department of Defense, and, as a result, no details of that part of the flight have been made public.

Unclassified payloads included a cryogenic heat pipe experiment; the Orbital Debris Radar Calibration Spheres; Cloud Logic to Optimize Use of Defense Systems (CLOUDS); the Cosmic Radiation Effects and Activation Monitor (CREAM); an experiment in fluid acquisition and resupply (FARE); the Hand-held, Earth-oriented, Real-time, Cooperative, User- friendly, Location-targeting and Environmental System (HERCULES); an experiment studying microcapsules in space; a study of space tissue loss; the visual function tester-2 (VFT-2); and radiation monitoring experiment (RME III).

Landing was originally scheduled for the Kennedy Space Center (KSC), but was diverted to Edwards Air Force Base in California because of heavy cloud cover at the Florida site. Utilizing its newly installed drag chute for the first time, Discovery landed at Edwards on December 9, 1992, after a flight of 7 days, 7 hours, 19 minutes and 47 seconds. An unexpected drama arose just after landing, when fuel leaked from a forward thruster. The crew remained safely aboard the shuttle while a fan dispersed the fuel and its fumes.

SPACE SHUTTLE MISSION 53: STS-54

(January 13–19, 1993)

U.S. Manned Spaceflight

John Casper commanded STS-54, which launched on January 13, 1993 aboard the shuttle Endeavour. He was accompanied by pilot Donald McMonagle and mission specialists Mario Runco, Gregory Harbaugh, and Susan Helms.

On the first day of the flight the crew deployed the fifth Tracking and Data Relay Satellite (TDRS-F), which was propelled into orbit by its attached Inertial Upper Stage (IUS).

The first of the six satellites in the TDRS network had been deployed during STS-6 in April 1983, and the sixth and final satellite in the series would be deployed during STS-70 in July 1995.

In a study of EVA procedures and techniques, Harbaugh and Runco made a spacewalk of 4 hours and 28 minutes in the shuttle's open payload bay on January 17. Their activities were videotaped for later study on the ground, and they also answered a lengthy list of questions about their experience after they returned to the shuttle, again for later study aimed at making future spacewalks easier and more productive.

Endeavour also carried the Diffuse X-ray Spectrometer (DXS) during STS-54. The DXS instrument was designed to study X-ray radiation emitted by diffuse sources in deep space.

Other payloads included a commercial bioprocessing instrument for life sciences experiments, and an experiment utilizing rodents to study skeletal system adaptation during spaceflight.

Foggy conditions at the Kennedy Space Center (KSC) in Florida delayed the scheduled landing by one orbit. The STS-54 crew landed at KSC on January 19, 1993 after a mission of 5 days, 23 hours, 38 minutes and 19 seconds.

SPACE SHUTTLE MISSION 54: STS-56

(April 8–17, 1993)

U.S. Manned Spaceflight

Launched aboard the shuttle Discovery on April 8, 1993, STS-56 featured the second flight of the Atmospheric Laboratory for Applications and Science (ATLAS-2).

Kenneth Cameron commanded STS-56. He was joined aboard Discovery by pilot Stephen Oswald and mission specialists Michael Foale, Kenneth Cockrell, and Ellen Ochoa.

A faulty instrument reading indicated a problem in the shuttle's main propulsion system during the first launch attempt on April 6. The night launch on April 8 suffered no delays.

Designed to study the interaction between the Sun and the Earth's middle atmosphere and the impact of that interaction on the Earth's ozone layer, the ATLAS-2 laboratory consisted of seven instruments, six of which were mounted on a Spacelab pallet in the shuttle's cargo bay. The Atmospheric Trace Molecule Spectroscopy (ATMOS) experiment, the Millimeter Wave Atmospheric Sounder (MAS), and the Shuttle Solar Backscatter Ultraviolet/A (SSBUV/A) spectrometer collected atmospheric data, while solar science observations were carried out by the Solar Spectrum Measurement (SOLSPEC) instrument, the Solar Ultraviolet Irradiance Monitor (SUSIM), the Active Cavity Radiometer (ACR), and the Solar Constant (SOLCON) experiments.

Because of poor weather at the Kennedy Space Center (KSC) in Florida, the originally scheduled landing was postponed, and the mission was extended for one day. Discovery landed at KSC on April 17, 1993, following a flight of 9 days, 6 hours, 8 minutes and 24 seconds.

SPACE SHUTTLE MISSION 55: STS-55

(April 26–May 6, 1993)

U.S. Manned Spaceflight

STS-55 was the second Spacelab mission dedicated to German scientific research. It launched aboard the shuttle Columbia on April 26, 1993 following the third abort of the space shuttle program (the first was a pad abort at T-9 seconds during Discovery's STS-41D mission in August 1984, and the second was an abort-to-orbit during the STS-51F launch of the shuttle Challenger in July 1985).

Steven Nagel served as commander of STS-55. He was accompanied by pilot Terence Henricks, mission specialists Jerry Ross, Charles Precourt, and Bernard Harris, and German payload specialists Ulrich Walter and Hans Schlegel.

Harris and Walter were medical doctors, specially chosen for the intense medical research to be conducted during the flight.

Originally scheduled to launch in February, STS-55 was first postponed while Columbia's three high-pressure oxidizer turbopumps were replaced. A new launch date of March 14 was scheduled, but that attempt was scuttled when a hydraulic hose in the shuttle's aft compartment burst during a Flight Readiness Test. As a precaution, all 12 of the hydraulic hoses in the aft compartment were removed and tested, and three were replaced. On March 21, the expected launch of the shuttle was put off by a delay in the launch of an unmanned Delta II rocket.

On March 22 the crew endured a pad abort, when Columbia's computers detected an incomplete ignition in the number three Space Shuttle Main Engine (SSME) just three seconds prior to launch. The fault was subsequently traced to a liquid oxygen leak in a valve inside the engine—a result of contamination during the manufacture of the valve. The nerve-wracking pad abort, as well as the prior difficulties that had already delayed the flight several times, led NASA officials to replace all three of the SSMEs as a precaution against further difficulties.

A faulty reading emanating from one of the shuttle's three inertial measurement units (IMUs) forced the postponement of a launch attempt on April 24; the IMU was replaced.

With launch finally accomplished on April 26, Launch Pad A at the Kennedy Space Center entered a period of scheduled refurbishment and modifications that would preclude its being used for another launch until February 1994.

Once in orbit, the crew worked in two shifts to conduct their Spacelab D-2 operations around the clock. They completed 88 experiments, in life sciences and materials science, technology applications, Earth observation, astronomy, and atmospheric physics.

One highlight of the life science experiments occurred when Dr. Bernard Harris administered the first intravenous drip (IV) ever used in space, as he injected fellow mission specialist Hans Schlegel and other crew members with a saline solution to study the efficiency of replacing fluids lost during the body's adaptation to weightlessness.

The IV experiment was part of a larger study of human adaptation to the space environment that included use of the Anthrorack, an advanced mini-diagnostic laboratory that returned a wealth of medical data.

Another major highlight of the mission was the German-manufactured Tests with Robotics Experiment (ROTEX). Ground controllers maneuvered the ROTEX robotic arm to perform the first remote control capture of a free-floating object in space.

The crew also performed several communications experiments. Using a Macintosh computer on board, crew members established an independent data link with ground controllers as part of the Crew Telesupport Experiment. As part of the Shuttle Amateur Radio Experiment (SAREX), the crew communicated with schoolchildren worldwide, and commander Steven Nagel communicated with the crew of the Russian space station Mir.

The extensive materials science experiments included use of the Holographic Optics Laboratory, the Material Science Autonomous Payload (MAUS), the Atomic Oxygen Exposure Tray, and crystal growth experiments. One result was the growth of a .78-inch gallium arsenide crystal—the largest grown in space up to that time.

Mission controllers opted to extend the flight by one day mid-way through the mission to allow the crew more time to continue the scientific experiments.

A frightening situation developed late in the flight, when a mistaken command emanating from Mission Control in Houston caused a loss of communications between the shuttle and mission managers on Earth. The communications breakdown lasted about an hour-and-a-half.

The crew also had to make a repair on a leaking nitrogen line in a tank designed to collect wastewater; and a malfunctioning refrigerator and freezer threatened to imperil their experiment samples, but crew members were able to transfer the samples to a backup unit.

Landing was originally scheduled for the Kennedy Space Center in Florida, but extensive cloud cover at the site forced mission controllers to divert the landing to Edwards Air Force Base in California, where the STS-55 crew returned to Earth on May 6, 1993, after a flight of 9 days, 23 hours, 39 minutes and 59 seconds.

With the completion of this fifty fifth mission, the total accumulated time in space for the space shuttle program passed one year: the five shuttles in the U.S. fleet (including the lost Challenger) had flown missions adding up to a total of 365 days, 23 hours and 48 minutes.

SPACE SHUTTLE MISSION 56: STS-57

(June 21–July 1, 1993)

U.S. Manned Spaceflight

Launched aboard the space shuttle Endeavour on June 21, 1993, STS-57 was the first shuttle mission to carry the SPACEHAB laboratory, a pressurized research and development module designed by Spacehab Inc., of Webster, Texas. The crew also retrieved the European Space Agency's EURECA European Retrievable Carrier, which had been deployed during STS-46 in 1992.

Ronald Grabe commanded STS-57. He was accompanied by pilot Brian Duffy, payload commander David Low, and mission specialists Nancy Sherlock, Peter Wisoff, and Janice Voss.

The launch had originally been scheduled for May, but was postponed to June 3, and was then further delayed when NASA officials decided to replace the

STS-57 payload commander David Low (top, on Endeavour's robotic arm) maneuvers mission specialist Peter Wisoff in a test of EVA procedures, June 25, 1993. [NASA/courtesy of nasaimages.org]

oxidizer turbopump on the number two Space Shuttle Main Engine (SSME) because of an improperly placed inspection stamp on the pump.

A truly strange phenomenon resulted in an additional delay when technicians heard an unexpected banging sound shortly after Endeavour arrived at the launch site. After a thorough investigation, the odd sound was traced to an assembly inside the 17-inch liquid hydrogen supply hose.

On June 20, another attempt was postponed five minutes before launch because of poor weather at the return-to-launch site at the Kennedy Space Center (KSC) and at the three overseas sites where the shuttle would have land in the event that an abort had been necessary.

The crew captured the European Retrievable Carrier (EURECA) on June 24, but the spacecraft would not respond to commands from ground controllers to stow its antennas. As a result, at the start of their scheduled EVA on June 25, David Low and Peter Wisoff manually folded the antennas before completing their assigned spacewalking exercises, a series of practice maneuvers which extended the total time of the EVA to 5 hours and 50 minutes.

Designed to greatly increase the pressurized workspace available to the crew for scientific research, the SPACEHAB module performed well on its first flight. The first such compartment to be privately developed by a commercial company and then flown on the shuttle, SPACEHAB facilitated 22 experiments by the crew, including materials and life sciences studies and a wastewater recycling experiment designed as a test of a system designed for potential use on the International Space Station (ISS).

On day two of the mission the crew spoke with President Bill Clinton.

Also carried aboard Endeavour during STS-57 were 11 Getaway Special experiments, including one designed by students and teachers from a school district in South Carolina that was identified by the acronym "CAN DO."

STS-57 was extended twice, for a total of two days, when poor weather at KSC forced mission controllers to put off landings on June 29 and June 30—the first time since 1986 (during STS-61C) that two attempted landings were waved off. Finally, after 9 days, 23 hours, 44 minutes and 54 seconds, Endeavour landed at KSC on July 1, 1993.

The STS-57 crew continued their mission activities even after they were on the ground, when they contacted the crew of STS-51, who were nearby in the shuttle Discovery on launch pad B. It was the first time that two crews communicated from inside their shuttles since 1985, when the STS-51D crew of Discovery chatted from space with the Challenger crew of STS-51B, which was on the ground at KSC.

SPACE SHUTTLE MISSION 57: STS-51

(September 12–22, 1993)

U.S. Manned Spaceflight

Launched aboard the space shuttle Discovery on September 12, 1993, STS-51 featured the successful deployment of an advanced communications satellite and the first flight of the ASTRO-SPAS instruments for astronomical observations.

The STS-51 crew endured the fourth pad abort of the space shuttle program, and at the end of their flight, they became the first shuttle crew to make a night landing at the Kennedy Space Center (KSC) in Florida.

The commander of STS-51 was Frank Culbertson. William Readdy served as pilot for the flight, and STS-51 mission specialists were James Newman, Daniel Bursch, and Carl Walz.